CLUSTER ANALYSIS FOR APPLICATIONS

Michael R. Anderberg

Air Force Systems Command
United States Air Force

ACADEMIC PRESS   New York and London   1973

A Subsidiary of Harcourt Brace Jovanovich, Publishers

ACADEMIC PRESS, INC.
111 Fifth Avenue, New York, New York 10003

United Kingdom Edition published by
ACADEMIC PRESS, INC. (LONDON) LTD.
24/28 Oval Road, London NW1

Library of Congress Cataloging in Publication Data

Anderberg, Michael R
  Cluster analysis for applications.

  (Probability and mathematical statistics)
  Originally presented as the author's thesis,
University of Texas at Austin, 1971.
  Bibliography: p.
  1. Cluster analysis. I. Title.
QA278.A5 1973   519.5'3   72-12202
ISBN 0–12–057650–3

AMS (MOS) 1970 Subject Classifications: 62-04, 62H30

This work is dedicated to our children,
Janel and Kenric,
the promise of the future

CONTENTS

PREFACE

Cluster analysis encompasses many diverse techniques for discovering structure within complex bodies of data. In a typical example one has a sample of data units (subjects, persons, cases) each described by scores on selected variables (attributes, characteristics, measurements). The objective is to group either the data units or the variables into clusters such that the elements within a cluster have a high degree of "natural association" among themselves while the clusters are "relatively distinct" from one another. The approach to the problem and the results achieved depend principally on how the investigator chooses to give operational meaning to the phrases "natural association" and "relatively distinct."

The need for cluster analysis arises in a natural way in many fields of study. Significant bodies of literature exist in each of the following areas:

1. Life sciences (biology, botany, zoology, ecology, paleontology).
2. Behavioral and social sciences (psychology, sociology, criminology, anthropology, linguistics, archaeology).
3. Earth sciences (geology, geography, regional studies, soil sciences, remote sensing).
4. Medicine (psychiatry, cytology, clinical diagnosis).
5. Engineering sciences (pattern recognition, artificial intelligence, systems science).
6. Information and policy sciences (operations research, information retrieval, political science, economics, marketing research).

The wide interest in this topic is evidenced by at least 600 relevant papers since 1960. Unfortunately, the literature is badly fragmented and only haphazard cross-fertilization between fields has occurred; judging by the citations in published papers, large bodies of researchers and analysts apparently are unaware of each other.

Unfortunately, the high level of journal activity has not been complemented with the publication of a good introductory text. The novice and casual user both find it necessary to read a staggering and varied array of journal articles to gain an adequate appreciation of the topic. And once the theory is in hand, there remains the computational problem of actually performing the cluster analysis, a problem addressed only indirectly in many papers.

There is an apparent and critical need for a book which distills the central elements of the literature into a single coherent presentation. The present work was written to be such a book. The first two chapters introduce the topic and establish a conceptual foundation for the further development of cluster analysis. Chapters 3–5 together give a comprehensive discussion of variables and association measures which should be of interest to a wide audience of analysts who use cluster analysis, factor analysis, or multidimensional scaling techniques; of particular interest are the strategies for dealing with problems containing variables of mixed types. Chapters 6 and 7 present the central techniques of cluster analysis with particular emphasis on computational considerations. Chapters 8 and 9 deal with techniques and strategies for making the most effective use of cluster analysis. Chapter 10 suggests an approach for the evaluation of alternative clustering methods, a topic which has received relatively little attention in the literature. The presentation is capped with a complete set of implementing computer programs listed in the Appendices. These programs have been carefully written to incorporate the following design features:

1. Wide variety of analysis alternatives
2. Simplicity of use
3. Economical and effective use of computer capacity
4. Large problem capabilities
5. Modular design
6. Extensive commentary on program operations
7. Machine independence

The book is written to be self-contained. Given a basic knowledge of univariate statistics and computer program usage, the reader should be able to perform penetrating and sophisticated analyses of real data. Analysts with applications for cluster analysis should find this book a useful reference work. Hopefully, the variety of clustering methods offered through the computer programs will tempt many experienced users of cluster analysis to experiment and discover the benefits of employing several methods rather than sticking with an old favorite alone. However, students and workers who have encountered the notion of cluster analysis only recently are intended to be the book's principal beneficiaries.

ACKNOWLEDGMENTS

My own education in this topic was through the laborious path of extensive reading in a wide variety of books and journals. At an early stage in this reading program, it became clear that the tools of cluster analysis should be available to much widened audiences and with substantially reduced individual effort. With the visionary consent of my supervisory committee, a Ph.D. program concentrating on applied cluster analysis was approved, and substantially this same text was submitted as a dissertation to the University of Texas at Austin in December 1971. Special thanks go to Professor Gerald R. Wagner who guided the program, chaired the supervisory committee and always provided a receptive ear for problems. Some of the best ideas to be found in these pages were prompted by the penetrating questions and comments from the other members of the committee: Professors Albert Shapero, Earl E. Jennings, and John J. Bertin. The special opportunity for doctoral study was provided by the United States Air Force as an integral part of my career assignment pattern. The Department of Aerospace Engineering at the University of Texas greatly enriched this opportunity by granting me the flexibility to design a unique program of study. The scope and depth of the book would have been much lessened without the support of the Defense Documentation Center which provided over one thousand government reports.

My most bountiful gratitude goes to my family for their encouragement, tolerance, and forbearance during the two and one half trying years of study. My wife, Lorna, has shared in full measure the agonies and triumphs while our children, Janel and Kenric, provided an incomparable incentive through the frequent query: "When is daddy going to finish his dissertation."

CHAPTER

I

THE BROAD VIEW OF CLUSTER ANALYSIS

This chapter gives an introduction to the subject of cluster analysis. The objective is to acquaint the reader with the broad nature of the subject. Computational details are postponed until later chapters in preference to establishing a strong intuitive grasp of key concepts first.

1.1 Category Sorting Problems

One of the most primitive and common activities of man consists of sorting like things into categories. The persons, objects, and events encountered in a single day are simply too numerous for mental processing as unique entities. Instead, each stimulus is described primarily in terms of category membership. For example, descriptions of people may involve items such as religion, ethnicity, political attitude, type of employment, style of dress, and so forth. Narrowly, each of these items is a specific attribute of the individual; but more broadly, category memberships tend to assign additional characteristics typical of the category but perhaps not specific to the individual. To illustrate, consider the following terms:

Mexican	teamster
Jew	priest
communist	homosexual
diplomat	football player

Each term evokes a class image which is a mixture of facts and emotions. Such categorical treatments in social settings are wellsprings of discrimination and

1

social injustice. On the other hand, without a scheme of categories each and every moment of the day would command the same degree of attention making life simply unmanageable. With a recognition of the fundamental role categories play in all human activity, consider some of the most important ways categories are used.

Classification, or identification, is the process or act of assigning a new item or observation to its proper place in an established set of categories. The essential attributes of each category are known, though there may be some uncertainty in assigning any given observation. Examples include the activities of Selective Service Boards classifying young men for the draft, a geologist identifying rocks, and a biologist cataloging flora and fauna. The classification problem may be complicated by imperfect class definitions, overlapping categories, and random variations in the observations. One way of dealing with these problems statistically is to find the probability of membership in each category for the new observation. The simplest classification criterion is to choose the most probable category. More sophisticated rules may be needed if the categories are not equally likely or if the costs of misclassification vary among the categories.

In discriminant analysis and pattern recognition the category structure is not so well known. The available information may vary from nearly complete descriptions down to knowing merely the number of categories. The distinctive element in this problem is the addition of a sample of observations from each class. The sample data may be used to estimate the missing information about the categories. The principal difference among various algorithms is the amount of structure assumed to be known *a priori*. To illustrate, suppose one wishes to predict whether individuals will be successful in some educational activity. If examples of past successes and failures are available, a classification procedure may be developed which will identify potential students with one class or the other, along with probabilities of membership. The same procedure could be applied to analyzing credit and insurance risks, screening people for disease during an epidemic, and recognizing handwritten letters or spoken words. Nagy (1968) gives a nice survey of the field, complete with 148 references. Dubes (1970) discusses several techniques and provides listings of computer programs for implementing them.

The cluster analysis problem is the last step in the progression of category sorting problems. In classification the category structure is known, while in discriminant analysis only part of the structure is known and missing information is estimated from labeled samples. The operational objective in both these instances is to classify new observations, that is, recognize them as members of one category or another. In cluster analysis little or nothing is known about the category structure. All that is available is a collection of observations whose category memberships are unknown. The operational objective in this case is to discover a category structure which fits the observations. The problem is frequently stated as one of finding the "natural groups." In a more concrete sense, the objective is to sort the observations into groups such that the degree of "natural association" is high among members of the same group and low between members of different groups. The essence of cluster analysis might be viewed as assigning appropriate meaning to the terms "natural groups" and "natural association."

1.2   Need for Cluster Analysis Algorithms

Even though little or nothing about the category structure can be stated in advance, one frequently has at least some latent notions of desirable and unacceptable features for a classification scheme. In operational terms the analyst usually is informed sufficiently about the problem that he can distinguish between good and bad category structures when confronted with them. Then why not enumerate all the possibilities and simply choose the most appealing?

The number of ways of sorting n observations into m groups is a Stirling number of the second kind (Abramowitz and Stegun, 1968)

$$\mathfrak{S}_n{}^{(m)} = \frac{1}{m!} \sum_{k=0}^{k=m} (-1)^{m-k} \binom{m}{k} k^n .$$

For even the relatively tiny problem of sorting 25 observations into 5 groups, the number of possibilities is the astounding quantity

$$\mathfrak{S}_{25}{}^{(5)} = 2{,}436{,}684{,}974{,}110{,}751 .$$

The problem is compounded by the fact that the number of groups is usually unknown, so that the number of possibilities is a sum of Stirling numbers. In the case of 25 observations

$$\sum_{j=1}^{j=25} \mathfrak{S}_{25}{}^{(j)} > 4 \times 10^{18} ,$$

a very large number indeed. It would take an inordinately long period of time to examine so many alternatives and the ability to make meaningful distinctions between cases would diminish rapidly. This approach might be likened to passing the entire population of New York City through a police lineup in the search for a single suspect. It is beyond human processing capacity.

Certainly most of the alternatives are inherently uninteresting or minor variations of some more appealing category structures. As a very mundane example, consider going to the grocery store and buying a month's worth of food for a family of four. The purchases likely will fill two carts and involve several hundred distinct pieces. Putting all these items into sacks for the trip home might be viewed as a cluster analysis problem in which each cluster is limited in size to the capacity of a sack. The most desirable arrangement is not obvious and the number of clusters will not be known until the last item is in a sack. The number of possible arrangements is enormous. But one can reject many as uninteresting because they put the milk on top of the bread or the cans on the fresh produce. A little thought might suggest putting the meat together in a sack, frozen foods in another, cans in a third category and so forth; or a wholly different scheme might come to mind. In any case, the human mind generates a series of short-cuts that permits a reasonable arrangement in little more time than it would take to stuff everything into sacks without regard for similarity or compatibility among the items. It is generally the intent of cluster analysis algorithms to emulate such human efficiency and find an acceptable solution while considering only a small number of the alternatives.

1.3   Uses of Cluster Analysis

Cluster analysis has been employed as an effective tool in scientific inquiry. One of its most useful roles is to generate hypotheses about category structure. An algorithm can assemble observations into groups which prior misconceptions and ignorance would otherwise preclude. An algorithm can also apply a principle of grouping more consistently in a large problem than can a human. Such methods can be particularly valuable in cases wherein the ultimate outcome is something like "Eureka! I never would have thought to associate X with A, but once you do the solution is obvious." A series of interconnected hypotheses may suggest models for the mechanisms generating the observed data. Put another way, cluster analysis may be used to reveal structure and relations in the data. It is a tool of discovery.

The results of a cluster analysis can contribute directly to development of classification schemes. Indeed, in biology and botany one of the principal applications of cluster analysis is to construct taxonomies. In other situations it may be possible to reduce a very large body of data to a relatively compact description through cluster analysis. If the grouping suggested by an algorithm is adopted for operational use, then it may become the *basis* for classifying new observations.

In a more theoretical vein, cluster analysis can be used to develop inductive generalizations. Strictly speaking, a set of results applies only to the sample on which they are based; but through appropriate modification they can be extended to describe adequately the properties of other samples and ultimately the parent population.

Since similar observations are grouped together, the individuals tend to assume class labels and the whole process may give names to things. This aspect of cluster analysis is most prominent in biology and botany, where the class name is part of the individual scientific name.

A most novel and unique application of cluster analysis is described by Bartels *et al.* (1970). These authors take quite literally the role of cluster analysis as being one of discovering "natural classes." They reason that if a suitable algorithm is applied to a set of data and the resulting clusters are only weakly differentiated, then the data probably consists of only one class. This test of homogeneity is applied to the screening of human tissue samples in the search for abnormal cells. A normal sample should be homogeneous while an abnormal one should result in significant grouping. This technique is a very clever way of using cluster analysis for classification.

The uses of cluster analysis described so far in this section are rather general in nature and may be applicable in almost any context. Adding another dimension to the discussion, it is useful to undertake a broad review of the various study disciplines where clustering has been used and to consider some of the entities subjected to cluster analysis. The many fields of study are grouped below into six major areas to help the reader organize this great variety of applications.

1. In the life sciences (biology, botany, zoology, entomology, cytology, microbiology, ecology, paleontology) the objects of analysis are life forms such as plants, animals, insects, cells, microorganisms, communities of interdependent members, and the fossil records of life in other times. The operational purpose of the analysis may range from developing complete taxonomies to delimiting the subspecies of a distinct but varied species.

2. Closely related to the life sciences are the medical sciences (psychiatry, pathology, and other specialties focusing on clinical diagnosis) where the objects of a cluster analysis may be diseases, patients (or their disease profiles), symptoms, and laboratory tests. The operational emphasis here is on discovering more effective and economical means for making positive diagnoses in the treatment of patients.

3. The behavioral and social sciences (psychology, sociology, criminology, education, anthropology, archeology) have provided the setting for an extraordinary variety of cluster analysis applications. The following entities have been among the many objects of analysis: training methods, behavior patterns, factors of human performance, organizations, human judgments, test items, drug users, families, neighborhoods, clubs and other social organizations, criminals and crimes, students, courses in school, teaching techniques, cultures, languages, artifacts of ancient peoples, and excavation sites. Factor analysis traditionally is a strong competitor to cluster analysis in these applications.

   4. The earth sciences (geology, soil science, geography, regional science, remote sensing) have included applications of cluster analysis to land and rock formations, soils, river systems, cities, counties, regions of the world, and land use patterns.

   5. Some engineering sciences (pattern recognition, artificial intelligence, systems science, cybernetics, electrical engineering) have provided opportunities for using cluster analysis. Typical examples of the entities to which clustering has been applied include handwritten characters, samples of speech, fingerprints, pictures and scenes, electrocardiograms, waveforms, radar signals, and circuit designs. Applications in engineering have been relatively few in number to date; any marked increase in the use of cluster analysis in this realm probably will require some dramatically new problem formulations.

   6. The information, policy, and decision sciences (information retrieval, political science, economics, marketing research, operations research) have included application of cluster analysis to documents and to the terms describing them, political units such as counties and states, legislators, votes on political issues, such issues themselves, industries, consumers, eras of good (and bad) times, products, markets, sales programs, research and development projects, political districting, investments, personnel assignments, credit risks, plant location, and floor plan designs. This realm of study probably is the source of the most innovative applications of cluster analysis in the last few years.

1.4   Literature of Cluster Analysis

   The literature of cluster analysis is both voluminous and diverse. Significant work may be found in almost any field of endeavor. Later chapters in this work cite and review numerous publications as they relate to particular topics. The purpose of this section is to illustrate the diversity of sources and give at least some general guidance for further reading.

   First of all, the terminology differs from one field to another. "Numerical taxonomy" is a frequent substitute for cluster analysis among biologists, botanists, and ecologists, while some social scientists may prefer "typology." In the fields of pattern recognition, cybernetics, and electrical engineering the terms "learning without teacher" and "unsupervised learning" usually pertain to cluster analysis. In information retrieval and linguistics "clumping" refers to a particular kind of clustering. In geography and regional sciences one may find the term "regionalization." Graph theorists and circuit designers prefer "partition" as a term describing a collection of clusters. Anthropologists have a problem they call "seriation" which consists of grouping ancient excavations into time strata. These terms will help identify a paper as being related to cluster analysis.

The diversity of terminology does not stop here. It extends throughout problem formulations, algorithm descriptions, and reporting of results. The cause is probably a mixture of professional jealousy, relative isolation among the fields, and genuine differences of viewpoint. For the novice, the disarray is bewildering and confusing; ultimately it is highly duplicative since the same idea is discovered repeatedly and published in a variety of journals.

Turning to books for a comprehensive survey is not very rewarding. Sokal and Sneath (1963) ranks as the classic work, at least in part because there is no other book that tries to be a complete reference work. But aside from its age, the book is diminished in value for the general user by the fact that a major portion of the text is devoted to philosophical issues unique to biological and evolutionary problems. Jardine and Sibson (1971) also direct their attention to the problems of the biological taxonomist. Their presentation is neither introductory nor elementary; the reader needs a working knowledge of the Sokal and Sneath book as well as considerable mathematical sophistication. Fisher (1968) describes a new method he devised and discusses its application to the aggregation problem of economics. Cole (1969) gives a collection of papers well worth reading. Tryon and Bailey (1970) describe their B C TRY system and illustrate its application with three well-studied data sets. Mathematically-oriented psychologists will be most comfortable with this book not only because of the psychological character of the applications but also because the methodologies are extensions of classical factor analysis. These five volumes exhaust the published books in the field. Only Sokal and Sneath try to provide anything approaching a general reference work. Any real education in cluster analysis must then involve extensive reading in learned journals.

As yet there is not a journal devoted exclusively to cluster analysis as a general topic. The most significant publications can be found in the following journals (listed alphabetically).

1. *Applied Statistics* (Series C, *Journal of the Royal Statistical Society*). Most papers come from authors in the United Kingdom. Computer programs are documented and listed in an algorithm supplement. The emphasis is on statistical methodology with illustrative applications.
2. *Biometrics*. An American journal with international contributors. The setting is one of life sciences but with many original contributions to statistical methodology.
3. *Biometrika*. Founded by Karl Pearson. One of the great statistical journals. Almost exclusively theoretical and with broadly international authorship.
4. *Computer Journal*. Published in the United Kingdom and dominated by contributions from the Commonwealth. Many of the papers provide a theoretical description and documentation for techniques proven in computer programs. Programs are listed in an algorithms supplement.
5. *Educational and Psychological Measurement*. An American journal which has published virtually all of L. L. McQuitty's work in cluster analysis. Computer programs are announced in an algorithm supplement.

6. *IEEE Transactions on Computers*. Cluster analysis is treated as a subfield of pattern recognition. The papers may range from quite readable to the abstract.
7. *Journal of Ecology*. Published in the United Kingdom and devoted to study of plants, animals, and other life forms. Cluster analysis is used frequently as a research tool leading to the reported results. Also includes methodological contributions.
8. *Multivariate Behavioral Research*. Problems in psychology provide the setting for a wide range of application papers. Methodology may also be found in this journal but at a lesser level of sophistication than in *Psychometrika*.
9. *Pattern Recognition*. Inaugurated in 1968, this journal is still very young and its ultimate personality is not yet clear. Primarily devoted to advancements in methodology.
10. *Psychometrika*. While formally devoted to problems of psychological measurement, this publication is an excellent statistical journal. Most papers are methodological.
11. *Systematic Zoology*. This journal is largely devoted to the development of cluster analysis for research in zoology and related life sciences. Many methodological innovations may be found in its pages. However much of the material is applicable only in the zoological setting.

This list of journals is by no means exhaustive. Most statistical journals have had a cluster analysis paper at one time or another. Survey and introductory articles are appearing in applications journals in increasing numbers. After becoming sensitized to the ideas and terminology of cluster analysis, it is surprising how frequently the subject is encountered in the literature.

In addition to the published literature, there is a wealth of contract reports, working papers, and other documents prepared under sponsorship of the United States Government. Such reports are announced and abstracted bimonthly in *Government Reports Announcements*. The reports may be purchased from the National Technical Information Service, Springfield, Virginia 22151. Reports of this kind form a major part of the cluster analysis literature. Government reports cited in the bibliography are identified with their AD, PB, or other order number so they can be acquired easily.

1.5   Purpose of This Book

The greatest deficiency in the current state of development of cluster analysis is the lack of a comprehensive and integrated approach to the problems of utilizing the methodology with real data. The literature is so diverse and rich that a person can read about hundreds of variations on cluster analysis; but he will also find it difficult to uncover practical guidance for making intelligent choices amidst this bewildering array of alternatives. A review of actual applications is of little help because authors typically give only their final selections and rarely any insight into the processes leading to the choices. An analyst who has a set of data to be clustered faces quite a number of questions, some of the most prominent being: How does one choose among similarity measures, clustering criteria, and algorithms? What considerations bear on each choice? How do the choices relate to each other? What strategies of utilization are most effective? How may the results be interpreted?

This book provides both an approach to answering such questions and the means to implement the resulting decisions quickly and efficiently. The necessary elements of data analysis, statistics, cluster analysis, and computer implementation are integrated vertically to cover the complete path from raw data to a finished analysis. A sizeable but manageable array of alternatives is provided for each stage of the analysis; this array is necessarily selective but the alternatives have been carefully chosen to include those which are best known and most universally applicable. To the extent possible this book is selfcontained; the reader should be able to use a wide variety of cluster analysis techniques without further reference to the literature. On the other hand, extensive citations to pertinent literature pinpoint sources of additional information. Finally, documented and tested computer programs are provided in the appendices to make the use of cluster analysis as painless and free of mechanical error as is possible.

2

CONCEPTUAL PROBLEMS IN CLUSTER ANALYSIS

The novice user of cluster analysis soon finds that even though the intuitive idea of clustering is clear enough, the details of actually carrying out such an analysis entail a host of problems. The foremost difficulty is that cluster analysis is not a term for a single integrated technique with well defined rules of utilization; rather it is an umbrella term for a loose collection of heuristic procedures and diverse elements of applied statistics. The actual search for clusters in real data involves a series of intuitive decisions as to which elements of the cluster analysis repertory should be utilized. Needed is a general framework for cluster analysis showing the steps involved, available alternatives, decision points, and relevant criteria for selecting among the options; but guidance of this scope simply is not to be found in the literature and this chapter falls short of completely filling the gap. However, in the following pages an outline of the major steps of a cluster analysis is proposed, the methodology is discussed as a tool of discovery, and various philosophical observations are offered pertinent to the questions of how cluster analysis should be used and interpreted.

2.1 Elements of a Cluster Analysis

When reading accounts of cluster analysis applications one sometimes has the impression that the investigator merely drops a batch of available data into a convenient computer code and interpretable results magically appear.

Typically, little discussion is devoted to all the choices leading up to the final results. Sometimes it is not clear whether any alternatives actually were considered or if the result was just forged directly.

What are the major choices in performing a cluster analysis? What considerations bear on these decisions? There appear to be at least nine major steps in clustering a set of data, all of which can shape the outcome profoundly.

2.1.1 CHOICE OF DATA UNITS

The objects of analysis are persons, animals, commodities, opinions, or other such entities. The actual mechanics of the analysis are performed on a sample of these entities, each member being denoted variously as a "data unit," "subject," "observation," "case," "element," "object," or "event."

There are two different situations of interest in the choice of data units. In the first instance the sample is the complete object of the analysis. The purpose is to discover a classification scheme for the given set of data units. It is not intended that the results should be applied to any additional data units outside the original sample. The principal consideration here is to make sure no important data units are omitted.

The second situation occurs more frequently and can be troublesome. In this instance the sample is a portion of a much larger population which is the true object of interest. Statistical training immediately suggests as relevant the familiar principles of random and independent selection. These principles are essential to any tractable testing of hypotheses, traditionally one of the principal concerns of statistical analysis. But cluster analysis generally is not involved with hypothesis testing; accordingly the necessity of imposing such principles should be carefully examined.

Randomization means that all data units are equally likely for selection as part of the sample. Under random selection, any groups that exist in the data will tend to be represented in the sample in proportion to their relative size in the population. Small or rare groups will tend to become lost amid the larger clusters. It may be necessary to be very selective about the data units to obtain any sizeable sample from such groups.

The independence assumption might also be violated systematically with profit. Independence means that the choice of each data unit is not influenced by the choice of any other. But, if selection of some data units promotes the candidacy of others, the effect should be exploited for the evidence of association rather than neutralized in deference to independence.

In a great many instances the data set is a product of another study or derives from routine administrative record-keeping. The existence of the data set itself may be an essential element of the decision to perform a cluster analysis. The problem in such cases is one of devising an effective procedure for screening data units to select the most relevant while culling out others inimical to the purpose of the analysis.

2.1.2   CHOICE OF VARIABLES

The data units cannot be merely arrayed for study like pieces of statuary. They must be consistently described in terms of their characteristics, attributes, class memberships, traits, and other such properties. Collectively, these descriptors are the variables of the problem. It is probably the choice of variables that has the greatest influence on the ultimate results of a cluster analysis.

Variables which are largely the same for all data units have little discriminating power whereas those manifesting consistent differences from one subgroup to another can induce strong distinctions. When a relevant discriminating variable is left out of the analysis some clusters may merge into an amorphous and confusing mass. It is not uncommon in cluster analysis to have one or more clusters that seem to make no sense and cannot be meaningfully subdivided in the chosen measurement space. On the other hand, inclusion of strong discriminators not particularly relevant to the purpose at hand can mask the sought-for clusters and give misleading results.

Zoologists take a unique approach to this problem through the "hypothesis of nonspecificity." Briefly it is assumed that the classification structure is dependent on many variables, any single one of which can be deleted or added without noticeable effect (Sokal and Sneath, 1963, pp. 85–91). As a consequence, numerical taxonomy investigations often involve huge numbers of variables. On the other hand, behavioral and social scientists, statisticians, and engineers strongly emphasize parsimony and seek to minimize the number of measured variables. This approach puts a premium on wise selection of variables both for relevance and discriminating power.

2.1.3   WHAT TO CLUSTER

The examples and much of the discussion up to this point deal with the clustering of data units as they are described by variables. However, it is just as reasonable to cluster variables according to their mutual behavior as manifested in the data units. Going a step further, there exist methods (Fisher, 1968; Hartigan, 1972) for the simultaneous clustering of data units and variables. It is even possible to consider an iterative approach in which the clustering is alternated between variables and data units until mutually harmonious classifications are achieved for both.

The clustering of variables has much in common with factor analysis. The two methodologies seek to explore the relationships among variables, though using vastly different techniques. Because of the differences, one should expect the diverse results to complement each other and provide greater insight when used together. More discussion of strategies for using cluster analysis, both alone and in conjunction with other methods, may be found in Chapter 9.

2.1.4 HOMOGENIZING VARIABLES

A common problem in real data is the lack of homogeneity among variables. In measuring association among variables different types of scales present difficult problems. For example, it is intuitively appealing that there should be a considerable degree of association between the variables "age" and "favorite movie star"; but defining a meaningful measure of this association may be frustrating.

When clustering data units it is necessary to amalgamate all the variables into a single index of similarity. The contribution of each variable to this composite depends on its scale of measurement and that of the other variables as well. The choice between grams and tons on one variable may be difficult enough, but when added to the choice between inches and miles on another it becomes not only a matter of relative weighting but one of finding meaning in the sum of diverse units.

Some investigators recommend reducing all variables to standard form (zero mean and unit variance) at the outset. Such suggestions simplify the mechanics of analysis but constitute a complete abdication of the analyst's responsibilities and prerogatives to a mindless procedure. There are ways of dealing with heterogeneous data without surrendering control of the analysis. This problem is given prominent consideration throughout this book. Chapter 3 is especially devoted to methods of reexpressing individual variables.

2.1.5 SIMILARITY MEASURES

Most cluster analysis methods require a measure of similarity to be defined for every pairwise combination of the entities to be clustered. When clustering data units the proximity of individuals is usually expressed as a distance. The clustering of variables generally involves a correlation or other such measure of association, many of which are discussed in the literature. Some have operational interpretations while others are rather difficult to describe. The measures interact with cluster analysis criteria so that some measures give identical results with one criterion and distinctly different results with another. The combined choices of variables, transformations, and similarity measures give operational meaning to the term "natural association." The choices are often made separately and should be reviewed for their composite effect to make sure the result is satisfactory. Further discussion of similarity measures may be found in Chapters 4 and 5.

2.1.6   CLUSTERING CRITERION

The term "cluster" is often left undefined and taken as a primitive notion in much the same manner as "point" is treated in geometry. Such treatment is fine for theoretical discussions. But when it comes to finding clusters in real data the term must bear a definite meaning. The choice of a clustering criterion is tantamount to defining a cluster. It may not be possible to say just what a cluster is in abstract terms, but it can always be defined constructively through statement of the criterion and an implementing algorithm.

Many criteria for clustering have been proposed and used. In some problems there is a natural choice, while in others almost any criterion might have status as a candidate. It should not be necessary to choose only a single criterion because clustering the data set several times with different criteria is a good way to reveal various facets of the structure. On the other hand, the expense of using cluster analysis prohibits trying out everything that is available. A reasonable selection of clustering criteria is presented in Chapters 6 and 7.

2.1.7   ALGORITHMS AND COMPUTER IMPLEMENTATION

Even after choosing data units, variables, what to cluster, transformations, a similarity measure, and a criterion, there still remains the problem of actually generating a set of clusters. Given all these prior choices there are frequently several alternative algorithms which ultimately give the same result. Some may be totally impractical with regard to computer storage requirements or time demands. Others may have special properties permitting trade-offs between the number of variables and number of data units. For example, in Chapter 6 two different algorithms are given for the Ward clustering method using euclidean distance. One will permit a practically unlimited number of variables but a maximum of about 250 data units on the CDC 6600. The other algorithm can trade space for variables to handle more data units. In the same core space needed for 250 data units above, the second algorithm could handle 1300 data units each measured on 25 variables. Magnetic disk and tape storage can extend capacity even further but at a cost in computing time.

The sequence of steps up to this point should have some apparent logic. The choice of an algorithm should come after the other choices because it is merely the means of implementation. Unfortunately, frequently all the choices are subordinated to the capabilities of an available computer program. The problem is made to fit the technique. It is possible that such an approach will give useful results; but a preferable state of affairs is to have a proven capacity to perform a variety of analyses in response to the demands of the problem. The computer programs in Appendixes E and F give a wide choice of methods.

2.1.8   NUMBER OF CLUSTERS

A substantial practical problem in performing a cluster analysis is deciding on the number of clusters in the data. Hierarchical clustering methods give a configuration for every number of clusters from one (the entire data set) up to the number of entities (each cluster has only one member). Other algorithms find a best fitting structure for a given number of clusters. Of course the latter can be repeated for different choices of the number of clusters to give a variety of alternatives. Some algorithms begin with a chosen number of groups and then alter this number as indicated by certain criteria with the objective of simultaneously determining both the number of clusters and their configuration.

The literature contains a lot of wishing for mechanical methods of determining the number of groups. It can be a difficult choice. Some investigators seem to have a preoccupation with an "optimal" number of groups, usually in conjunction with a search for a single "best" classification scheme for the data set. The possibility of several alternative classifications, each reflecting a different aspect of the data, is rarely entertained.

2.1.9   INTERPRETATION OF RESULTS

Having finally produced a set of clusters it is time to make some use of the results. At the least sophisticated level the clusters are summary descriptive statistics much like the mean and variance. For a given similarity measure, criterion, and algorithm the production of clusters can be a straightforward mechanical procedure. There might even be a strong temptation to apply standard tests of significance on the differences between group means or the ratio of between group to within group sums of squares. The trouble with making such tests is that they are hardly relevant. The whole focus of the clustering criterion and algorithm is to give a set of clusters that are well differentiated from each other. "Highly significant" results then should occur with monotonous regularity.

At the next higher level of sophistication the set of choices described in the last eight sections may be so well defined that the clusters necessarily have the desired properties. The results are inherently interesting because they are the product of a procedure which satisfies all relevant requirements. Such is the case in political districting problems (Garfinkel and Nemhauser, 1970) and floor plan design problems (Grason, 1970). Long experience with a particular method in a special application can produce much the same effect; the U.S. Air Force has used a job clustering program (named CODAP) to analyze job inventories in many Air Force specialties. The program has been in use since at least 1964 and now constitutes a routine part of occupational surveys. Archer (1966) describes the basic methodology used in CODAP with a detailed example.

At the highest level of sophistication the results of a cluster analysis are an aid to reasoning from the data to explanatory hypotheses about the data. Indeed, a set of clusters may be viewed as a proposition concerning the organization of the data. The proposition may contain elements of sheer falsehood and nonsense; it may also contain the seeds of novel interpretation for what is already known, as well as suggestions about previously unnoticed regularities and relations. Most persons using cluster analysis probably employ it in such an exploratory fashion, but sometimes with an excess willingness to accept the gospel as pronounced by the computer. The tendency to ascribe truth to numbers produced mechanically is well known in the field of factor analysis and no less prevalent in cluster analysis. Perhaps such misuse of the methodology stems from the lack of a general philosophical framework for its utilization and interpretation. Section 2.3 offers for consideration some partial contributions to such a framework which hopefully will prove useful and intelligent in practical problems.

2.2 Illustrative Example

The foregoing discussion includes some uncommon ideas about cluster analysis which can be illustrated conveniently with an example from a familiar context. Take a deck of ordinary playing cards and set aside the jokers. Now sort the 52 cards into clusters. What do you find? The most probable answer is that it is hard to choose among all the groupings that come to mind. Figure 2.1 shows six possibilities which will suffice for the present discussion.

Classification A is not very natural but is a probable choice if one is forced to generate 26 clusters. Somewhat more appealing are the 13 groups of classification B. Notice that B can be obtained from A by using the groups of A as building blocks. A sequence of classifications in which larger clusters are obtained through merger of smaller ones is a *nested or hierarchical classification*. Classification C is a hierarchical outgrowth of both A and B. It is also a bridge player's classification: honors versus trash. In many card games the players arrange their hands according to suits; an application of this simple principle is embodied in classification D. If a small child was asked to sort the deck he likely would come up with something like classification E. The bridge player might interrupt with a preference for a division on the basis of minor versus major suits rather than black versus red; of course both of these divisions are related hierarchically to classification D. Classification F is the only meaningful organization for an avid hearts player. Quite obviously it is not hierarchically related to any of the other five classifications. Together these six classifications illustrate a variety of conditions that might occur in any problem.

1. There are *many meaningful groupings.* Each reflects a different aspect of how a deck of cards can be used. Notice that each classification embodies a number of groups tailored to its own character rather than conforming to any prior ideas of how many groups there should be.

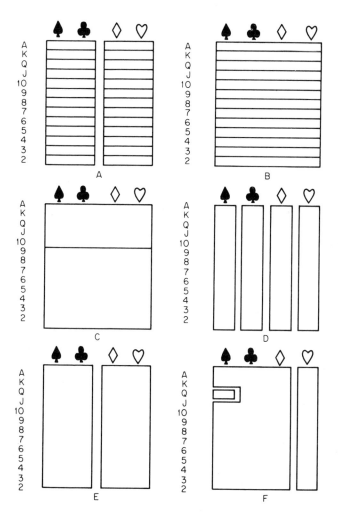

Fig. 2.1. Playing card example.

2. *A single classification may give a distorted view* of a multifaceted data set. Classification F is very meaningful in the game of hearts, but it is so specialized as to be meaningless in any other context.

3. When using a hierarchical method of cluster analysis, *early decisions can preclude certain meaningful groupings* at later stages. If classification B is the next step after A, then certainly classification E can never be achieved in this sequence.

4. On the other hand, *the same classification may be achieved in several different ways.* Classification E may be obtained hierarchically from both A and D, which are not hierarchically compatible with each other.

5. If there are several meaningful groupings, *a variety of cluster analysis techniques will be needed* to reveal them. Certainly classifications C and E could not both come from a single set of clustering decisions.

2.3   Some Philosophical Observations

> The use of any knowledge reaches into three areas of the mind: the search for truth, the skill of forecasting, and the gift to imagine a future different from the present. There will never be clear-cut rules of procedure. The best we can hope for is to sort out the different strands and look for regularities as they are combined into cords. The more we succeed the more energy and freedom will be left for the creative and innovating element indispensible in all utilization. (Lazarsfeld and Reitz, 1970, p. 14.)

Cluster analysis is one of the few systematic techniques that can help one "look for regularities." Other such aids are factor analysis (Harman, 1960; Rummel, 1970), multidimensional scaling (Greene and Carmone, 1970), and methods seeking the "intrinsic dimensionality" of a data set, the latter best exemplified by the paper of Shepard and Carroll (1966). The intelligent use of such techniques to explore data sets can open new vistas. Traditional models can be temporarily set aside while taking a fresh look. Questions can be asked that would have been decidedly impractical to study before. To gain such advantages it helps to have some guiding philosophy of utilization for cluster analysis. This section is devoted to some fundamental aspects of such a guiding philosophy.

2.3.1   A BASIC QUESTION

One of the first questions faced by a novice user of cluster analysis is "how do you know when you have a good set of clusters?" Such a question frequently proves to be difficult to answer; before reading further, the reader should pause and try to answer it himself. The way the analyst deals with this question depends on his reasons for using cluster analysis. The three interpretative uses introduced in Section 2.1.9 will provide a convenient framework for discussion.

In the first case, the set of clusters is used like a descriptive summary statistic. There is no room to assess whether the clusters are "good" or "bad," because they are simply statements of fact: this set of clusters is the mechanical result of submitting the data to the chosen procedure. The only question that can properly arise is whether the procedure was correctly applied to the data, just as with the computation of other descriptive statistics such as the mean and variance.

In the second case, the clustering procedure is so well defined as to insure that the clusters have the desired properties simply because of their method of generation. The clusters themselves cannot be subject to question as long as the procedure is accepted as valid.

In the third case, the cluster analysis procedure is an exploratory device. Too little is known about the data set to say whether a given clustering procedure should produce a set of clusters with any relevance at all. The clusters merely constitute a proposal about the underlying structure of the data set. Now this structure may not be unique, or at least not uniquely determinant from the given set of data. Several different plausible explanations may account for all the known information. And the original question arises again: how do you know when you have a good set of clusters? The answer is that the clusters are not interesting in themselves at all; the point of interest is in inferences about the structure of the data. The observation "All entities in this cluster are similar to each other" is not an explanation of the data structure; it is a consequence of this structure. This explanatory structure is the object of the search. Its description is in terms of principles and ideas, not individual data units. Once a satisfactory structure is known and defended on its own merits, any cluster analysis that contributed to its discovery is only of historical interest. If desired, the original data units can be classified in accordance with this structure to obtain "a good set of clusters." But without such a structure in abstract there is no criterion for judging "goodness" of clusters in numerical data.

2.3.2 THE ROLE OF THE ANALYST

The value of exploratory cluster analysis is primarily in the tendency for new arrangements of data units or variables to suggest relationships and principles previously unnoticed. The substantive results are not the output of the computer but the new ideas prompted in the analyst's mind. These results may take the form of hypotheses to be suitably tested or they may be so selfevident as to be adopted immediately without further ado. The clustering that suggests such ideas may itself prove to be only weakly consistent with the notions it prompts. The clustering obtained by heuristically and intuitively reorganizing according to the new ideas may be far more appealing than anything obtainable by a systematic clustering procedure. Cluster analysis is a tool for suggestion and discovery. It is not in itself a wellspring of either truth or falsehood. A consequence of these views is that the mechanical result of a clustering procedure is itself empty of all meaning.

The use of exploratory cluster analysis is conditioned by (at least) three essential elements: (1) the context of the problem, (2) the analyst's knowledge of the context, and (3) the analyst's research objective. Consider each of these elements in turn.

The context of the problem includes virtually anything that may influence the observation or interpretation of the data. The most pertinent aspects of context include the actual (versus intended) circumstances of data collection, related matters of fact, and relevant theory. Context is a collective term for all those elements of knowledge that pertain to the object of study. Any explanatory structure proposed for the observed data needs to be consistent with the context.

The analyst's knowledge and understanding of the problem context has long been recognized as an important element of all scientific inquiry. At the very least, he cannot insure that his explanations, hypotheses, and conclusions are consistent with any elements of context unknown to him. But in a more direct sense, all such knowledge conditions not only how events are observed, but also whether they are observed at all. As Hanson (1958, p. 51) notes: "To one ignorant of what happens as a rule in bridge, 'finesse' will explain nothing. Even though nothing escapes his view while the finesse is made, he will not see the finesse being made." Some prior knowledge is needed to make sense of observation. But the same prior facts and dispositions can block recognition or prompt outright rejection of the explanations being sought. Medawar (1969, p. 52) observes: "We scientists often miss things that are 'staring us in the face' because they do not enter into our conception of what might be true, or alternatively, because of a mistaken belief that they could not be true." The latter point is especially troublesome because much of one's knowledge is taken as true by force of authority and not as a matter of personal understanding. Reputed facts known as matters of faith cannot be rejected without questioning the authority that presents them as true. On the other hand, understanding why the "fact" is thought to be true implies the flexibility to adapt to new evidence.

In addition to the ability to evaluate observations, the knowledge of context is fundamental to the conceptual power to explain causes and findings. Again, Hanson (1958, p. 54) makes the point well:

> We have had an explanation of x only when we can set it into an interlocking pattern of concepts about other things, y and z. A *completely* novel explanation is a logical impossibility. It would be incomprehensible...; it would be imponderable, like an inexpressible or unknowable fact.

An explanation cannot stand by itself unrelated to any other knowledge. Indeed more often than not an explanation consists of linking familiar ideas into a new ensemble which accounts for more of observed reality than could the ideas taken individually.

The analyst's research objectives permeate the entire investigation. They motivate the enterprise in the first place and very effectively shape both the evaluation of observations and explanation of facts as they are perceived. These objectives are the determinants of relevance. The analyst does not simply go out and browse in the world like cattle grazing on the Great Plains of the 1800s. He proceeds with purpose and ignores or prunes away the unnecessary and distractive elements as he sees them. Neither is his objective unitary in nature. He may seek a solution with many objectives in mind. Some objectives may be solidly bounded and act as constraints as well. The ideal solution is one which dominates all other alternatives in regard to every objective. The more usual case will involve some intuitive compromise which depends at least in part on how well the various objectives are fulfilled in the alternatives encountered on the way to choosing a tentative solution. The process is much like buying a house in a new city. At the outset the objectives include nice neighborhood, close to schools and shopping, attractive design, and so forth. In addition to the objectives there are some constraints: four bedrooms, basement, two car garage, fenced yard, all brick, 2000 square feet of living space, under $20,000, and so forth. After the shattering revelation that no such thing has existed since the Korean War, old objectives fall into the "nice to have" category, rigid constraints become very flexible, and the would-be house buyer tries to make out as best he can in the circumstances. Likewise, context and objectives interact in cluster analysis, the latter adjusting to accommodate a new appreciation of the former.

2.3.3 SOME REMARKS ON UTILIZATION

It would be appropriate at this point to present a coherent integrated philosophy of discovery in which cluster analysis (and other such methods) plays the role of a hypothesis generator. Such a philosophy would be a major contribution in its own right but is not presently available. Indeed the cluster analysis literature is quite devoid of philosophy for the utilization of cluster analysis in general. Sokal and Sneath (1963) devote a major portion of their text to philosophical problems in zoology and the influence of numerical taxonomy; but most of these remarks apply only to the zoological context. The literature in the philosophy of science is not much more helpful except for Hanson's (1958) *Patterns of Discovery*. Neither cluster analysis nor any other systematic aid to discovery is mentioned in the book. However, Hanson examines the means by which scientists (physicists in particular) reason from data to hypotheses about the data. Such reasoning is neither inductive nor deductive; it is "retroductive." Kepler's discovery that the orbit of Mars is elliptical rather than circular (as universally held by astronomers of the time) is recounted in Hanson's Chapter IV as a classic illustration of retroductive reasoning. While Kepler could have made no use of cluster analysis even if it was available, it seems clear from the account that retroductive reasoning and exploratory cluster analysis are different aspects of the same kind of intellectual activity. Hanson's remarks about retroductive reasoning apply with equal force to cluster analysis. Perhaps the relationships can be formalized to produce progress in both areas.

Even though a complete guiding philosophy of cluster analysis utilization is not available, some individual principles can be stated and discussed. Hopefully these principles will be seen as useful and intelligent advice for applications.

1. Any given set of data may admit of many different but meaningful classifications. Each classification may pertain to a different aspect of the data. It is unnecessarily narrowing to seek a single "right" classification. Several different clustering procedures will be needed to discover multiple classifications. Such was the message of the card example. Incidentally this principle has interesting implications for data sets that are thought to be well understood. New insight and understanding might result from alternative classifications suggested through a cluster analysis. Totally unexpected aspects of structure might be revealed in the process.

2. Cluster analysis is a device for suggesting hypotheses. The classification of data units or variables obtained from a cluster analysis procedure has no inherent validity. The analyst should not feel any pressure to embrace a particular classification, nor should he feel bound to the details of a classification he finds generally interesting. The worth of a particular classification and its underlying explanatory structure is to be justified by its consistency with known facts and without regard to the manner of its original generation. Ideally, the set of clusters generated by a cluster analysis procedure will produce combinations of entities which otherwise might never be considered for examination but reveal aspects of the data which are self-evident in retrospect. Hanson (1958, p. 87) puts it a little differently,

> Perceiving the pattern in phenomena is central to their being "explicable as a matter of course." Thus the significance of any blob or line in [a diagram] eludes one until the organization of the whole is grasped; then this spot or that patch, becomes understood as a matter of course.

3. A set of clusters is not itself a finished result but only a possible outline. It follows then that there is little justification for using excessively detailed and expensive algorithms when results of the same general character can be

achieved with cheap and intuitive procedures. Of course, some procedures are both elegant and low in cost and may be preferred for this combination. But one of the most striking things about the many methods in the literature is the high degree of redundancy when applied to a set of data. The ideal would be to have a small stable of algorithms minimally duplicative among themselves but collectively representative of all the general types of classifications that might be produced by all other algorithms put together.

4. Cluster analysis methods involve a mixture of imposing a structure on the data and revealing that structure which actually exists in the data. The notion of finding "natural groups" tends to imply that the algorithm should passively conform like a wet teeshirt. Unfortunately, practical procedures involve fixed sequences of operations which systematically ignore some aspects of structure while intensively dwelling on others. Such properties are often quite inadvertent and therefore discovered only by observing their effects on data. To a considerable extent a set of clusters reflects the degree to which the data set conforms to the structural forms embedded in the clustering algorithm.

5. In view of the preceding four points it should come as no surprise that the results of a cluster analysis method rarely suggest a satisfactory structure for the total set of data. More commonly one or more interesting clusters lead to inferences about part of the data. Along this line, certain clusters may be so natural and selfevident (once discovered) as to constitute "features" of the data likely to be revealed by almost any method. Rather than continue to recover the obvious, it is economical and relatively riskless to remove such features from the data set as they are found and concentrate attention on the more confused residue. Of course, adding or deleting variables alters the measurement space, and consequently such features may lose their distinctive character or alter their composition in response to such a change.

6. In the quest for clusters two possibilities are often overlooked.

a. *The data may contain no clusters.* When clustering variables, nearly complete independence or orthogonality would lead to such a result. When clustering data units, an absence of discriminating variables and a more or less uniform distribution of points in the measurement space would lead to a distinct lack of cohesion.

b. *The data may contain only one cluster.* When clustering variables, nearly complete dependence or colinearity would lead to such a result. In the clustering of data units, the absence of discriminating variables combined with meaningful mutual association among all data units would give only one cluster.

Clearly these two possibilities are opposite extremes with all other possibilities falling between them.

7. Prior knowledge about the population can be grossly misleading when applied to sample data, especially when the circumstances of data collection are not completely known. For example, suppose it is known that there are five groups in the population. If one group is rare or has been excluded systematically in data collection there likely will be only four real groups in the sample. But forcing the clustering algorithm to produce five groups will create clusters which should be nonsense in this same framework of prior knowledge. The same result easily can arise when the chosen variables inadequately distinguish among the various classes or promote distinctions not relevant to the immediate purpose of the analysis.

2.4   A Note on Optimality and Intuition

The search for the "optimal solution" seems to be a persistent element wherever mathematical methods are applied. It is comforting to formulate a problem as the search for optimality because when the solution is found it can be eagerly adopted without further ado. Of course, the ability to proceed thusly depends on the analyst being able to specify an analysis method which assigns a meaningful value to every alternative. In the effort to choose a suitable objective function and adequately constrain the problem there is a well-known tendency to replace the original problem with another that can be solved. The transformed problem may be a far cry from the original, but it is frequently preferred because the formulation is explicit.

An alternative to optimality is to use heuristics and intuition to find solutions "by inspection." Let it be agreed that such an approach may be totally infeasible for many kinds of problems. Nevertheless, there are some problems where it may be a useful data analysis strategy to heuristically rearrange the data in such a way as to improve the chances of finding good solutions "by inspection." The approach enhances the efficacy of subjective judgement rather than surrendering to a completely specified procedure before the first candidate solutions are seen. Rather than remove human judgement altogether, the idea is to focus judgement on plausible alternatives. An added bonus comes in the form of being able to use qualitative information as a natural part of the analysis. A few trial solutions based on likely criteria or inherently meaningful methods may provide very good results with little effort. A possible outcome might even include the insight to state an appropriate optimal formulation when none seemed available before.

CHAPTER

3

VARIABLES AND SCALES

Most discussions of statistical theory are based on the convenient assumption that the variables are of a single type, usually continuous and on an interval scale. In this context the most powerful and sophisticated of mathematical techniques may be applied to good advantage. However, in real world problems, natural formulations frequently entail a mixture of variable types. Guidance as to how to handle such cases is exceedingly rare; there appears to be no published account of alternative analysis techniques which approaches the scope of this chapter. The first part of the chapter is devoted to the careful construction of a classification scheme for variables. To describe adequately the important aspects of variables it is necessary to use a cross-classification based on the size of the range set and the scale of measurement. Each possible variable type is illustrated with an example to show how the formal distinctions have force in practice. Hopefully the use of this classification scheme will help to reduce the incidence of ambiguities and other sources of confusion in this book. The second and largest portion of the chapter is devoted to techniques for converting variables from one scale to another. A typical problem of this type is that of categorizing interval data. Because this topic is treated only in bits and pieces in the literature, a comprehensive approach is adopted, and many alternative schemes are examined. Accordingly the discussion is rather lengthy. As an aid to effective reading it is suggested that the chapter be approached as follows:

1. Scan the table of contents to appraise briefly the range of material in this chapter.

2. Read the chapter, but skip all four digit subsections (e.g., Section 3.2.1.1).

3. Read the chapter in its entirety.

This approach will help to put the many details into proper perspective.

3.1 Classification of Variables

The application of any statistical technique ultimately involves a confrontation with real data and the special considerations attendant to particular kinds of data. A systematic and comprehensive classification of variables provides a convenient structure for identifying essential differences among data elements. This section presents a cross-classification of variables based on two familiar descriptive schemes.

3.1.1 CLASSIFICATION ACCORDING TO SIZE OF THE RANGE SET

From a mathematician's point of view it is natural to distinguish among variables on the basis of the number of elements in the range set, that is, the number of distinct values the variable may assume. To explore this approach adequately it is useful to have available the following concepts about counting the number of elements in a set.

1. A set is *finite* if its elements may be put in one-to-one correspondence with a subset of the positive integers, the latter containing a largest number. Less formally, a set is finite if its elements can be counted and some definite integer is given as the number of elements in the set.

2. A set is *countably infinite* if its elements may be put into one-to-one correspondence with the set of positive integers. The latter set is infinite since there is no largest integer (the claim that some number, say k, is the largest integer is refuted by exhibiting $k + 1$, a larger integer). Less formally, the set may be counted, at least in principle, but it would take infinitely long to do so. However, between any two given elements of the set there is a finite number of other elements in the set.

3. A set is *uncountably infinite* if its elements cannot be put into one-to-one correspondence with the positive integers. Fundamentally, between any two real numbers there are infinitely many real numbers as opposed to the finiteness characterizing an interval in a countably infinite set. Given a starting point in the set, it is not meaningful to speak of "a next number." If the set consists of all real numbers between 1.5 and 2.0, what is "the next number" after 1.5? It is not 1.5000001 since 1.50000009 lies between the two numbers. Indeed, any candidate for "next number" fails since infinitely many numbers may be found between 1.5 and the candidate. Hence, the set is uncountable.

With these concepts, a familiar classification scheme for variables is as follows:

1. A *continuous variable* has an uncountably infinite range set. Typically such a variable may assume any value in an interval (say 1.5 to 2.0) or a collection of such intervals.

2. A *discrete variable* has a finite, or at most countably infinite range set.

3. A *binary or dichotomous* variable is a discrete variable which may take on only two values.

3.1.2   CLASSIFICATION ACCORDING TO SCALE OF MEASUREMENT

In the social and behavioral sciences one frequently encounters a classification of variables based on their scales of measurement. It will be convenient to illustrate this scheme with a variable X and two objects, say A and B, whose scores on X are x_A and x_B, respectively.

1. A *nominal scale* merely distinguishes between classes. That is, with respect to A and B one can only say $x_A = x_B$ or $x_A \neq x_B$.

2. An *ordinal scale* induces an ordering of the objects. In addition to distinguishing between $x_A = x_B$ and $x_A \neq x_B$, the case of inequality is further refined to distinguish between $x_A > x_B$ and $x_A < x_B$.

3. An *interval scale* assigns a meaningful measure of the difference between two objects. One may say not only that $x_A > x_B$ but also that A is $x_A - x_B$ units different than B.

4. A *ratio scale* is an interval scale with a meaningful zero point. If $x_A > x_B$ then one may say that A is x_A/x_B times superior to B.

These scale definitions are ordered hierarchically from nominal up to ratio. Each scale embodies all the properties of all the scales below it in the ordering. Therefore, by giving up information one may reduce a scale to any lower order scale. Considerations involved in such scale transformations are discussed in the second half of this chapter.

Frequently variables on nominal and ordinal scales are referred to as *categorical* variables or *qualitative* variables, often with ambiguity as to whether any order relation exists. For contrast, variables on interval or ratio scales are then referred to as *quantitative* variables.

3.1.3   CROSS-CLASSIFICATION AND EXAMPLES

Table 3.1 summarizes the definitions for the two classification schemes and illustrates the resulting cross-classification with examples. Several of these examples bring out subtle distinctions worthy of brief discussion. The zero point of the Kelvin temperature scale (denoted °K) is absolute zero, whereas the zero point of the Celsius scale (°C) is the freezing point of water. It makes sense to say 200°K is twice as hot as 100°K since the heat content of some substance is twice as great. However, a similar statement using the Celsius scale would be nonsense. Hence the Kelvin scale may be treated as a ratio scale while the Celsius scale is at best an interval scale.

The example for the continuous–ordinal case involves quantities which may be measured very precisely by appropriate equipment and on a continuous scale. Now suppose human judges are asked to make judgements of the same quantities. The judges probably will be able to make very good pairwise distinctions as to which of two stimuli is the more intense. But in a full array of paired comparisons they might not be expected to make very precise judgements as to the amount of difference for each pair. Now if the judge records his response by assigning a score reflecting his appreciation of the difference, the resulting set of scores would be essentially on a continuous scale but would only contain ordinal information.

The case of a continuous–nominal variable is very difficult to illustrate since

TABLE 3.1
CROSS-CLASSIFICATION OF VARIABLES WITH EXAMPLES

Scale of measurement	Size of range Set		
	Continuous: May assume an uncountably infinite number of values	*Discrete:* May assume a finite (or at most countably infinite) number of values	*Binary:* May assume only two values
Ratio: If $x_A > x_B$, A is x_A/x_B times greater than B and $x_A - x_B$ units greater than B	Temperature in °K, weight, height, age	Counts such as numbers of children, hospitals, cars	Unit price of soft drinks in vending machines—bottles or cups: 10¢, cans: 15¢
Interval: If $x_A > x_B$, A is $x_A - x_B$ units greater than B	Temperature in °C, specific gravity	Serial numbers, TV channel numbers	How many wives do you have (assuming the only legal answers are 0 or 1)?
Ordinal: Either $x_A > x_B$, $x_A = x_B$, or $x_A < x_B$	Human judgements of texture, brightness, sound intensity	Military rank, models in a line of cars (Ford, Mercury, Lincoln), wide, medium, narrow	Tall–short, good–bad, big–small, wide–narrow
Nominal: Either $x_A = x_B$ or $x_A \neq x_B$	Absurd—requires an uncountably infinite number of distinct classes	Eye color, place of birth, favorite actor	Yes–no, present–absent, dead–alive, on–off, true–false

an uncountably infinite set of unordered categories is required, whereas examples of nominal variables which come to mind inherently seem to involve a finite number of categories. The examples in the discrete and binary columns seem to be selfevident and not requiring further explanation.

3.1.4   A SPECIAL FEATURE OF BINARY VARIABLES

A binary variable is simply a discrete variable which may assume only two values and so may not seem worthy of separate consideration. However, the property of having only two values gives binary variables a very special flexibility when using analysis techniques which are invariant under linear transformations.

Let u and v be a pair of distinct scores for a binary variable. By application of a linear transformation

$$f(u) = a + bu,$$

any other pair of scores x and y may be obtained as

$$x = a + bu, \quad \text{and} \quad y = a + bv.$$

The constants a and b are

$$b = \frac{(x - y)}{(u - v)}, \quad a = x - bu = y - bv.$$

Thus there is an equivalent linear transformation for any mapping from u and v to x and y. The result is that the scores for the two classes of a binary variable are arbitrary when the analysis technique is invariant under linear transformations. In effect, the class scores are used merely as class labels; hence only nominal scaling is assumed for binary variables.

Among the more familiar analysis techniques which are invariant under linear transformations are:

1. simple correlation,
2. linear regression,
3. discriminant analysis,
4. factor analysis,
5. canonical correlation,
6. any other method based on a linear model.

These techniques usually are employed with data measured on interval scales. But in the case of binary variables, the scale of definition is irrelevant by virtue of the fact that only nominal scaling is needed. Thus qualitative variables defined as dichotomies may be included directly in such analyses based on interval scales.

3.2   Scale Conversions

Most analysis techniques assume a homogeneity of scale types, whereas real data sets often feature mixed scales. One approach to handling such problems is to choose a particular scale type and then suitably transform variables to achieve homogeneity.

Scales usually are ordered in the sequence nominal, ordinal, interval, and ratio, with the progression reflecting increasing information demands for scale definition. Hence promoting a variable implies the utilization of additional information or acceptance of a new assumption. Likewise, demotion of a variable involves relinquishing some information (and sometimes use of additional information as well). In all the techniques for scale transformation discussed below, one must use critical judgement as to whether the technique is appropriate to the particular problem. Indeed, informed judgement is an alternative to all these techniques and sometimes provides the only means of taking account of important but unquantifiable considerations.

Ratio scales are excluded from the following discussion of scale conversions. None of the analysis techniques employed in the subsequent discussion of cluster analysis will require ratio scales. When ratio variables are encountered, they are treated as interval variables by simply failing to make any use of available origin information.

3.2.1   INTERVAL TO ORDINAL

The problem here is to define contiguous categories over the interval scale such that objects within a category are treated as being of equal rank while the ordinal order relation among objects in different categories is maintained. In a typical case categories are constructed by aggregating distinct objects. Information loss takes two forms:

    1. Distinctions between objects within the same class are lost or ignored.
    2. Distinctions between objects in different classes retain order properties but magnitudes of the distinctions are lost or ignored.

Most techniques are directed at minimizing information loss given some number of categories as the desired result.

It is generally not necessary to distinguish between continuous and discrete variables when considering scale conversions. In the typical case, the raw data will represent the sum total of quantitative information about the variable. The number of data points is always finite and so may be considered as discrete points extracted from an underlying continuum.

The problem of converting from interval to ordinal scales is rarely of interest in itself. However, this conversion forms a necessary step in most changes from interval to nominal scale, a very interesting case.

3.2.1.1 No Substantive Change Necessary

In some cases, transition from an interval to an ordinal scale involves nothing more than a change of viewpoint. The notion is articulated most easily through use of examples.

  1. Age is a ratio variable but is typically measured in discrete units (such as years) which naturally form ordered, common-length classes of 1, 2, 3, etc., units of age.

  2. Education is a ratio variable when measured in terms of years of education or number of highest grade completed. As with age, the use of discrete units of measurement naturally defines ordered classes.

In both cases the key element is the acceptability of the discrete unit of measurement as a uniform definition of class size. Possible objections that may arise include:

  1. The number of classes may be numerous. If children are the object of analysis the number of age classes would be small, whereas analysis of the entire population would necessarily involve perhaps 100 classes.

  2. Equal length classes may not be satisfactory. One may desire some classes to be only a fraction of the unit of measurement (e.g., under 6 months of age) while others are multiples (e.g., 17–21 years) or even of unspecified length (e.g., over 65 years).

3.2.1.2 Substitution

An occasionally useful device is to delete a variable and replace it by another variable which is related closely but measured on the desired kind of scale. The primary considerations in using this method are: (1) retain as much as possible of the relevant information included in the original variable, and (2) minimize the introduction of extraneous and perhaps misleading information.

Some illustrative examples may help to establish the idea.

  1. Education measured in terms of highest grade completed could be replaced with the ordered categories:
    a. no grammar school,
    b. some grammar school,
    c. eighth grade completed,
    d. high school graduate,
    e. bachelor's degree, etc.

2. Suppose the object of analysis is conspicuous consumption by American families. A variable of interest might be the original purchase price of the family car as obtained by direct survey or through tax records. This variable could be replaced with the ordered categories

    a. compact: Pinto, Gremlin, VW;
    b. low priced: Chevrolet, Ford, Plymouth;
    c. medium priced: Chrysler, Oldsmobile, Mercury;
    d. luxury: Rolls Royce, Cadillac, Imperial.

In both these cases the interval and ordinal variable are reasonably parallel.

3.2.1.3   Equal Length Categories

If there is no satisfactory substitute variable available, one is faced with the dual problems of deciding the appropriate number of categories and the cut points on the continuum separating classes. One fairly simple approach is to choose some number of classes and then divide the range of the variable into

	2	3	4	5	6
0					
5					
7					
9					
10					
15				19.0	15.8
20			23.7		
25					
26					
27					
29		31.6		38.0	31.6
42	47.5		47.5		47.5
48					
51					
55				57.0	
61					
62		63.3	71.1		63.3
73				76.0	
78					
79					79.2
81					
84					
85					
95					

Fig. 3.1. Partition of 24 numbers into equal length classes.

equal length intervals. In case the relevant number of categories is not obvious, a reasonable extension of this approach is to consider 2, 3, 4, etc., categories and choose the most satisfactory grouping. To illustrate, consider Fig. 3.1, where 24 random numbers have been ordered and cut into 2–6 equal length classes. From a purely structural standpoint, the three group arrangement has some appeal. It has cut points in the longest empty intervals; and even if none of the partitions are themselves entirely satisfactory, some of the subgroups may be worthy of adoption or suggest a useful overall partition.

This method amounts to constructing a series of histograms. The typical approach to constructing a histogram is to divide the range of the variable into equal classes and then count the number in each class by identifying the class membership of each observation. Using this latter approach, any questions about what would happen with some other number of classes may be answered only by repeating the entire process. However, by ordering the observations first, one can determine very quickly the location of class boundaries among the observations so that many different histograms can be examined with little more effort than just one.

3.2.1.4   Equal Membership Categories

The preceding scheme of equal length categories can result in a gross imbalance of membership among the categories. Actually only the largest and smallest values in the entire sample determine the cut points; the other values are ignored completely. Another simple alternative is to partition the data into classes with equal (as near as possible) membership. This approach might be described through adoption of a model in which any relevant classes are *a priori* equally probable.

The 24 number example used to illustrate equal length partitions is shown in Fig. 3.2. Four groups might have some appeal in this partition. Three groups seem to provide somewhat less attractive cut points. Again, this sort of attack on the problem may be most useful for suggesting some subgroups or otherwise contributing to the organization of one's thinking.

3.2.1.5   Assumed Distribution for Proportion in Each Class

A degree of apparent statistical sophistication can be achieved through distributional assumptions. The naive methods described in preceding sections have distributional explanations, as will be seen. However, the mere existence of such explanations should not be a source of enhancement for the validity or status of the method. An association with the normal distribution or any other common form is not to be confused with the relevance of a particular technique to the purpose of the particular study.

Let X be a random variable with cumulative distribution function (CDF) $F(x)$. Let X be defined over the interval $[a,b]$ such that $F(a) = 0$ and $F(b) = 1$.

0					
5					
7					
9					4
10				5	
15			6		
20					
25		8			8
26					
27				10	
29					
42	12		12		12
48					
51					
55				15	
61		16			16
62					
73			18		
78					
79				20	20
81					
84					
85					
95					
	2	3	4	5	6

Fig. 3.2. Partition of 24 numbers in equal membership classes.

Now divide the interval $[a, b]$ into n equal segments, each of length

$$L = (b - a)/n.$$

The separation points between segments are then

$$y_0 = a,$$
$$y_i = y_{i-1} + L, \qquad i = 1, \ldots, n - 1,$$
$$y_n = b.$$

The proportional of the population falling in the ith segment from y_{i-1} to y_i is $F(y_i) - F(y_{i-1})$.

These results can then be used to construct a model of the following form:

The observed data are from an underlying distribution of known form. The classes in the data are defined by dividing the abscissa into n equal length segments. The fraction of observations in the ith class is then equal to the area under the density function between the end points of the class, i.e., $F(y_i) - F(y_{i-1})$.

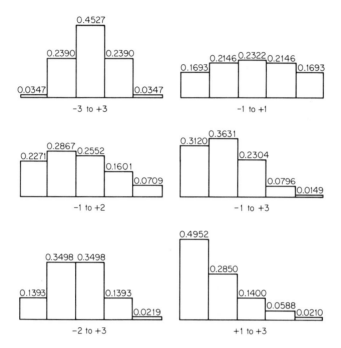

Fig. 3.3. Distribution among normal classes for various choices of truncation points.

If $F(x)$ is taken to coincide with the empirical distribution in the sample, the result is the method of equal length categories in Section 3.2.1.3. If $F(x)$ is taken as the uniform distribution, the result is the method of equal membership categories in Section 3.2.1.4. In both of these cases the distribution has a finite domain which is taken to range from the smallest to the largest of the actual observations.

Distributions with infinite domains such as the gamma and normal result in some complications. To obtain correspondence with the finite range of values in the sample it is necessary to truncate the distribution. The ultimate distribution of observations among classes is very sensitive to the choice of these truncation points. To illustrate, Fig. 3.3 shows the results of partitioning a normal distribution into 5 equal length classes for several choices of truncation points.

The numbers used in these examples may be determined from tables as follows:

    1. Calculate class boundaries as standard (zero mean, unit variance) scores.

    2. Look up the CDF at each boundary.

```
RANGE FROM-1.63 TO 2.00 DIVIDED INTO 2 TO  7 CLASSES

  2 GROUPS
CUT POINTS
  .183
PROPORTION IN EACH CLASS
  .5632   .4368
CUMULATIVE PROPORTIONS
  .5632 1.0000

  3 GROUPS
CUT POINTS
 -.422     .789
PROPORTION IN EACH CLASS
  .3081   .4843   .2077
CUMULATIVE PROPORTIONS
  .3081   .7923 1.0000

  4 GROUPS
CUT POINTS
 -.725     .183   1.092
PROPORTION IN EACH CLASS
  .1977   .3655   .3129   .1239
CUMULATIVE PROPORTIONS
  .1977   .5632   .8761 1.0000

  5 GROUPS
CUT POINTS
 -.906   -.180    .547   1.273
PROPORTION IN EACH CLASS
  .1416   .2660   .3014   .2061   .0850
CUMULATIVE PROPORTIONS
  .1416   .4076   .7090   .9150 1.0000

  6 GROUPS
CUT POINTS
 -1.028  -.422    .183    .789   1.394
PROPORTION IN EACH CLASS
  .1089   .1992   .2552   .2291   .1441   .0635
CUMULATIVE PROPORTIONS
  .1089   .3081   .5632   .7923   .9365 1.0000

  7 GROUPS
CUT POINTS
 -1.114  -.595   -.076    .443    .962   1.481
PROPORTION IN EACH CLASS
  .0879   .1547   .2093   .2175   .1737   .1066   .0503
CUMULATIVE PROPORTIONS
  .0879   .2426   .4519   .6694   .8431   .9497 1.0000
```

Fig. 3.4. Computer program output for normal classes.

3. Calculate the proportion in each class as the difference in the CDF between the upper and the lower boundaries of the class.

4. Normalize the proportion by the total area under the density between the truncation points, i.e., the difference in values of the CDF at the upper and lower truncation points.

Since this procedure can be tedious and a source of error, a short FORTRAN IV computer program for the normal case is provided in Appendix B. The user need only specify the upper and lower truncation points (as z scores) and the maximum number of classes of interest. The results include the cut points, proportion in each class and cumulative proportions. Example output is shown in Fig. 3.4.

The cumulative proportions [i.e., the $F(y_i)$] are the most convenient to use in dealing with real data. If there are m observations in the sample, then the first class has $mF(y_1)$ observations and in general $mF(y_i)$ in the first through ith classes. Using the sample output of Fig. 3.4, the 24 number example is shown grouped in Fig. 3.5.

	2	3	4	5	6
0					
5					
7				0.1416	0.1089
9					
10			0.1977		
15					
20		0.3081			0.3081
25					
26					
27				0.4076	
29					
42					
48					
51	0.5632		0.5632		0.5632
55					
61					
62				0.7090	
73					
78		0.7923			0.7923
79					
81			0.8761		
84				0.9150	
85					0.9365
95					

Fig. 3.5. Partition of 24 numbers using normal class proportions with truncation points −1.633 and 2.000.

Good estimates of the truncation points may be known from earlier analyses or as a result of theoretical considerations. In many problems there will be no such prior information and the truncation points will have to be estimated from the sample itself. The largest observation of a sample of n observations, x_n say, is a possible estimator of the upper truncation point but suffers from an obvious bias to be too small. Robson and Whitlock (1964) show that a much more satisfactory estimator is $x_n + (x_n - x_{n-1})$, where x_{n-1} is the next largest observation. The corresponding estimator for the lower truncation point is $x_1 + (x_1 - x_2)$, where x_1 is the smallest observation and x_2 the next smallest. These estimators are unbiased for the uniform distribution and much less biased than x_n and x_1 for other distributions.

The example collection of 24 numbers has a mean of 44.46 and a standard deviation of 30.28. The standardized truncation points estimated from the sample are then

$$z = \frac{0 + (0 - 5) - 44.46}{30.28} = -1.633,$$

$$z = \frac{95 + (95 - 85) - 44.46}{30.28} = 2.000.$$

These values were used for Figs. 3.4 and 3.5.

3.2.1.6   Assumed Distribution for Cut Points between Classes

In the preceding section classes were constructed by dividing the abscissa into equal intervals. Another possible way to use a distributional assumption is to divide the distribution into equal probability classes, that is, each class has the same area under the density function as every other class. In the notation of the preceding section

$$F(y_i) - F(y_{i-1}) = 1/n, \qquad i = 1, \ldots, n.$$

If the sample of m observations is taken to approximate the population closely, then the method is to assign the first m/n observations to the first class, the second m/n to the second class, and so forth. This technique is the same as the method of equal membership categories in Section 3.2.1.4.

An alternative way of using equal probability classes is as follows:

1. Assume a distributional form.
2. Estimate the distributional parameters from the sample.
3. For each candidate number of classes, say k, calculate the cumulative fraction of the sample in each class and predecessors, i.e.,

$$F(y_i) = i/k$$

for the ith class.

4. Using tables find y_i from $F(y_i)$.

5. If y_i is in standard form, use the parameters estimated in step 2 to transform y_i to a cut point in the sample. To illustrate, the 24 number example is assumed to come from an underlying normal distribution. The mean and standard deviation are $\bar{x} = 44.46$ and $s_x = 30.28$. The cut points between classes for a standard normal distribution are given in Table 3.2. These cut points (z_i say) are transformed to cut points in the sample (y_i say) by

$$y_i = \bar{x} + s_x z_i.$$

The results are shown in Fig. 3.6.

TABLE 3.2
CUT POINTS FOR EQUAL PROBABILITY STANDARD NORMAL CLASSES[a]

i \ n	2	3	4	5	6	7	8	9
1	0.0000	0.4303	0.6742	0.8415	0.9674	1.0676	1.1504	1.2208
2			0.0000	0.2529	0.4303	0.5656	0.6742	0.7645
3					0.0000	0.1797	0.3182	0.4303
4							0.0000	0.1394

i \ n	10	11	13	14	15	16	17	18
1	1.2817	1.3354	1.3832	1.4263	1.4655	1.5014	1.5344	1.5651
2	0.8415	0.9083	0.9674	1.0201	1.0676	1.1108	1.1504	1.1869
3	0.5240	0.6042	0.6742	0.7361	0.7914	0.8415	0.8870	0.9288
4	0.2529	0.3483	0.4303	0.5020	0.5656	0.6226	0.6742	0.7213
5	0.0000	0.1139	0.2100	0.2929	0.3657	0.4303	0.4884	0.5410
6		0.0000	0.0000	0.0963	0.1797	0.2529	0.3182	0.3769
7				0.0000	0.0000	0.0834	0.1570	0.2226
8							0.0000	0.0736

[a] For n classes, there are $n - 1$ cut points, symmetrically located about 0.0.

3.2.1.7   *One-Dimensional Hierarchical Linkage Methods*

The central problem in converting from interval to ordinal scale is a matter of organizing observations into groups. The techniques of cluster analysis and discriminant analysis are well suited to this problem as will be demonstrated in Sections 3.2.1.7–3.2.1.11. The application of these techniques to the one-dimensional problem of scale conversion also provides a simple introduction to the details of actually using cluster analysis and discriminant analysis with numerical data.

	2	3	4	5	6
0					
5					
7					
9					
10					11.2
15				19.0	
20			24.1		
25					
26					
27					27.4
29		31.5		36.8	
42	44.5		44.5		44.5
48					
51				52.2	
55		57.5			
61					61.6
62			64.9	70.0	
73					76.8
78					
79					
81					
84					
85					
95					

Fig. 3.6. Partition of 24 numbers using equal probability normal classes.

The first step in using cluster analysis is to choose a measure of distance or similarity between the observations. Because of the one-dimensional nature of the problem many measures collapse to a common expression. For example, consider the family of Minkowski metrics

$$D = \left[\sum_{i=1}^{i=n} |x_{ij} - x_{ik}|^L \right]^{1/L},$$

where n is the number of dimensions, j and k identify different observations, and L is the order of the metric. Since n is 1, D reduces to

$$D = |x_j - x_k|$$

for all values of L. This measure seems to be the most natural and will be adopted for use throughout this section.

The simplest clustering method is single linkage. In this one-dimensional context the method is even simpler.

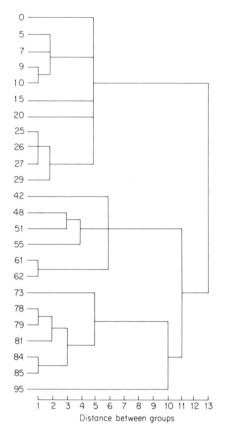

Fig. 3.7. Single-linkage clustering.

1. Order the observations in ascending sequence and treat each observation as a group with one member.

2. Examine all pairs of adjacent groups and find those two which are closest together, the distance between them being the distance between their nearest members.

3. Repeat step 2 until there is only one group.

The result of using single linkage clustering on the 24 number example is shown in Fig. 3.7. Notice that this method identifies the largest gaps in the sequence of observations. The method is sufficiently simple to be performed by hand.

A closely related clustering method is complete linkage. The algorithm for finding the cluster structure is the same as single linkage except that the distance between two groups is the distance between their farthest members,

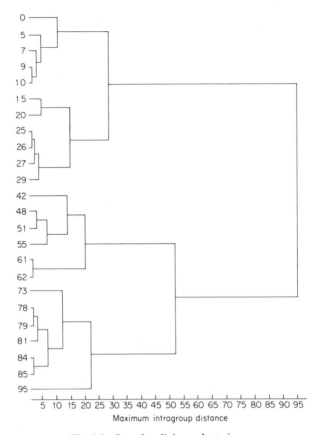

Fig. 3.8. Complete-linkage clustering.

i.e., the largest distance between two observations in the group which would result from merger of the candidate groups. Again only pairing of adjacent groups need be considered. The result of applying this method is shown in Fig. 3.8.

It is interesting to note that the single and complete linkage results give identical sequences of merges (except for ties). Such similar results may appear often in the one-dimensional case. It is even tempting to speculate that the two methods always give equivalent results in these cases. These thoughts are dispelled easily by deleting the observation 73; the change in the complete linkage case is minor, while that in the single linkage case is substantial.

3.2.1.8   Ward's Hierarchical Clustering Method

Ward (1963) and Ward and Hook (1963) describe another hierarchical clustering method based on within group variance rather than linkage. A

TABLE 3.3
WARD HIERARCHICAL CLUSTERING OF 6 NUMBERS

Number of groups	Groups	W
6	(1) (2) (5) (7) (9) (10)	0.0
5	(1, 2) (5) (7) (9) (10)	0.5
4	(1, 2) (5) (7) (9, 10)	1.0
3	(1, 2) (5, 7) (9, 10)	3.0
2	(1, 2) (5, 7, 9, 10)	17.75
1	(1, 2, 5, 7, 9, 10)	67.33

brief description of the one-dimensional case will suffice for the present as the method and some variations are described fully in Chapter 6.

Suppose a sample of n observations has been partitioned into g groups, the ith group containing n_i observations with mean $\bar{x}_i$. The within group sum of squared deviations for this g-group partition is

$$W = \sum_{i=1}^{i=g} \sum_{j=1}^{j=n_i} (x_{ij} - \bar{x}_i)^2, \tag{3.1}$$

where x_{ij} is the jth observation in the ith group. The value of W may be calculated for any partition. The quantity W/n is the pooled within group variance for the partition.

The Ward clustering technique proceeds as follows:

1. Begin with n groups, each group consisting of one observation. At this stage $W = 0$.

2. At each stage reduce the number of groups by one through merger of those two groups whose combination gives the least possible increase in W.

3. Continue for a total of $n - 1$ mergers until there is only one group.

This technique tends to give minimum W partitions for each number of groups from n to 1. The nesting of groups to form larger groups sometimes precludes finding the minimum W partition as in the following example. Suppose a sample of six observations gives the scores 1, 2, 5, 7, 9, and 10. The Ward clustering technique will give the sequence of groupings shown in Table 3.3.

It may be readily verified that for two groups the partition (1, 2, 5) (7, 9, 10) gives $W = 13.0$, a somewhat better result than obtained with the Ward method. The problem arises because the formation of three groups involves the three alternatives:

$$(1, 2) \ (5, 7) \ (9, 10) \quad W = 3.0,$$
$$(1, 2) \ (5) \ (7, 9, 10) \quad W = 5.67,$$
$$(1, 2, 5) \ (7) \ (9, 10) \quad W = 9.33.$$

Either of the last two alternatives will lead to the optimum partition for two groups, but they are inferior to the first and are rejected at the three group stage. Most hierarchical clustering methods are subject to this sort of difficulty, the single linkage method being a notable exception.

It could be embarrassing to offer the partition (1,2) (5,7,9,10) and have a casual observer simply move one observation and produce a substantial improvement. Such shortcomings of the Ward technique may be countered effectively by using the Ward result as a trial partition and subjecting it to the iterative improvement technique described in Section 3.2.1.9.

In most cases the Ward technique involves sufficient computation to discourage a pencil and paper approach. However, the computer program for multidimensional clustering given in Appendix E performs very nicely on one-dimensional problems. The result of using this program on the 24 number example gives a sequence of merges differing only slightly from that obtained with complete linkage. In much larger problems there likely would be substantial differences between the products of the two methods.

3.2.1.9   Iterative Improvement of a Partition

Among the principal nonhierarchical clustering methods is a class of iterative techniques which begin with an initial partition and try to make improvements through reassigning observations from one class to another. Various versions of this general approach are described in Chapter 7. In a one-dimensional form many of these methods reduce to the particular technique described in this section or to some equivalent procedure.

The principle of least squares is the most common statistical approach to fitting a model to observed data. In the present one-dimensional grouping problem this principle is applied in the form of minimizing the sum of squared deviations about the group means, the W function of Eq. (3.1). Fisher (1958) and Ericson (1964) prove that one-dimensional least-squares partitions are necessarily contiguous, i.e., each group corresponds to a single interval disjoint from all other groups. For a sample of n observations to be divided into g groups, this result reduces the number of possible partitions from a Stirling number of the second kind (Abramowitz and Stegun, 1968)

$$\mathfrak{S}_n^{(g)} = \frac{1}{g!} \sum_{k=0}^{k=g} (-1)^{g-k} \binom{g}{k} k^n$$

to a binomial coefficient (Fisher, 1958),

$$_{n-1}C_{g-1} = \binom{n-1}{g-1}.$$

For 24 observations and 5 groups the reduction is from 4.85×10^{14} to 42,504. While the number of possibilities is reduced by 10 orders of magnitude, it is still rather unappealing to enumerate the remainder just to find the best least-squares partition of 24 observations into 5 groups! Fisher (1958) refers to a computer program which is somewhat more efficient than total enumeration, but he gives too few details to permit writing another program with comparable capabilities.

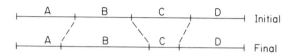

Fig. 3.9. Movement of interclass boundaries.

Because a least-squares partition is contiguous, a partition into g groups will involve $g - 1$ intergroup boundaries. Beginning with some partition of the sample into g groups, the optimum partition then can be reached simply by moving intergroup boundaries. For example, in Fig. 3.9 an initial partition of equal length segments is modified by moving boundaries. Note that a group with only one member makes zero contribution to the W function; consequently no group will vanish in such a process. The question then is what procedure should be used to move boundaries? A simple and intuitively appealing approach is as follows:

1. Choose an interclass boundary to test (for definiteness begin with the leftmost boundary).

2. Try moving the boundary one observation to the right. If that move fails to reduce W, try a move to the left. If movement in either direction reduces W, then continue with single moves in the successful direction until no further improvement is possible.

3. Repeat step 2 for each of the remaining boundaries.

4. Repeat steps 1–3 until every boundary is at a minimum, that is, movement of one observation in either direction fails to improve the W function.

The simplicity of this algorithm lies in the consideration of only one observation at a time and making the obvious improvements. Herein also lies its weakness: the single move technique insures only a local minimum. Further improvements may be possible by considering simultaneous movement of multiple observations at possibly several boundaries. But, explicit consideration of these possibilities involves another combinatorial explosion and tends toward total enumeration. Even though a global minimum cannot be guaranteed by using single moves, improvement of any local minimum involves sufficiently complex changes to the partition that it would be most remarkable for a casual observer to spot a better partition.

The computational aspects of this approach may be carried out using the W function of Eq. (3.1). However, in this one-dimensional case a more convenient computational procedure is available. The fundamental sums of squares identity in the analysis of variance is

$$\sum_{i=1}^{i=g} \sum_{j=1}^{j=n_i} (x_{ij} - \bar{x})^2 = \sum_{i=1}^{i=g} \sum_{j=1}^{j=n_i} (x_{ij} - \bar{x}_i)^2 + \sum_{i=1}^{i=g} n_i(\bar{x}_i - \bar{x})^2$$

or

$$T = W + B.$$

The notation is the same as used in Eq. (3.1) except for the addition of $\bar{x}$ as the mean of the entire sample. Obviously T is a constant for any partition of a given sample. It follows then that minimizing W is equivalent to maximizing B, the criterion used in this development. Since only a single boundary is moved at each iteration only two terms (1 and 2 say) in the B sum of squares are affected. It is easy to show then

$$n_1(\bar{x}_1 - \bar{x})^2 + n_2(\bar{x}_2 - \bar{x})^2 = (n_1 \bar{x}_1)^2/n_1 + (n_2 \bar{x}_2)^2/n_2 - (n_1 + n_2) \bar{x}^2.$$

The last term on the right is unaffected by the partition of the $n_1 + n_2$ observations between the two groups. Hence, moving an observation x^* from group 1 to group 2 gives an increase in B if

$$\frac{(n_1 \bar{x}_1 - x^*)^2}{n_1 - 1} + \frac{(n_2 \bar{x}_2 + x^*)^2}{n_2 + 1} > \frac{(n_1 \bar{x}_1)^2}{n_1} + \frac{(n_2 \bar{x}_2)^2}{n_2}.$$

The numerator of each term is simply the squared sum of all observations in the group, while the denominator is the number of observations in the group. These quantities conveniently characterize the groups and make for particularly simple implementation of the single move technique. A FORTRAN IV version of the algorithm is given in Appendix B along with instructions for its use. The results of applying this technique to the 24 number example are shown in Fig. 3.10. The initial grouping for each case was an equal membership partition.

3.2.1.10   *The Linear Discriminant Function*

The most commonly described case of multiple group discriminant analysis may be stated as follows (Anderson, 1958, Chap. 6): Let $\mu_1, \mu_2, \ldots, \mu_g$ be the mean vectors of g multinormal populations with common covariance matrix Σ. The populations are present in the grand ensemble in proportions $p_1, p_2, \ldots, p_g$. If an observation is from population j and is classified wrongly as being from population i, then a loss $L(i:j)$ is incurred. With these definitions,

	2	3	4	5	6
0					
5					
7					
9				4	4
10					
15			6	2	2
20					
25					
26					
27					
29		11	5	5	5
42	12				
48					
51					
55					4
61					
62		6	6	6	
73					3
78					
79					
81					
84					
85					
95	12	7	7	7	6
g	2	3	4	5	6
W	4186.9	1617.9	760.2	671.7	454.1

Fig. 3.10. Improving an equal membership partition (showing size of each group).

the optimal Bayes classification rule is to assign a new observation x to that class i such that

$$x^T \Sigma^{-1}(\mu_i - \mu_j) - \tfrac{1}{2}(\mu_i + \mu_j) \Sigma^{-1}(\mu_i - \mu_j) \geqslant \ln \frac{p_j L(i:j)}{p_i L(j:i)}, \qquad (3.2)$$

for all $j = 1, \ldots, g, j \neq i$. If x is unidimensional and sample estimates replace population parameters, the inequality (3.2) becomes

$$\frac{x(\bar{x}_i - \bar{x}_j) - \tfrac{1}{2}(\bar{x}_i + \bar{x}_j)(\bar{x}_i - \bar{x}_j)}{s_x^2} \geqslant \ln \frac{p_j L(i:j)}{p_i L(j:i)}. \qquad (3.3)$$

The rule in effect says to make all pairwise comparisons between populations and choose that one which dominates all the others. Comparing two populations, say i and j, in the one-dimensional case may be accomplished by calculating the cut point between classes. In this case, population i is preferred to population j for all x such that

$$x \geqslant C_{ij} = \frac{s_x^2}{(\bar{x}_i - \bar{x}_j)} \ln \frac{p_j L(i:j)}{p_i L(j:i)} + \frac{(\bar{x}_i + \bar{x}_j)}{2}, \qquad (3.4)$$

assuming $\bar{x}_i > \bar{x}_j$.

When prior estimates for the p_i are not available, one might consider the hypothesis of equally likely groups ($p_i = 1/k$). An alternative to the latter is to estimate the p_j/p_i from the data as

$$p_j/p_i = n_j/n_i \qquad (3.5)$$

where n_i is the number of observations in the ith class. Another frequent assumption is that all misclassifications are equally costly, that is

$$L(i:j) = L, \qquad \text{for all } i \text{ and } j.$$

In case the *expected* costs of misclassification are equal, that is

$$p_j L(i:j) = p_i L(j:i),$$

the first term in Eq. (3.4) is zero. The cut point between a pair of classes simply reduces to the midpoint between their means in this case.

This specialization of discriminant analysis provides a fairly simple means of categorizing an interval variable. All that is required is a series of matching measurements on both the interval variable of interest and some categorical variable. The latter variable might be chosen to have some close association with the interval variable or it may be chosen so as to have characteristics which should be reflected in the categorized version of the interval variable. For example, the objects of analysis might be some segment of the population, the interval variable annual income, and the ordinal variable education level (not a high school graduate, high school graduate, BA, MA, PhD). Then, the discriminant analysis technique would partition the income scale so as to predict education level with minimal loss.

The now familiar 24 number example has been augmented with a three group ordinal variable as shown in Fig. 3.11. The class means and class sizes are

$$\bar{x}_A = 14.6, \qquad \bar{x}_B = 52.0, \qquad \bar{x}_C = 73.4,$$
$$n_A = 10, \qquad n_B = 5, \qquad n_C = 9.$$

The sample variance is 917.1. Assuming equal expected costs of misclassification, Eq. (3.4) gives

$$C_{BA} = (52.0 + 14.6)/2 = 33.3,$$
$$C_{CB} = (73.4 + 52.0)/2 = 62.7,$$
$$C_{CA} = (73.4 + 14.6)/2 = 44.0.$$

The resulting rule is to assign the following intervals to each class:

$$A: (0, 33.3) \qquad B: (33.3, 62.7) \qquad C: (62.7, 95).$$

The cut point $C_{CA} = 44.0$ is suppressed since the B class effectively separates the A and C classes.

Now suppose the costs of misclassification are all equal and the proportions for the populations are estimated using Eq. (3.5). Then Eq. (3.4) gives

$$C_{BA} = \frac{917.1}{52.0 - 14.6} \ln \frac{10}{5} + 33.3 = 50.3,$$

$$C_{CB} = \frac{917.1}{73.4 - 52.0} \ln \frac{5}{9} + 62.7 = 37.5,$$

$$C_{CA} = \frac{917.1}{73.4 - 14.6} \ln \frac{10}{9} + 44.0 = 45.6.$$

In this case only two groups result because there is no region in which B is preferred to *both* A and C; B is preferred to A only if x is greater than 50.3 while B is preferred to C only if x is less than 37.5. The elimination of one group may be either acceptable or objectionable depending on the particular purpose of the study and the role the categorized variable is to play.

The reference variable in the example might be taken to have its classes ordered as $C > B > A$. This assumption might seem reasonable since the class means have this sequence. However, any other ordering (or even no ordering at all) would give exactly the same results. Therefore the reference variable need not fulfill any requirements other than the researcher's satisfaction with the kind of partition it induces.

Notice that perfect discrimination is not achieved. If it could be achieved then the reference variable might be substituted directly. The cut points resulting from this procedure depend on the assumptions of normality, equal variance, equal misclassification costs and the approach for estimating the proportional representation of the populations in the grand ensemble. These cut points may not perform well on the sample, especially when the sample is small, as in the example. As a consequence, it might be more satisfactory in some instances to identify a minimum misclassification partition by inspection. One such partition is shown in the last column of Fig. 3.11.

3.2.1.11   Cochran and Hopkins Method

In the preceding section the classes constructed from the interval variable were in one-to-one correspondence with the classes of the reference categorical variable. Cochran and Hopkins (1961) provide another technique based on discriminant analysis. They were concerned with discrimination using qualitative variables and found it necessary to consider the problem of categorizing a normal variate so it could fit with the rest of the variables.

Interval value	Ordinal class	Equal expected losses	Estimated proportions	By inspection
0	A			
5	A			
7	A			
9	A	A-10	A-10	A-9
10	A	B-1	B-1	B-0
15	A	C-0	C-1	C-0
20	A			
25	A			
26	A			BA
27	B			
29	A	BA 33.3		
42	C		CA 45.6	
48	B	A-0		
51	B	B-3		A-1
55	C	C-3		B-5
61	B			C-3
62	C	CB 62.7		
73	B			CB
78	C	A-0	A-0	
79	C	B-1	B-4	A-0
81	C	C-6	C-8	B-0
84	C			C-6
85	C			
95	C			
Misclassifications		5	6	4

Fig. 3.11. Discriminant based partitions.

Suppose an interval variable is normally distributed in each of two classes with means $\bar{x}_1$ and $\bar{x}_2$ and common variance s_x^2. The question which Cochran and Hopkins consider is: Assuming equal costs and probabilities of misclassification, how should the interval variable be categorized into n groups to retain maximum discriminability? Their results may be stated in the following manner (different from that used in their paper): Calculate the ith cut point between classes as

$$C_i = (\bar{x}_1 + \bar{x}_2)/2 + c_i s_x,$$

where c_i is given in Table 3.4. Cochran and Hopkins consider no more than six classes because this level of division retains over 94% of the discriminating power.

TABLE 3.4
VALUES OF c_i AND RELATIVE POWER

i \ n	2	3	4	5	6
1	0.0	−0.6	−1.0	−1.2	−1.4
2		0.6	0.0	−0.4	−0.7
3			1.0	0.4	0.0
4				1.2	0.7
5					1.4
Power	0.636	0.810	0.882	0.920	0.942

To illustrate use of the technique, the 24 number example has been augmented with a two-class interval variable and categorized as shown in Fig. 3.12.

Interval value	Ordinal class	2	3	4	5	6
0	A					−0.7
5	A				5.6	
7	A					
9	B					
10	A			11.7		
15	A					
20	B		23.9			20.8
25	A					
26	B					
27	B					
29	A	41.98		41.98	29.9	41.98
42	B					
48	A					
51	A				54.1	
55	B		60.1			
61	B					
62	B			72.3		63.2
73	B					
78	B				78.4	
79	B					
81	A					
84	B					84.7
85	B					
95	B					
		2	3	4	5	6

Fig. 3.12. Categorization using the Cochran and Hopkins method.

3.2.2   Interval to Nominal

The problem in this case is very similar to that of converting from interval to ordinal scales as in Section 3.2.1. In fact, every one of the techniques described in the latter section may be employed on the present problem; since their usage is identical in both cases the descriptions of Section 3.2.1 will not be repeated.

When the techniques of interval to ordinal conversion are used the resulting classes are constrained to satisfy some order restriction. The strongest manifestation of an order restriction is that each class is connected and contiguous, i.e., it consists of a single interval in the one-dimensional case. However, Fisher (1958) and Ericson (1964) prove that one-dimensional least squares partitions are necessarily contiguous; as a consequence it becomes fairly difficult to state rational principles which fail to give connected and contiguous classes.

At least one really distinctive principle gives noncontiguous classes. In many instances it is quite meaningful to consider some observations as central while others are extreme. Alternative descriptions include routine–exceptional and near–far. The extremes typically occur in both positive and negative or high and low senses. For example, persons might be measured in terms of how much they exceed or fall short of an ideal weight for their age and height. From the viewpoint of the need for professional medical or dietary help, the following classes might be of interest:

−20 to +20	Normal
−35 to −21 and +21 to +35	In need of counseling
less than −35 and more than +35	Remedial action indicated

These classes have an order property, but it is quite different from that in the original variable. The classes definitely are not contiguous or connected on the original interval scale.

3.2.3   Ordinal to Nominal

Given a set of ordered classes, a natural way to obtain nominal classes is to forget the order properties and keep the same classes. However, the absence of the order restriction permits use of the central-extreme principle as in Section 3.2.2. It may also be useful to consider degrees of extremes. Thus if class C is the central class in the sequence $A > B > C > D > E$, one might adopt the revised classifications:

C	normal
B ∪ D	moderately deviant
A ∪ E	extreme.

Even though no example seems readily apparent, there may exist situations in which discrete ordinal variables might be reorganized by using an odd–even principle; that is, the odd numbered elements are assigned to one class while the even numbered elements are assigned to the other. The principle of substituting a closely related variable of the desired type may also produce reorganizations prohibited by the prior order restriction.

Aside from these few instances it is quite difficult to identify principles of reorganization which capitalize on the easing of order restrictions. However, in any given situation the analyst may be able to marshall supplementary information or apply his judgement to define classes which are suited uniquely to the problem at hand but defy description in any abstract terms.

3.2.4 NOMINAL TO ORDINAL

The essential problem here is to impose an ordering on nominal classes. Aside from simply substituting a related ordinal variable or applying informed judgement, the only available approaches involve use of a reference variable. In retrospect it is not at all surprising that the reference variable is a necessary ingredient in solving this scale transformation problem. Unordered categories can be given an order only by introduction of new information which links the categories with an external ordering. If the ordering is manifested in the form of a variable, it is most natural to link it with the nominal variable through their joint responses in the empirical sample.

When the order restriction was relaxed in conversions from ordinal to nominal scales, there was the opportunity to merge disconnected classes which previously were separated in deference to the order requirement. This same principle is applicable in reverse when making a nominal to ordinal transformation. That is, it may be necessary to break down or decompose classes before they are suitable for ordering. For example, if the classes are central and extreme, then the extreme class will have to be split into low and high before the ordering low, central, high can be achieved. Identification of such cases is dependent primarily on the particular purpose of the analysis, additional information about the data units and the vigilance of the analyst.

3.2.4.1 *Correlation with an Interval Variable*

Suppose scores are assigned to nominal classes such that every item in a given class receives the same score. These scores may be used as interval values to calculate the ordinary product-moment correlation with some interval variable. What scores will give the maximum correlation? As shown in Section 3.2.6.2 the optimal scores are the class means on the reference variable. This technique both orders and spaces the categories so as to provide an effective means of making a nominal to interval scale transformation. In the present case only order information is of interest and the spacing may be ignored. See Section 3.2.6.2 for an example illustrating use of the technique.

3.2.4.2   Rank Correlation and Mean Ranks

If N data units have been measured on variables X and Y, both of which have been transformed to ranks so that the ith data unit is characterized by ranks x_i and y_i, then the Spearman rank correlation between X and Y is

$$r = 1 - 6 \frac{\sum_{i=1}^{i=N} (x_i - y_i)^2}{N(N^2 - 1)}.$$

Now let X be a nominal variable of g groups, n_i observations in the ith group, and Y be an ordinal reference variable. Define x_i as the common rank of all elements in the ith nominal class and y_{ij} as the Y rank of the jth element in the ith class. The rank correlation may then be written as

$$r = 1 - 6 \frac{\sum_{i=1}^{i=g} \sum_{j=1}^{j=n_i} (x_i - y_{ij})^2}{N(N^2 - 1)}.$$

Now consider the problem of ordering the nominal classes through assignment of class ranks so as to maximize the rank correlation with the ordinal reference variable Y. Clearly maximizing the rank correlation is equivalent to minimizing

$$\sum_{i=1}^{i=m} \sum_{j=1}^{j=n_i} (x_i - y_{ij})^2$$

or

$$\sum_{i=1}^{i=m} \left[n_i x_i^2 - 2n_i x_i \bar{y}_i + \sum_{j=1}^{j=n_i} y_{ij}^2, \right]$$

where $\bar{y}_i$ is the mean Y rank in the ith nominal class. The last term in brackets is a constant for all orderings since the Y ranks are fixed; only the x_i values are in question. Therefore the problem may now be stated as: assign the integers $1, 2, \ldots, m$ to $x_1, x_2, \ldots, x_m$ so as to minimize

$$C = \sum_{i=1}^{i=m} n_i x_i (x_i - 2\bar{y}_i). \tag{3.6}$$

One way of finding the ordering giving the highest rank correlation is to simply enumerate all $m!$ distinct orderings and choose the one with the smallest value of the reduced criterion C given in Eq. (3.6). An alternative approach is to begin with some ordering and attempt to improve the criterion value through iterative changes of the order. In particular consider the case of swapping the ranks of two classes, say j and k. Since a swap of ranks will affect only two of the m terms in Eq. (3.6), the swap is advantageous if

$$n_j x_j(x_j - 2\bar{y}_j) + n_k x_k(x_k - 2\bar{y}_k) > n_j x_k(x_k - 2\bar{y}_j) + n_k x_j(x_j - 2\bar{y}_k),$$

or after rearranging, if

$$(x_j - x_k)[2(n_j \bar{y}_j - n_k \bar{y}_k) - (n_j - n_k)(x_j + x_k)] < 0. \tag{3.7}$$

Notice that $n_j \bar{y}_j$ is merely the total of the Y ranks for all elements in the jth class. If $n_j = n_k$ then the swap criterion reduces to

$$(x_j - x_k)(\bar{y}_j - \bar{y}_k) < 0,$$

which may occur only if the two factors are of opposite sign, a condition that may occur only if

$$x_j > x_k \quad \text{and} \quad \bar{y}_j < \bar{y}_k \quad \text{or} \quad x_k < x_j \quad \text{and} \quad \bar{y}_j > \bar{y}_k.$$

It follows then that no pairwise swaps can improve the rank correlation if the x_i ranks are assigned in the same sequence as the $\bar{y}_i$ means, provided the classes are all the same size. This simple device of ordering the X classes by the $\bar{y}_j$ values will give a very good initial ordering when the classes vary in size. Indeed, the result may be more appealing than the ordering which gives the highest rank correlation. Consider the example of 9 items shown in Table 3.5.

TABLE 3.5
RANK CORRELATION EXAMPLE

Item	Nominal Class	Y rank
1		1
2		2
3		3
4	A	4
5		5
6		6
7		7
8	B	8
9		9

Intuitively (and by the mean Y rank criterion), class A should be assigned rank 1 and class B rank 2. However, the criterion (3.7) for rank swap is

$$(2 - 1)[2(17 - 28) - (2 - 7)(2 + 1)] = -7 < 0,$$

which indicates that the opposite ordering gives a higher rank correlation. Indeed, the rank correlation corresponding to $x_A = 1$ and $x_B = 2$ is 0.756, while that corresponding to the assignment $x_A = 2$ and $x_B = 1$ is 0.765. The problem here is the size disparity between the classes. This example should serve as a warning to examine critically the reasonableness of an ordering that departs from the order of the mean ranks.

3.2.5   ORDINAL TO INTERVAL

Each ordinal class consists of a collection of data units indistinguishable from each other, at least in regard to the ordinal variable. The classes themselves are ordered and may be thought of as being ranked. All the members of a given class are tied in rank and so are treated identically. With this collective view in mind, the problem is to assign a score to each class which preserves the rank order and exhibits magnitude differences among the classes, these differences being meaningful in some relevant sense. From another view, the problem is to locate on an interval scale as many points as there are classes (each point representing the coincidence of possibly many data units) with the sequence of points respecting the prior order.

3.2.5.1   Class Ranks

A simple and intuitive approach to assigning scores to ordered classes is simply to use class ranks. One possible interpretation of this approach is that the classes are spaced equally along the interval scale. This approach may also be viewed as the inverse of the method of equal length categories described in Section 3.2.1.3.

While the use of ranks as interval scores may seem a little naive, it turns out that some statistical techniques are remarkably robust to their use. In particular, the product moment correlation coefficient is quasi-invariant to order-preserving (monotone) transformations. That is, given some set of "true scores" the correlation between the "true scores" and ranks is likely to be very high. In turn, the correlation with some other variable is likely to be little dependent on whether ranks or "true scores" are used in the calculation.

Labovitz (1967) suggests the use of ranks, and Labovitz (1970) attempts to justify empirically the technique with correlations between some real data and ordered sequences of random numbers. In his examples ranks correlate with the real data and the random numbers very highly indeed, 0.97 in the worse case. The approach is criticized by Mayer (1970, 1971b), Schweitzer and Schweitzer (1971), and Vargo (1971). Labovitz (1971) replies to these objections.

The critics point out that the use of ranks can seriously underestimate the correlation if the "true scores" are exponential or logarithmic transformations

TABLE 3.6
CORRELATIONS BETWEEN x AND e^x

N	5	10	15	20	50	100
r	0.88628	0.71687	0.60992	0.53870	0.35252	0.25203

of ranks. That is, x and $\ln x$ correlate poorly as do x and e^x, especially as the number of ranks becomes large. Schweitzer and Schweitzer give the correlation between ranks (x) and the exponential function of ranks (e^x) for selected values of N (the number of ranks) as shown in Table 3.6. Labovitz's critics consider these results to be a telling indictment of the use of ranks. However, anytime it might be suspected that the use of ranks will underestimate correlation, a simple remedy is to try $\ln x$ and e^x as alternative scores and choose the set giving the largest correlation. The stability of the correlation coefficient (as demonstrated by Labovitz) combined with the trial of three radically different scoring systems should result in only small errors of underestimation.

To illustrate the approach, suppose each observation in a set of 24 is scored on both an ordinal variable with 6 classes and an interval variable (Y), the scores on the latter being the same as in previous examples. Using the interval variable for comparison, the correlation is calculated for ranks, logarithms of ranks and exponentials of ranks as shown in Table 3.7. The correlation of class ranks with Y is remarkably high, so high that one would not normally bother to check $\ln x$ and e^x. The correlations for these latter two transformations are also rather high, primarily because so few ranks are involved.

In contrast to the purely empirical approach of Labovitz, Stuart (1954) gives a mathematical treatment of the effects of using ranks. His results may be stated briefly as follows. Suppose N samples of size n are drawn from a continuous distribution with finite variance and the members of each sample are ranked from 1 to n. The correlation between variate values and ranks may be calculated for the entire set of nN observations. As the number of samples N increases without bound, this correlation approaches a limit which Stuart dubs "the correlation between variate values and ranks for samples of n" and denotes as C_n. He further shows that for any continuous distribution

$$C_n = \left(\frac{n-1}{n+1}\right)^{1/2} C,$$

where C is the limit of C_n as the sample size n increases without bound. For the uniform distribution the value of C is 1, while for the normal distribution C is $(3/\pi)^{1/2} = 0.977$. For the gamma distribution with density function

$$f(x) = \frac{1}{\Gamma(m)} e^{-x} x^{m-1},$$

TABLE 3.7
CORRELATIONAL COMPARISON OF RANKS,
LOGARITHMS, AND EXPONENTIALS AS SCORES

Y	Ranks (x)	$\ln x$	e^x
0	1	0	2.718
5	1	0	2.718
7	1	0	2.718
9	1	0	2.718
10	1	0	2.718
15	1	0	2.178
20	2	0.693	7.389
25	2	0.693	7.389
26	2	0.693	7.389
27	3	1.099	20.086
29	3	1.099	20.086
42	3	1.099	20.086
48	3	1.099	20.086
51	4	1.386	54.598
55	4	1.386	54.598
61	5	1.609	148.41
62	5	1.609	148.41
73	5	1.609	148.41
78	5	1.609	148.41
79	5	1.609	148.41
81	6	1.792	403.43
84	6	1.792	403.43
85	6	1.792	403.43
95	6	1.792	403.43
r with Y	0.978	0.946	0.846

the value of C depends on m as

$$C = (3m/\pi)^{1/2}\,\Gamma(m + \tfrac{1}{2})/\Gamma(m + 1).$$

Stuart illustrates the effect of varying m with the values (error for $m = 2$ corrected) shown in Table 3.8 and the following limits:

$$\lim_{m \to \infty} C = (3/\pi)^{1/2} = 0.977, \qquad \lim_{m \to 0} C = 0.$$

Evidently C is degraded severely by small values of m but is fairly stable for large values.

The value of C is important primarily for establishing the value of C_n. For samples of size n the correlation between variate values and ranks is asymptotically C_n. The values of C for the normal and uniform distributions

TABLE 3.8
EFFECT OF m ON C

m	$\frac{1}{2}$	1	2	3	4
C	0.779	0.866	0.917	0.936	0.946

indicate that ranks will be very satisfactory substitutes for variate values. The values of C_n for various values of n make the point as shown in Table 3.9.

On the other hand, the results with the gamma distribution show that the use of ranks is not without its pitfalls. Indeed, by suitable choice of m the correlation between ranks and variate values may be made arbitrarily small.

TABLE 3.9
EFFECT OF SAMPLE SIZE FOR THE UNIFORM AND NORMAL DISTRIBUTIONS

n	5	10	15	20	25
C_n (uniform)	0.816	0.904	0.935	0.951	0.960
C_n (normal)	0.797	0.884	0.914	0.930	0.939

This latter possibility is another manifestation of the phenomenon observed earlier as the poor correlation between ranks and logarithmic or exponential transformations. Aitken (1966) and Aitken and Hume (1966) extend Stuart's results to truncated normal distributions.

3.2.5.2   Expected Order Statistics

A classical method of assigning scores to ordinal data is to use expected order statistics. Let $x_1, x_2, \ldots, x_n$ be a random sample from a population with continuous cumulative distribution function $F(x)$. Because $F(x)$ is continuous, the observations are all distinct and so may be ordered uniquely. The ith order statistic for the sample is defined as the ith largest of the n observations. The expectation or mean of the ith order statistic is a well-defined quantity, though it may be somewhat tedious to calculate. Their calculation for the normal distribution is treated by Harter (1961). A concise summary of statistical theory for order statistics is provided by Gibbons (1971, Chapter 2).

In the case of the normal distribution the computational procedures need not be a matter of concern since tables have been compiled and published.

The expected order statistics are known as normal scores and tabulated in the following four references:

1. Harter (1961). Tabulates normal scores for up to 400 classes with accuracy to five decimal places.
2. David *et al.* (1968). Duplicates the table in Harter (1961) and discusses both background theory and computational considerations.
3. Fisher and Yates (1953). See Table XX, "Scores for Ordinal (or Ranked) Data," pp. 58–59. Tabulates normal scores for up to 50 classes with accuracy to two decimal places.
4. Pearson and Hartley (1954). See Table 28, "Mean Positions of Ranked Normal Deviates (Normal Order Statistics)," p. 175. Gives normal scores with three decimal place accuracy for up to 20 classes, two decimal place accuracy for up to 50 classes.

In the case of the uniform distribution, expected order statistics are simply ranks or linear transformations of ranks. The use of ranks was discussed in Section 3.2.5.1.

While expected normal order statistics may have some appeal because of their distributional association, they give only slightly different results from those obtained with ranks, at least in the case of correlational analyses. To illustrate, the correlation between normal scores and ranks was computed under the assumption all classes are of the same size. The computed correlations are presented in Table 3.10. It appears that the additional effort of using normal scores will produce few rewards over those obtained with ranks.

TABLE 3.10
CORRELATIONAL COMPARISON
OF NORMAL SCORES AND RANKS

Number of Classes	Correlation
2	1.00000
3	1.00000
4	0.99917
5	0.99813
6	0.99708
7	0.99609
8	0.99518
9	0.99435
10	0.99359
11	0.99289
12	0.99226
13	0.99168
14	0.99114
15	0.99065

3.2.5.3 *Assuming Underlying Distribution for the Data Units*

Using ranks or normal scores makes no use of the relative sizes of the classes. Even realizing that relative class sizes might be a sampling artifact, it is intuitively appealing that very large classes should be associated with long segments of the derived interval scale, while the small classes should be associated with short segments. This notion can be put into practice by supposing that the data units were drawn from an underlying distribution and that the observed categories are sections from the distribution. Class scores are then obtained by sectioning the assumed distribution using the proportions observed in the sample and calculating for each section a summary statistic such as the mean or median.

Let n_k be the number of sample observations, out of the total of n, falling in the first k classes. The proportion of the population in the first k classes is estimated as $p_k = n_k/n$. The variate value x_k corresponding to the upper boundary of the kth class is the solution to $p_k = F(x_k)$, where $F(x)$ is the CDF of the assumed distribution. The mean for the kth class is then

$$\bar{x}_k = \int_{x_{k-1}}^{x_k} xf(x)\,dx/[F(x_k) - F(x_{k-1})], \qquad (3.8)$$

where $f(x)$ is the density corresponding to $F(x)$. The choice of x_0 depends on the assumed distribution. The median for the kth class, m_k, is the solution to

$$(p_k + p_{k+1})/2 = F(m_k). \qquad (3.9)$$

In all cases $p_0 = 0$. As a matter of convenience, it will usually suffice to choose $F(x)$ as being in standard form. This method may be thought of as the inverse of the method described in Section 3.2.1.5. Incidentally, Hamdan (1971a) proves that the x_k solutions to $p_k = F(x_k)$ are maximum likelihood estimators for the class boundaries in the assumed distribution; the x_k are therefore consistent, asymptotically efficient, and asymptotically normal.

For the uniform distribution over the unit interval

$$F(x) = x, \qquad 0 \leqslant x \leqslant 1,$$

it is easy to show that

$$\bar{x}_k = m_k = \tfrac{1}{2}(p_k + p_{k-1}).$$

For the standard normal distribution

$$f(x) = \frac{1}{(2\pi)^{1/2}} e^{-x^2/2},$$

TABLE 3.11
CLASS BOUNDARIES AND SCORES FROM DISTRIBUTIONAL ASSUMPTIONS

Class			Uniform		Normal		
k	n_k	p_k	x_k	Means	x_k	Means	Medians
1	13	0.13	0.13	0.065	−1.13	−1.62	−1.51
2	5	0.18	0.18	0.155	−0.92	−1.01	−1.02
3	6	0.24	0.24	0.210	−0.71	−0.81	−0.81
4	8	0.32	0.32	0.280	−0.47	−0.59	−0.58
5	12	0.44	0.44	0.380	−0.15	−0.31	−0.31
6	16	0.60	0.60	0.520	0.25	0.05	0.05
7	7	0.67	0.67	0.635	0.44	0.35	0.35
8	15	0.82	0.82	0.745	0.92	0.67	0.66
9	8	0.90	0.90	0.860	1.28	1.07	1.08
10	10	1.00	1.00	0.950	∞	1.86	1.65

the class means assume the simple form

$$\bar{x}_k = [f(x_{k-1}) - f(x_k)]/[F(x_k) - F(x_{k-1})].$$

To determine class medians it is necessary to resort to Eq. (3.9). Both the means and medians may be easily determined from standard normal tables. Special tables for normal means and medians are provided by David *et al.* (1968). Tables for normal section means alone are given by Sandon (1961, 1962).

The use of this technique is illustrated in Table 3.11. Ten classes involving a total of 100 observations are assigned scores using the methods just developed. It is interesting that normal means and normal medians coincide closely except for the stronger tendency of the normal means to isolate the extreme classes.

3.2.5.4   *Correlation with a Reference Variable*

Another approach to choosing class scores is to select values which maximize the correlation of the ordinal variable with some reference variable. For an interval reference variable the optimal score (disregarding order) for each class is the mean score of the class members on the interval variable as shown in Section 3.2.6.2. If the class means for the reference variable follow the prior ordering then this technique will be useful. Otherwise it becomes necessary to search for an optimal set of scores which maximize correlation subject to the order constraint. Fortunately this problem has been considered by at least one group of researchers.

Bradley *et al.* (1962) formulate the problem as one of maximizing a variance ratio subject to constraints that the class scores have the desired order and zero mean. A quadratic programming problem is the result. The solution technique is not unduly complicated though the authors report that their computer program is very long due to a profusion of contingencies and special cases. Fortunately it seems unnecessary to be concerned with implementing this technique because one of the principal empirical results is that the variance ratios for optimal scores are little different from those obtained using ranks. Indeed, the authors find that when there are real and appreciable differences between classes "the scale values that are optimal tend to approach equally spaced scale values" (Bradley *et al.*, 1962, p. 368). They further see these results as a source of confidence for equally spaced scales. Their testimonial for the use of ranks is all the more remarkable considering the apparent extensive effort that went into development and test of the optimal technique.

3.2.6 NOMINAL TO INTERVAL

As in the case of ordinal to interval conversions the objective is to find for each class a score to be applied uniformly to every member of the class. In the present case the classes are unordered and any technique adopted must induce an ordering as well as a spacing of the classes. This dual aspect of the problem makes it the most difficult of all those discussed in this chapter.

3.2.6.1 *Two-Step Composition of Methods*

Since the nominal to interval scale conversion problem has two major aspects, it may be a useful device to decompose the problem to two steps: first find some satisfactory ordering of the classes, and second assign scores to the ordered classes. Techniques for these steps are discussed in Sections 3.2.4 and 3.2.5. An advantage of using such a decomposition is that the analyst's informed judgement may play a large role in either step. Information sufficient to describe a partial ordering of the classes conceivably could be used in defining a complete ordering whereas such information might not contribute at all to the method of the next section. Another consideration is the almost overwhelming evidence of Section 3.2.5 that once an ordering is achieved simple ranks are likely to be as good as most other scores. This result is a strong indication that the analyst's effort should be concentrated most heavily on ordering the classes rather than making delicate adjustments to the spacing between classes.

3.2.6.2 *Correlation with an Interval Variable*

The concept of a reference variable to guide scale conversions has had an important role throughout this chapter. In the present case the reference variable can provide an effective means of representing the full complement of information needed to assign interval scores to nominal classes.

Let X be a nominal variable with g classes and let Y be an interval variable. How can X be used most effectively to predict Y? The well-known answer is to find the b coefficients which maximize R^2 for the regression equation

$$Y = \sum_{i=1}^{i=g} b_i X_i. \tag{3.10}$$

Since X is a nominal variable, the X_i conventionally are treated as "dummy variables" such that $X_i = 1$ if an observation falls into the ith class and 0 otherwise. Since only one X_i variable is nonzero for each observation, Eq. (3.10) decomposes into g simple regression equations

$$Y_i = b_i X_i, \tag{3.11}$$

each equation concerned with predicting Y within the associated X class. It is well known that maximizing R^2 for Eq. (3.11) is achieved by choosing b_i as the mean Y value for all observations in the ith class, that is,

$$b_i = \sum_{j=1}^{j=n_i} y_{ij}/n_i, \tag{3.12}$$

where n_i is the number of observations in the ith class. Now using the class means of Eq. (3.12) as class scores for the X variable gives the maximum R^2 for Eq. (3.10). These class scores may be applied to the class members and used to calculate the product moment correlation with the reference variable. The squared correlation will be the same as R^2. Consequently the class scores which maximize the correlation between the nominal variable and the ordinal reference variable are the class means of Eq. (3.12).

To illustrate this technique, suppose a group of persons have been surveyed on a number of questions, among them education level. Some methods of quantifying education level were discussed earlier; but in this case suppose it is desired to quantify education level in terms of annual income. Now the mean annual income for each education class can be calculated. But, if annual income was not part of the survey, there need not be a problem since this quantity is figured for the country as a whole, by educational class. According to the *Information Please Almanac, Atlas and Yearbook* (1970), p. 131, annual mean income in 1967 by education level is as shown in Table 3.12. The maximal correlation property is maintained by using the actual mean income, or any linear transformation thereof, because correlation is invariant to linear transformations. The last column in Table 3.12 shows a transformed score obtained by dividing each of the income values by $3606.

Since the class scores can be determined by mean data from either the sample or the larger population, the classes can be scaled or scored with reference variables that could not be measured on the sample. For example, education level could have been scored using mean lifetime income, a quantity which is certainly unknown for the sample respondents until they actually die.

This technique can be especially useful for quantifying variables such as ethnicity, state or country of birth, religion, and such other clearly qualitative variables. Of course, each quantification is the result of a particular reference variable chosen to reflect a definite aspect of the nominal variable. For example, scores for religious denominations using total national membership as a reference variable would be quite different from those based on the average annual income of individual members. Care and thought are essential to achieving any meaningful results with this approach.

TABLE 3.12
MEAN ANNUAL INCOME BY EDUCATION CLASS

Years of School	Income	Transformed Score
Elementary		
less than 8 years	$3606	1.000
completed 8 years	$5189	1.435
High School		
1–3 years	$6335	1.755
4 years	$7629	2.110
College		
1–3 years	$8843	2.450
4 years or more	$11924	3.320

3.2.7   DICHOTOMIZATION

Many similarity coefficients and some clustering methods are specialized to deal with binary or dichotomous variables exclusively. Further, since binary variables may be treated as interval variables (see Section 3.1.4), dichotomization provides another means of attributing interval characteristics to nominal and ordinal variables. These considerations make dichotomization methods worthy of some serious thought.

For interval variables dichotomization is merely a special case of interval to ordinal (Section 3.2.1) or interval to nominal (Section 3.2.2) scale conversion. The case of two classes neither requires unique considerations nor permits unusual approaches when dealing with interval variables. On the other hand, none of the techniques discussed up to this point may be used alone to dichotomize ordinal and nominal variables.

3.2.7.1   Ordinal to Dichotomous

With a set of g ordered classes the problem is to choose among the $g - 1$ possible splits which preserve order. For example, with five ordered classes

$$A > B > C > D > E$$

the possible splits are

$$\text{(A) (B, C, D, E)}$$
$$\text{(A, B) (C, D, E)}$$
$$\text{(A, B, C) (D, E)}$$
$$\text{(A, B, C, D) (E).}$$

From the context of the problem it may be possible to rule out some alternatives or even identify a uniquely appropriate choice. On the other hand the choices may seem bewildering.

As earlier, the notion of a reference variable can be employed to advantage. Suppose an interval variable can be found which should be reflected in the split. Then the interval scores of the reference variable may be used to calculate the W function of Eq. (3.1) for the grouping of observations defined by each split. The split with the smallest value of W gives the strongest division of the reference variable. This method is very similar to the Automatic Interaction Detection (AID) technique of Sonquist and Morgan (1964). It is also very similar to the iterative improvement method described in Section 3.2.1.9; in the latter case single observations were moved, while in the present instance whole groups are moved.

As shown in Section 3.2.1.9 minimizing W is equivalent to maximizing the between group sum of squares which may be written as

$$B = (n_1 \bar{x}_1)^2/n_1 + (n_2 \bar{x}_2)^2/n_2 - (n_1 + n_2) \bar{x}^2. \tag{3.13}$$

The last term is a constant for all splits, so the maximum B can be found by maximizing the first two terms in Eq. (3.13). The quantity $n_1 \bar{x}_1$ is readily recognized to be the total of scores for observations in group 1. Thus each split may be tested using only the total score and number of observations for each group.

To illustrate, the 24 number sample of earlier examples is taken as the reference variable for 6 ordered groups as shown in Table 3.13. The valuation of each split is shown in Table 3.14. The split (A, B, C) (D, E, F) is the best in terms of the maximum B function criterion. However, the split (A, B, C, D) (E, F) is almost as good and easily might be preferred on the basis of supplementary information.

This same technique may be employed by replacing the reference variable scores with group scores derived by any of the methods of Section 3.2.5. For example, take ranks as group scores. Using the previous example, the 6 observations in group A each score 1, the 3 observations in group B each score 2, and so forth. The evaluation of the splits is then as shown in Table 3.15.

TABLE 3.13
GROUPS FOR BINARY SPLITS USING A
REFERENCE VARIABLE

Groups	n_i	$n_i \bar{x}_i$
A = (0, 5, 7, 9, 10, 15)	6	46
B = (20, 25, 26)	3	71
C = (27, 29, 42, 48)	4	146
D = (51, 55)	2	106
E = (61, 62, 73, 78, 79)	5	353
F = (81, 84, 85, 95)	4	345

TABLE 3.14
EVALUATION OF BINARY SPLITS USING A
REFERENCE VARIABLE

Split	$(n_1 \bar{x}_1)^2/n_1 + (n_2 \bar{x}_2)^2/n_2$
(A) (B, C, D, E, F)	58,500
(A, B) (C, D, E, F)	61,700
(A, B, C) (D, E, F)	63,900
(A, B, C, D) (E, F)	63,200
(A, B, C, D, E) (F)	54,800

TABLE 3.15
EVALUATION OF BINARY SPLITS USING CLASS
RANKS

Split	$(n_1 x_1)^2/n_1 + (n_2 x_2)^2/n_2$
(A) (B, C, D, E, F)	318
(A, B) (C, D, E, F)	324
(A, B, C) (D, E, F)	339
(A, B, C, D) (E, F)	335
(A, B, C, D, E) (F)	306

Surprisingly the splits have the same relative worth in both examples. This result is not to be expected in general. By proper choice of the reference variable, any of the potential splits can dominate the others.

3.2.7.2   Nominal to Dichotomous

The problem in this case is to cluster the nominal classes into two groups. However, since the classes are scaled only nominally, external information is needed to define intergroup similarities.

One approach is to use an interval reference variable and find the grouping of classes which minimizes the W function of Eq. (3.1), or equivalently maximizes the B function of Eq. (3.13). Since there is no ordering, the number of different ways g groups can be clustered into two supergroups is $2^{g-1} - 1$. For small values of g it is feasible to enumerate the possibilities. However, it is unnecessary to consider all these cases. This approach may be stated alternatively as a one-dimensional least-squares partition of the reference variable, subject to the constraint that the nominal groups may not be subdivided. As mentioned previously, Fisher (1958) and Ericson (1964) prove a one-dimensional least-squares partition is contiguous. However, the values of the reference variable may overlap the class boundaries and make the meaning of contiguity rather unclear. Since the B function depends only on the mean and size of each group, all observations in each group may be thought of as being located at the group mean. It is clear then that a contiguous partition of the reference variable necessarily respects the ordering defined by the group means. Thus, the first approach of Section 3.2.7.1 may be followed exactly, except that the original classes are ordered by class means rather than by a defined ordering.

Instead of using the same reference variable to both order and group, one might choose different reference variables for each task. Or, the ordering might be found intuitively and then employed in conjunction with the ordinal to dichotomous methods of Section 3.2.7.1. Ordering nominal classes by use of an ordinal reference variable is discussed in Section 3.2.4 and provides yet another approach to this problem.

3.3   The Application of Scale Conversions

A great many ideas for scale conversions have been offered in this chapter. One purpose behind presenting such a wide variety is to allow the analyst to choose a technique well suited to the particular problem at hand rather than feel compelled to use some marginally relevant "standard" technique. But in spite of its scope this chapter should not be considered exhaustive; additional ideas are to be welcomed and judged on their merits.

These techniques for scale conversion will be met repeatedly in the remainder of the text. They will play a central role in the formulation of various strategies for analysis of mixed variable data sets, a very real and common problem. Even at this stage it is obvious that a data set can be "homogenized" by choosing a dominant variable type and converting other variables as necessary. Use of this approach often involves both special cautions and unique opportunities peculiar to the specific conversion problem. Consequently, mixed variable strategies are discussed in detail for each of several rather broad problem classes. As will be seen, a sound capacity to deal simultaneously with variables of several types will permit increased flexibility of problem formulation and will promote a more comprehensive analysis.

4

MEASURES OF ASSOCIATION AMONG VARIABLES

In order to cluster variables it is necessary to have some numerical similarity measurements to characterize the relationships among the variables. The conventional approach to this requirement is to compute a measure of association for every pairwise combination of the variables; in a problem with N variables there are $\binom{N}{2} = \frac{1}{2}N(N-1)$ different pairs. A basic working assumption of all cluster analysis methods is that these numerical measures of association are all comparable to each other; that is, if the measure of association for one pair is 0.72 and for another pair 0.59, then the first pair is associated more strongly than the second. Of course, each measure reflects association in only a particular sense, and some care is needed to choose a measure appropriate to the problem and its context.

This chapter is a broad review of many different measures of association. Careful attention is given to describing the statistical background and operational interpretation (if any) for each measure. Measures between interval variables are developed first and followed by measures between nominal variables. Many different measures between binary variables are introduced and explored. Finally several strategies are offered for dealing with mixed data sets. Measures between ordinal variables are not included for two reasons: first it is most unusual in practice to find a multivariable data set consisting predominantly of ordinal variables; second, an adequate discussion of ordinal measures would require space and time quite out of proportion to the potential for application in most clustering problems. The author is not aware of any cluster analysis in which ordinal measures were used. In case the topic does arise in an application, the classic works of Kendall (1955) and Kruskal (1958) should be quite adequate sources of information.

The measures of association uniformly will be required to be symmetric. That is, if $A(X, Y)$ is the association between X and Y, then $A(X, Y)$ is symmetric if $A(X, Y) = A(Y, X)$. Measures based on conditional probabilities are often asymmetric because in general $P(X|Y) \neq P(Y|X)$. However, by employing arithmetic or geometric means of the conditional probabilities such measures may be made symmetric so that the limitation is not critical. The requirement for symmetry is an inherent element of most clustering methods including all those treated in this book.

4.1   Measures between Ratio and Interval Variables

This section is devoted primarily to the classical product moment correlation and a related measure for ratio variables. Rather than merely writing down these measures and stating their properties, a tutorial development is presented in which many statistical and algebraic concepts are introduced and related to each other.

4.1.1   THE ANGLE BETWEEN VECTORS

The behavior of two variables, say X and Y, in a data set may be represented by vectors of scores†

$$X^T = (x_1, \ldots, x_n) \quad \text{and} \quad Y^T = (y_1, \ldots, y_n).$$

The ith component of each vector is the score of the ith data unit (out of n) measured on the variable. In linear algebra the inner product (or scalar product) of two vectors is

$$\langle X, Y \rangle = X^T Y = \sum_{i=1}^{i=n} x_i y_i.$$

In statistics this quantity is known as the sum of cross products between X and Y. The inner product of a vector with itself, $X^T X$, is known as the sum of squares for X. The square root of the sum of squares is the Euclidean norm or length of the vector and is conventionally written as $|X|$ or $\|X\|$. With this notation, an alternative expression for the inner product between X and Y is

$$X^T Y = |X||Y| \cos \alpha, \qquad (4.1)$$

---

† The T superscript indicates transpose. All vectors are henceforth taken as column vectors; therefore row vectors are represented as transposed column vectors.

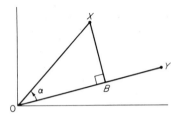

Fig. 4.1. Illustration of the inner product.

where α is the angle between X and Y. In a two-dimensional space this relation can be depicted as in Fig. 4.1. The distance from 0 (the origin) to B is $|X|\cos\alpha$ as is well known from elementary geometry; this quantity is also the orthogonal projection of X onto Y. The inner product then is seen to be the product of the length of Y and the length of the projection of X onto Y. Solving Eq. (4.1) for the cosine of the angle gives

$$A(X, Y) = \cos\alpha = \frac{X^\mathrm{T}\, Y}{|X|\,|Y|} = \frac{\sum\limits_{i=1}^{i=n} x_i y_i}{\left(\left[\sum\limits_{i=1}^{i=n} x_i^2\right]\left[\sum\limits_{i=1}^{i=n} y_i^2\right]\right)^{1/2}}. \tag{4.2}$$

The cosine of the angle is a measure of similarity between X and Y; the more nearly parallel the two vectors the greater is the cosine. The product of norms, $|X|\,|Y|$, is the maximum value of $X^\mathrm{T}\, Y$. Hence the cosine may also be viewed as $\cos\alpha = X^\mathrm{T}\, Y/(X^\mathrm{T}\, Y)_{\max}$. This measure is independent of the length of the vectors as is obvious from the geometry of Fig. 4.1. Algebraically, let b and c be two scalar constants and define $\hat{X} = bX$ and $\hat{Y} = cY$. Then

$$\frac{\hat{X}^\mathrm{T}\, \hat{Y}}{|\hat{X}|\,|\hat{Y}|} = \frac{\sum\limits_{i=1}^{i=n} (bx_i)(cy_i)}{\left(\left[\sum\limits_{i=1}^{i=n} b^2 x_i^2\right]\left[\sum\limits_{i=1}^{i=n} c^2 y_i^2\right]\right)^{1/2}}$$

$$= \frac{bc \sum\limits_{i=1}^{i=n} x_i y_i}{|bc|\left(\left[\sum\limits_{i=1}^{i=n} x_i^2\right]\left[\sum\limits_{i=1}^{i=n} y_i^2\right]\right)^{1/2}}$$

$$= \mathrm{sgn}\,(bc)\frac{X^\mathrm{T}\, Y}{|X|\,|Y|},$$

where $\mathrm{sgn}\,(bc)$ is the sign ($+$ or $-$) of the product of b and c. In a more compact notation

$$A(X, Y) = \mathrm{sgn}\,(bc)\, A(bX, cY). \tag{4.3}$$

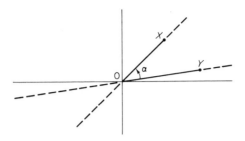

Fig. 4.2. Geometric interpretation of cosine invariance.

Thus the cosine of the angle between X and Y is invariant to uniform multiplicative scaling except for a possible change of sign. The geometric interpretation of this relation is shown in Fig. 4.2. Every point on the line $0X$ and its projections in both directions (except for the origin, 0, itself) is equivalent to X under this measure and likewise for the line $0Y$ and Y. In this particular example the cosine of the angle between X and Y is positive; it is also positive for any pair of points (on the two lines) which lie in the same quadrant and negative if the points lie in different quadrants. Thus, the cosine is a many-to-one transformation which effectively ignores the relative magnitudes between the vectors.

4.1.2   THE PRODUCT MOMENT CORRELATION COEFFICIENT

The *means* for the variables X and Y are computed from sample data as

$$\bar{x} = \sum_{i=1}^{i=n} x_i/n \qquad \text{and} \qquad \bar{y} = \sum_{i=1}^{i=n} y_i/n.$$

Now if the variable means are subtracted from the original scores, the vectors of centered scores are obtained as

$$\hat{X}^{\mathrm{T}} = [(x_1 - \bar{x}), \ldots, (x_n - \bar{x})], \qquad \hat{Y}^{\mathrm{T}} = [(y_1 - \bar{y}), \ldots, (y_n - \bar{y})].$$

Note that $\hat{X}^{\mathrm{T}}$ and $\hat{Y}^{\mathrm{T}}$ have zero means. The inner product of two centered vectors is called the scatter of X and Y. The inner product of $\hat{X}$ with itself is the scatter of X or the sum of squared deviations about $\bar{x}$. If the scatter is divided by n then the covariance and variance are recognized as

$$\mathrm{cov}\,(X, Y) = \frac{\hat{X}^{\mathrm{T}}\hat{Y}}{n} = \frac{1}{n}\sum_{i=1}^{i=n}(x_i - \bar{x})(y_i - \bar{y}),$$

$$\mathrm{var}\,(X) = \frac{\hat{X}^{\mathrm{T}}\hat{X}}{n} = \frac{1}{n}\sum_{i=1}^{i=n}(x_i - \bar{x})^2.$$

The covariance of X and Y is also known as the product moment of X and Y; likewise, var(X) is the product moment of X. The (sample) product moment correlation between X and Y is then defined as

$$r = r(X, Y) = \frac{\text{cov}(X, Y)}{[\text{var}(X)\text{var}(Y)]^{1/2}} = \frac{\sum\limits_{i=1}^{i=n}(x_i - \bar{x})(y_i - \bar{y})}{\left(\left[\sum\limits_{i=1}^{i=n}(x_i - \bar{x})^2\right]\left[\sum\limits_{i=1}^{i=n}(y_i - \bar{y})^2\right]\right)^{1/2}} \quad (4.4)$$

Comparing Eqs. (4.2) and (4.4) it is apparent that r is equal to the cosine of the angle between the centered vectors $\hat{X}$ and $\hat{Y}$.

An alternative view of correlation can be had in terms of standardized vectors in which the components are transformed as

$$x_i^* = (x_i - \bar{x})/[\text{var}(X)]^{1/2}.$$

The vector X^* with components x_i^* has zero mean and unit variance. The inner product of X^* and Y^* is then

$$X^{*T} Y^* = \frac{\sum\limits_{i=1}^{i=n}(x_i - \bar{x})(y_i - \bar{y})}{[\text{var}(X)\text{var}(Y)]^{1/2}},$$

so $r = X^{*T} Y^*/n$. The correlation is actually the covariance of the standardized vectors.

Since the product moment correlation is the cosine of the angle between the centered vectors, it has the multiplicative invariance of the cosine. In addition, it is also invariant under the uniform addition of a constant to each element of X or Y. Let $x_i^+ = x_i + b$ for all i. Then the mean of the transformed variable is $\bar{x}^+ = \bar{x} + b$. Therefore,

$$x_i^+ - \bar{x}_i^+ = (x_i + b) - (\bar{x} + b) = x_i - \bar{x}.$$

Thus, the added constant is subtracted out in the process of centering the scores. The combined effect of these two forms of invariance is that the product moment correlation coefficient is invariant to any linear transformation, except for a possible change in sign, that is

$$r(X, Y) = \text{sgn}(ce)r(b + cX, d + eY). \quad (4.5)$$

The correlation coefficient has a stronger form of invariance than the cosine because it is unaffected by the uniform addition of a constant to each element of a score vector. But this same property means that the correlation coefficient is less discriminating than the cosine because for given X and Y there are many more members in the equivalence class of all linear transformations of X and Y than in the equivalence class of all multiples of X and Y.

The essential difference between the two measures is that the cosine is based on the original scores (deviations from the origin) while the correlation coefficient is based on centered scores (deviations about the mean). If the origin is well established and meaningful, then the original scores have meaning in an absolute sense and the cosine is an appropriate measure of association. If the origin is arbitrary or chosen for convenience, then the original scores are meaningful relative to each other and to their mean but not relative to the origin; in this case the correlation coefficient is an appropriate measure of association. Recalling the definitions of Section 3.1, the distinction between these two cases is precisely the difference between ratio and interval scales, respectively. Thus, the cosine makes use of ratio scale information while the correlation coefficient utilizes only interval scale information.

4.2   Measures between Nominal Variables

It is conventional and convenient to represent the joint distribution of two categorical (either ordinal or nominal) variables in a contingency table as in Table 4.1. An n_{ij} entry in the table is the number of data units which fall in

TABLE 4.1
GENERAL FORM OF THE CONTINGENCY TABLE

Variable A \ Variable B	1	2	$\cdots$	q	Totals
1	n_{11}	n_{12}	$\cdots$	n_{1q}	$n_{1.}$
2	n_{21}	n_{22}	$\cdots$	n_{2q}	$n_{2.}$
$\vdots$	$\vdots$	$\vdots$	$\cdots$	$\vdots$	$\vdots$
p	n_{p1}	n_{p2}	$\cdots$	n_{pq}	$n_{p.}$
Totals	$n_{.1}$	$n_{.2}$	$\cdots$	$n_{.q}$	$n_{..}$

both the ith class of variable A and the jth class of variable B. If all entries and marginal totals are divided by $n_{..}$ (the total number of data units) then the table entries are in terms of frequencies, f_{ij}. Ordinal measures of association depend on the sequence of rows and columns (the prior ordering) whereas the nominal measures of this section are invariant to any permutation of rows and columns.

4.2.1 CHI-SQUARE BASED MEASURES

Some old and traditional measures of association are based on the chi-square statistic, a familiar quantity in the analysis of contingency tables. The chi-square statistic itself usually is used to test the hypothesis of independence between the variables in the table. Taking o_{ij} as the observed count in cell ij and e_{ij} as the corresponding expected count under the hypothesis of independence, the sample chi-square statistic is defined as

$$\chi^2 = \sum_{i=1}^{i=p} \sum_{j=1}^{j=q} (o_{ij} - e_{ij})^2/e_{ij}.$$

Under the hypothesis of independence the expected count in cell ij is,

$$e_{ij} = f_{i.} f_{.j} n_{..} = \frac{n_{i.} n_{.j}}{n_{..}}.$$

The observed count is $o_{ij} = f_{ij} n_{..} = n_{ij}$. Then,

$$\chi^2 = \sum_{i=1}^{i=p} \sum_{j=1}^{j=q} (n_{ij} - n_{i.} n_{.j}/n_{..})^2/(n_{i.} n_{.j}/n_{..}), \tag{4.6a}$$

$$\chi^2 = n_{..} \left[\sum_{i=1}^{i=p} \sum_{j=1}^{j=q} (f_{ij} - f_{i.} f_{.j})^2/f_{i.} f_{.j} \right], \tag{4.6b}$$

$$\chi^2 = n_{..} \left[\sum_{i=1}^{i=p} \sum_{j=1}^{j=q} \frac{f_{ij}^2}{f_{i.} f_{.j}} - 1 \right] = n_{..} \left[\sum_{i=1}^{i=p} \sum_{j=1}^{j=q} \frac{n_{ij}^2}{n_{i.} n_{.j}} - 1 \right], \tag{4.6c}$$

which are all familiar computing forms of the chi-square statistic.

The value of χ^2 itself increases without bound as $n_{..}$ increases and so is not a suitable measure of association even though it is useful as a test statistic. A partial remedy is found in

$$\phi^2 = \chi^2/n_{..},$$

which is known as the mean-square contingency. But this quantity itself is dependent on the size of the table. To see why, suppose $p = q$ so the table is square, and further suppose that the variables are associated perfectly, i.e., $n_{i.} = n_{.i} = n_{ii}$ for all i. Substituting these values in Eq. (4.6c) and noting there are only p nonnull cells, the value of χ^2 is $n_{..}(p - 1)$ so that ϕ^2 has a maximum of $p - 1$ in a square table. In a rectangular table ($p \neq q$) perfect association is attained by concentrating all entries in the longest diagonal as in the case of a square table. The number of cells in such a diagonal is $\min(p,q)$, so that χ^2 is $n_{..} \min\{(p - 1), (q - 1)\}$ and ϕ^2 is $\min\{(p - 1), (q - 1)\}$. With these facts in mind various attempts have been made to norm ϕ^2 to have the conventional range of 0 to 1. Tschuprow suggested the geometric mean of $(p-1)$ and $(q-1)$ as a norming factor to give the measure

$$T = \left\{ \frac{\chi^2/n_{..}}{[(p-1)(q-1)]^{1/2}} \right\}^{1/2}. \tag{4.7}$$

Cramér (1946) usually is credited with suggesting the maximum of ϕ^2 as a norming factor to give the measure

$$C = \left\{ \frac{\chi^2/n_{..}}{\min\,[(p-1),(q-1)]} \right\}^{1/2}. \tag{4.8}$$

Actually Maung (1941, p. 195) preceded Cramér with this suggestion and showed that C^2 is the mean square of the nonzero canonical correlations between the two sets of categories. That is, if r_i is the ith of k nonzero canonical correlations, then

$$C^2 = \sum_{i=1}^{i=k} r_i^2/k.$$

Pearson suggested another measure based on ϕ^2,

$$P = \left(\frac{\phi^2}{1+\phi^2} \right)^{1/2} = \left(\frac{\chi^2}{n_{..} + \chi^2} \right)^{1/2}. \tag{4.9}$$

This measure is known as the coefficient of contingency. Assuming that the distribution underlying the contingency table is bivariate normal with correlation parameter r and the classes are constructed very finely so as to conform closely to the correlation surface, then P^2 approaches r^2 as $n_{..}$ increases without bound (Kendall and Stuart, 1961, Vol. II, p. 557). However, in typical tables categories are rather broad, the number of observations relatively small, and the assumption of bivariate normality unwarranted.

The chi-square statistic and measures derived from it are useful as tests of hypotheses. But they may not be so useful as measures of association. Goodman and Kruskal (1954, p. 740) pinpoint the major problem:

> One difficulty with the use of the traditional measures, or of any measures that are not given operational interpretation, is that it is difficult to compare meaningfully their values for two cross classifications. Suppose that C [the coefficient of contingency in their notation] turns out to be 0.56 and 0.24 respectively in two cross-classification tables. One wants to be able to say that there is a higher association in the first table than the second, but investigators sometimes restrain themselves, with commendable caution, from making such a comparison. Their restraint may stem in part from the noninterpretability of C.

In cluster analysis meaningful comparisons among all pairwise combinations of variables are essential. All the methods discussed in this book rely on the consistency of the association measure across all comparisons. Now the traditional measures are not necessarily lacking in this regard; but their lack of operational interpretation makes it impossible to determine in what sense they are consistent. The alternative measures discussed in the following sections are interpretable more directly and generally are to be preferred for this reason.

4.2.2   TWO MEASURES BASED ON OPTIMAL CLASS PREDICTION

One of the most appealing ways to give meaning to the association between variables is through measuring the power of one as a predictor for the other. For the moment take an asymmetric point of view and suppose the objective is to guess the B-variable membership class for an observation. If the observation's A class is unknown, the best one can do is to choose the B class with the largest marginal total, that is, the value of m satisfying

$$f_{.m} = \max\{f_{.1}, \ldots, f_{.q}\}.$$

The probability of error in this case is $P_1 = 1 - f_{.m}$. Now, suppose the A class for an observation is known to be a. Then only row a of the contingency table is of interest and the best guess is the B class corresponding to the largest entry in row a, that is, the value of m_a satisfying

$$f_{am_a} = \max\{f_{a1}, \ldots, f_{aq}\}.$$

The a subscript on the m is added to stress that the optimal B class may vary from row to row; if it did not vary the additional information about the A class would be useless. Given that a is known, the probability of error is $P_1 = 1 - f_{am_a}/f_{a.}$. But there are p rows each of which occurs with frequency $f_{i.}$. Therefore the unconditional probability of error is

$$P_2 = \sum_{i=1}^{i=p} f_{i.}(1 - f_{im_i}/f_{i.}) = 1 - \sum_{i=1}^{i=p} f_{im_i}.$$

The last term is the sum of the maximum entries for each row of the table. Goodman and Kruskal (1954) suggest as a measure of the predictive power of A for B the relative decrease in error probability due to knowledge of the A class; that is,

$$L_{\mathrm{B}} = \frac{P_1 - P_2}{P_1} = \frac{\sum\limits_{i=1}^{i=p} f_{im_i} - f_{.m}}{1 - f_{.m}},$$

where the B subscript indicates prediction of the B class. The analogous expression for prediction of the A class is

$$L_A = \frac{\sum\limits_{j=1}^{j=q} f_{m_j j} - f_{m.}}{1 - f_{m.}}.$$

If the prediction of A from B is taken as being equally important as the prediction of B from A, then a symmetric relationship is obtained by considering the prediction of A half the time and B the other half. The probability of error when a predictor class is unknown is then

$$P_1 = 1 - \tfrac{1}{2}(f_{.m} + f_{m.}),$$

and when a predictor class is known the probability of error is

$$P_2 = 1 - \tfrac{1}{2}\left(\sum\limits_{i=1}^{i=p} f_{im_i} + \sum\limits_{j=1}^{j=q} f_{m_j j}\right).$$

The "lambda" measure is then

$$L = \frac{P_1 - P_2}{P_1} = \frac{\sum\limits_{i=1}^{i=p} f_{im_i} + \sum\limits_{j=1}^{j=q} f_{m_j j} - f_{.m} - f_{m.}}{2 - f_{.m} - f_{m.}}, \tag{4.10a}$$

or

$$L = \frac{\sum\limits_{i=1}^{i=p} n_{im_i} + \sum\limits_{j=1}^{j=q} n_{m_j j} - n_{.m} - n_{m.}}{2n_{..} - n_{.m} - n_{m.}}. \tag{4.10b}$$

Thus, L is the relative decrease in error probability due to the use of predictor classes when the directions of prediction are equally important. Goodman and Kruskal (1954, p. 743) note that L is indeterminant only if the entire population lies in a single cell; otherwise L lies between 0 and 1 inclusive, in particular between L_A and L_B inclusive. The value of L is 1 if and only if the entire population lies in isolated cells, i.e., cells which are the only nonnull cells in both the row and column. In the case of statistical independence (the products of the marginal frequencies equal the cell frequencies throughout the table) L is 0; but a 0 value for L need not imply independence.
A measure closely related to L is

$$D = P_1 - P_2 = \tfrac{1}{2}\left[\sum\limits_{i=1}^{i=p} f_{im_i} + \sum\limits_{j=1}^{j=q} f_{m_j j} - f_{.m} - f_{m.}\right], \tag{4.11a}$$

or

$$D = \frac{\sum_{i=1}^{i=p} n_{im_i} + \sum_{j=1}^{j=q} n_{m_j j} - n_{.m} - n_{m.}}{2n_{..}}.$$

(4.11b)

This quantity is the actual reduction in the error probability (also the actual increase in the probability of correct prediction) as a consequence of using predictor information. The value of D is always determinant and lies between 0 and 1 inclusive. Also, $D \leqslant L$ since $D = LP_1$ and $P_1 \leqslant 1$.

The primary distinction between these two measures is that D depends on the difference $P_1 - P_2$, whereas L depends on the ratio, P_2/P_1. Table 4.2 illustrates three equivalent cases for each measure.

TABLE 4.2
COMPARISON OF MEASURES D AND L

Case	1	2	3	4	5	6
P_1	0.20	0.40	0.60	0.45	0.60	0.75
P_2	0.00	0.20	0.40	0.15	0.20	0.25
D	0.20	0.20	0.20	0.30	0.40	0.50
L	1.00	0.50	0.33	0.67	0.67	0.67

Obviously the two measures treat association in different senses. As a result of using predictor information one expects to reduce the number of errors by $100D\%$ and the error rate by $100L\%$.

4.2.3   CANONICAL CORRELATION

An entirely different approach to measuring association between nominal variables is to assign a score to each class for each variable and then calculate the ordinary product moment correlation using these scores. The question is: what scores should be chosen? An appealing answer is to choose those scores which give the maximum correlation between the two variables. The computation of these scores is a problem of canonical correlation.

In the ordinary canonical correlation problem two sets of variables $\{x_1, \ldots, x_p\}$ and $\{y_1, \ldots, y_q\}$ are used to form the linear composites $X = \sum_{i=1}^{i=p} a_i x_i$ and $Y = \sum_{j=1}^{j=q} b_j y_j$. The object of the analysis is to find two sets of coefficients $\{a_1, \ldots, a_p\}$ and $\{b_1, \ldots, b_q\}$ which maximize the correlation between the composites X and Y. In the case of the contingency table, let a_i be the score to be assigned to the ith A class and b_j the score for the jth B class; further, let x_i be 1 if the observation falls in the ith A class, y_j be 1 if the observation falls in the *j*th **B** class, and each be 0 otherwise. Then for any sample observation, X is simply the score for the **A** class into which the observation falls and Y is the score for the corresponding **B** class. Thus, the correlation between X and Y is the correlation between two vectors of class scores for the sample of observations.

Since canonical correlation is not an elementary subject in statistics, the derivation of the a_i and b_j scores is developed directly as a problem of maximizing product moment correlation. The development is rather lengthy, though not difficult to follow, and accordingly is presented in Appendix A. For the remainder of this discussion it will suffice to say that such class scores may be determined and that no other set of scores can give a larger value of correlation. A computer program in Appendix C may be used to compute scores and correlations with a minimum of effort.

There is no lack of interpretability for a canonical correlation based on a contingency table; it is the largest possible correlation that may be obtained by assigning scores to classes and using the scores in the ordinary product moment correlation formula. The question in any particular case is whether this kind of association between nominal variables is relevant to the problem at hand and consistent with other known information. The principal features of this approach are that class labels are replaced by class scores and prediction of class membership on the basis of error probabilities is replaced by prediction of class scores against a least squares criterion. If the classes can be viewed meaningfully as lying on an underlying continuum, then the class scores provide a spacing and ordering of the classes along this continuum. However, a different set of scores is obtained each time a variable is paired with another. The scores for variable A in the comparison of variables A and B may be quite different than in the comparison of variables A and C. The variables appear as chameleons changing their colors to match their surroundings; they mutually adapt through their scores to achieve the maximum possible harmony in the single pairwise comparison; there is no way to do better in a correlational sense. When comparing correlations from different tables, the comparison is meaningful if it is reasonable (at least potentially) to talk about class scores and their prediction; the unique element is that the scores are specialized to the individual prediction rather than remaining constant across all predictions as in the case of correlation with interval variables.

Allowing the order and spacing between classes to float may have enlightening consequences, especially in regard to order. In some cases it might be a useful heuristic to leave the order free but require the scores to be spaced equally once the order is determined. In short, the optimal scores are replaced by their ranks. Section A.3 discusses the details of this approach; an implementing computer program is given in Appendix C. At least a partial motivation for using ranks as scores is provided by the material in Section 3.2.5, especially the robust nature of the product moment correlation coefficient under monotone transformations. In particular, if some hypothetical "true" scores have the same order as the optimal scores produced by the canonical correlation method, then the use of ranks should result in only relatively small decrements in correlation from that which might be achieved with the "true" scores. The use of ranks is also more in keeping with the original nominal character of the variables since it only attempts to induce order, but not distinctive spacing.

Before leaving the topic of canonical correlation, it should be noted that there are usually several canonical correlations associated with two sets of variables. These correlations usually are determined in order of magnitude with the fiıst representing the largest possible association between the two sets of variables, the second representing the largest possible association between the residuals after removal of the variation accounted for by the first canonical correlation, and so forth, ·until the set of canonical correlations collectively account for the entire variation of one variable in terms of the other. The focus in this section has been on the largest canonical correlation alone. Srikantan (1970) and Stewart and Love (1968) deal with measures involving the entire set of canonical correlations. And as noted earlier, the measure of Eq. (4.9) is a composite of the canonical correlations. However, operational interpretations for these latter measures have not been developed and consequently it is not clear just what they do measure.

4.2.4   OTHER MEASURES BETWEEN NOMINAL VARIABLES

A number of other measures between nominal variables exist in the literature. They are mentioned here to complete the picture for nominal measures but are not discussed in detail either because they lack operational interpretations or involve very special assumptions.

An important traditional measure is tetrachoric correlation. The assumed model in this case is a bivariate normal distribution between two interval variables which have been dichotomized to give a 2×2 table. The approach is to find the point of dichotomy on each of the marginal univariate normal distributions and then the correlation parameter in the bivariate distribution which is consistent with these points of dichotomy. Carroll (1961) is an advocate of this measure. Castellan (1966) gives a brief summary of the principal theoretical concepts involved and discusses a number of computational approximations. Froemel (1971) provides additional information on the computational aspect. Lancaster and Hamdan (1964) develop a polychoric correlation coefficient for the general $p \times q$ table, and Hamdan (1971b) gives a simple example to illustrate the computational approach.

Next, there are two measures based on the information theoretic concept of entropy. Linfoot (1957) proposes a measure which is a function of the difference between the observed entropy of the table and the expected entropy under the hypothesis of independence (similar to chi-square). The measure has the range 0 to 1 and is offered as a correlation measure for contingency tables. Pearson (1966) considers Linfoot's approach too complex and offers another measure. Basically the variance and covariance terms in Eq. (4.4) are replaced with entropy based analogs. In actual computations with test data Pearson's measure tends to be somewhat larger than Linfoot's and both sometimes exceed the canonical correlation obtained with optimal scores. This latter fact denies the authors' contentions that the measures have correlational meaning, because no set of scores can give a larger value from Eq. (4.4) than those associated with the canonical correlation.

Finally, there is intraclass correlation which is detailed fully in Haggard's (1958) monograph. In substance the method uses analysis of variance techniques to estimate certain within and between group variance components for the computation of a correlation ratio. Linscheid and Stone (1971) offer a computer program to calculate this measure.

4.3   Measures between Binary Variables

The 2×2 contingency table has received a great deal of attention in the statistical literature. It is frequently the case that variables can be expressed meaningfully as dichotomies such as high–low, present–absent, good–bad, yes–no, and so forth. Section 3.2.7 gives some suggestions for dichotomizing variables so they can fit this mold. As will be seen the information loss suffered in conversion to binary variables can be somewhat compensated by additional advantages.

It is certainly possible to continue utilizing the contingency table of Table 4.1. However, the widely adopted alternative shown in Table 4.3 permits a much simplified notation for the 2×2 case. Note that $n = a + b + c + d$.

TABLE 4.3
SPECIAL FORM OF 2×2 CONTINGENCY TABLE

Variable A \ Variable B	1	0	Totals
1	a	b	$a + b$
0	c	d	$c + d$
Totals	$a + c$	$b + d$	n

A common convention for 2 × 2 tables is to employ the class labels 1 and 0 to represent the positive and negative sides of the dichotomy. These labels often are used as class scores as well; for example an observed data unit would score 1 if it bore a particular attribute and 0 otherwise.

4.3.1 BINARY VERSIONS OF EARLIER MEASURES

An easy introduction to binary measures can be had in terms of the measures already introduced in this chapter. Since each of these measures has a direct counterpart in the 2 × 2 case, binary variables may be included routinely in any of the preceding analyses.

4.3.1.1 *The Angle between Vectors*

If the 1–0 labels in Table 4.3 are taken as scores, then the factors in Eq. (4.2) are computed as

$$\sum_{i=1}^{i=n} x_i y_i = a, \quad \sum_{i=1}^{i=n} x_i^2 = a+b, \quad \sum_{i=1}^{i=n} y_i^2 = a+c.$$

Then the cosine of the angle between the vectors A and B is

$$\frac{a}{[(a+b)(a+c)]^{1/2}} = \left[\left(\frac{a}{a+b}\right)\left(\frac{a}{a+c}\right)\right]^{1/2}. \tag{4.12}$$

Sokal and Sneath (1963, p. 130) attribute this measure to Ochiai, a Japanese zoologist.

In some cases it is not clear that one category has a definite positive character relative to the other and consequently the choice of which class to assign the score of 1 is arbitrary. But it makes a difference which class is assigned the unit score since, in general, a reversal of scoring gives a different result. Therefore in ambiguous situations it may be desirable to take the geometric mean of the cosines obtained both with the scoring of Table 4.3 and with reverse scoring. The geometric mean would involve the fourth root of a number in this case; the square of the geometric mean usually is adopted instead to give

$$\left[\left(\frac{a}{a+b}\right)\left(\frac{a}{a+c}\right)\left(\frac{d}{b+d}\right)\left(\frac{d}{c+d}\right)\right]^{1/2}. \tag{4.13}$$

It is worth noting that each of the terms in expressions (4.12) and (4.13) are conditional probabilities. For example, $a/(a+b)$ is the conditional probability that a data unit scores 1 on variable B given that it scored 1 on variable A. Then expression (4.12) is the geometric mean of conditional probabilities associated with cell a while expression (4.13) is the geometric mean of conditional probabilities associated with the diagonal.

Recalling the multiplicative invariance of the cosine measure, the number 1 may be replaced in the scoring of Table 4.3 by any other number but 0 and the expressions (4.12) and (4.13) will remain unaffected. However, the cosine is not invariant under addition of a constant to all scores; therefore, the expressions (4.12) and (4.13) depend on one class being scored with 0 in each dichotomy.

4.3.1.2   *The Product Moment Correlation Coefficient*

The squares and products in Eq. (4.4) may be expanded to give the alternative form

$$r = \frac{\sum\limits_{i=1}^{i=n} x_i y_i - \dfrac{1}{n}\left(\sum\limits_{i=1}^{i=n} x_i\right)\left(\sum\limits_{i=1}^{i=n} y_i\right)}{\left\{\left[\sum\limits_{i=1}^{i=n} x_i^2 - \dfrac{1}{n}\left(\sum\limits_{i=1}^{i=n} x_i\right)^2\right]\left[\sum\limits_{i=1}^{i=n} y_i^2 - \dfrac{1}{n}\left(\sum\limits_{i=1}^{i=n} y_i\right)^2\right]\right\}^{1/2}}. \tag{4.14}$$

Using the 1–0 scoring system of Table 4.3 it follows that

$$\sum_{i=1}^{i=n} x_i = a + b \qquad \text{and} \qquad \sum_{i=1}^{i=n} y_i = a + c.$$

Substituting these expressions into Eq. (4.14) along with the sum of products and sums of squares expressions found in Section 4.3.1.1 gives

$$r = \frac{a - (a + b)(a + c)/n}{\{[a + b - (a + b)^2/n][a + c - (a + c)^2/n]\}^{1/2}}$$

$$= \frac{an - (a + b)(a + c)}{\{(a + b)[n - (a + b)](a + c)[n - (a + c)]\}^{1/2}}$$

$$= \frac{ad - bc}{\{(a + b)(c + d)(a + c)(b + d)\}^{1/2}}. \tag{4.15}$$

By substituting a, b, c, and d into Eq. (4.6a) and noting that $p = q = 2$, it is also possible to show that $r^2 = \phi^2 = \chi^2/n = T^2 = C^2$. In the psychological literature expression (4.15) is frequently called ϕ because of this equivalence. Also note that removing the bc term in the numerator of Eq. (4.15) gives expression (4.13).

Since r is invariant under a linear transformation, the two scores 0 and 1 are arbitrary because they can be linearly transformed to any other pair of scores as discussed in Section 3.1.4. Further, since any pair of scores gives the same result as any other pair, every pair is optimal and the canonical correlation method of Section 4.2.3 also reduces to Eq. (4.15).

Finally, when Yates' correction for continuity is applied to the chi-square statistic,

$$\chi_c^2 = \sum_{i=1}^{i=p} \sum_{j=1}^{j=q} (|o_{ij} - e_{ij}| - 0.5)^2/e_{ij},$$

which in this 2×2 case reduces to

$$\chi_c^2 = \frac{n(|ad - bc| - 0.5n)^2}{(a+b)(c+d)(a+c)(b+d)}.$$

Stiles (1961) has used $\log_{10}\chi_c^2$ as an association factor in information retrieval.

4.3.1.3   *Optimal Prediction Measures*

Using the conventions of Table 4.3, the terms of Eqs. (4.10b) and (4.11b) are expressed as follows:

$$n_{.m} = \max\{(a+c),(b+d)\}, \qquad n_{m.} = \max\{(a+b),(c+d)\},$$

$$\sum_{i=1}^{i=p} n_{im_i} = \max(a,b) + \max(c,d), \qquad \sum_{j=1}^{j=q} n_{m_j j} = \max(a,c) + \max(b,d).$$

Substitution of these quantities in Eqs. (4.10b) and (4.11b) gives complex appearing (but easy to use) forms which do not yield to efforts of simplification.

Kendall and Stuart (1961, Vol. II, p. 547) point out one special way in which a classical measure on the 2×2 table can be obtained from lambda [Eq. (4.10b)]. The marginal totals in the table reflect the relative frequency of occurrence in the population for each of the classes; but sometimes it is desirable to construct an artificial population with specified marginal proportions. Consider multiplying the rows and columns of Table 4.3 by the following factors:

$$\text{first row:} \quad (cd)^{1/2}/k,$$
$$\text{second row:} \quad (ab)^{1/2}/k,$$
$$\text{first column:} \quad (bd)^{1/2}/k,$$
$$\text{second column:} \quad (ac)^{1/2}/k,$$

where $k = (abcd)^{1/2}$. Then Table 4.3 is transformed to Table 4.4. Assuming that $ad > bc$ the terms in the lambda measure computed on Table 4.4 are:

$$n_{.m} = n_{m.} = \tfrac{1}{2}M$$

$$\sum_{i=1}^{i=p} n_{im_i} = \sum_{j=1}^{j=q} n_{m_j j} = 2(ad)^{1/2}.$$

TABLE 4.4
TRANSFORMED 2 × 2 TABLE

$(ad)^{1/2}$	$(bc)^{1/2}$	$\frac{1}{2}M$
$(bc)^{1/2}$	$(ad)^{1/2}$	$\frac{1}{2}M$
$\frac{1}{2}M$	$\frac{1}{2}M$	M

Substituting these expressions into Eq. (4.10b) gives

$$L = \frac{4(ad)^{1/2} - M}{2M - M} = \frac{2(ad)^{1/2} - \frac{1}{2}M}{\frac{1}{2}M}.$$

Since $\frac{1}{2}M = (ad)^{1/2} + (bc)^{1/2}$, L becomes

$$L = \frac{(ad)^{1/2} - (bc)^{1/2}}{(ad)^{1/2} + (bc)^{1/2}}. \tag{4.16}$$

Proceeding in a similar manner it is easy to show that the D measure of Eq. (4.11b) is

$$D = \frac{4(ad)^{1/2} - M}{2M} = \frac{1}{2}L.$$

Therefore there is no point in distinguishing between D and L in the 2 × 2 case. Expression (4.16) originally was proposed by Yule (1912) and is sometimes called the *coefficient of colligation*, which traditionally is denoted by the symbol Y. Yule also proposed a related measure of association

$$Q = \frac{ad - bc}{ad + bc}. \tag{4.17}$$

Since Y was constructed to be independent of the marginal totals, Q also shares this property because $Q = 2Y/(1 + Y^2)$. Kendall and Stuart (1961, Vol. II, p. 545) and Goodman and Kruskal (1954, p. 750) also note that Q is the 2 × 2 version of an ordinal measure which the latter authors call "gamma."

Edwards (1963) argues that a measure of association on a 2 × 2 table should be a function of the cross ratio

$$X = bc/ad. \tag{4.18}$$

He notes in particular that the cross ratio itself and its logarithm have been used in studies of genetic linkage. It is seen easily that Y and Q are functions of the cross ratio

$$Y = \frac{1 - X^{1/2}}{1 + X^{1/2}} \qquad \text{and} \qquad Q = \frac{1 - X}{1 + X},$$

provided $ad \neq 0$. These latter two expressions lead to one more interesting form. The hyperbolic tangent is defined as

$$\tanh u = \frac{e^u - e^{-u}}{e^u + e^{-u}} = \frac{1 - e^{-2u}}{1 + e^{-2u}} \cdot$$

Comparing the functional forms, it is easy to verify that

$$Y = \tanh\left(-\tfrac{1}{4}\ln X\right) \qquad \text{and} \qquad Q = \tanh\left(-\tfrac{1}{2}\ln X\right).$$

Edwards (1963) omits the logarithms and negative signs so that his results in this case are quite in error.

4.3.2   MATCHING COEFFICIENTS

A natural and intuitively appealing approach to assessing similarity in the 2×2 table is simply to count the total number of relevant matches between the variables. This sort of device spawns a surprising number of measures because there are several ways of interpreting the phrase "relevant matches." Two factors can account for all the variations. First there is the question of what to do with 0–0 matches. If the dichotomies are of the present–absent type, the data units in cell d of Table 4.3 do not possess either attribute. For example, suppose the data units are animals and the variables are "has feathers," and "has webbed feet." Dogs and cats and many other animals would fall into cell d because there is no way they could have such attributes. It would be misleading to allow these 0–0 matches to contribute to the measure of association between dogs and cats. Second is the question of how to weight matches and mismatches. Subject to the preceding remarks about 0–0 matches, cells a and d represent matches, while cells b and c represent mismatches between the variables. Various reasons for weighting one diagonal over the other will be discussed in the context of the individual measures. Table 4.5 provides a summary of the various measures along with names traditionally associated with them. Every mechanically derived combination is included in the table even though five possibilities apparently are worthless. The fourteen measures are discussed individually as follows:

1. The value of this measure is the probability that a randomly chosen data unit will score 1 on both variables. It excludes 0–0 matches as irrelevant in counting the number of times the two variables match (the numerator) but does count 0–0 matches in determining the number of possibilities for a match (the denominator).

2. The value of this measure is the probability that a randomly chosen data unit achieves the same score on both variables. The 0–0 matches are given full weight.

3. The value of this measure is the conditional probability that a randomly chosen data unit will score 1 on both variables, given that data units with 0–0 matches are discarded first. The 0–0 matches are treated as totally irrelevant.

4, 8, and 12. These measures treat the 0–0 matches as relevant in the numerator but exclude such matches from the denominator. Since the numerator usually can be viewed as the number of relevant possibilities fulfilled, it is nonsense to include in the numerator that which is excluded specifically from the denominator.

5 and 9. These two measures are analogous to measure number 1 since they exclude 0–0 measures in the numerator while including them in the denominator. They have not appeared in the clustering literature and no interpretation seems readily available. However, they do not seem to have obvious faults which might prompt summary rejection as with numbers 4, 8, and 12.

TABLE 4.5
MATCHING COEFFICIENTS

Weighting of matches, mismatches	0–0 matches in denominator	0–0 matches in numerator	
		Excluded	Included
Equal weights	Included	1. Russell and Rao $$\frac{a}{a+b+c+d}=\frac{a}{n}$$	2. Simple matching $$\frac{a+d}{a+b+c+d}=\frac{a+d}{n}$$
	Excluded	3. Jaccard $$\frac{a}{a+b+c}$$	4. Nonsense $$\frac{a+d}{a+b+c}$$
Double weight for matched pairs	Included	5. Not recommended $$\frac{2a}{2(a+d)+b+c}$$	6. Unnamed $$\frac{2(a+d)}{2(a+d)+b+c}$$
	Excluded	7. Dice $$\frac{2a}{2a+b+c}$$	8. Nonsense $$\frac{2(a+d)}{2a+b+c}$$
Double weight for unmatched pairs	Included	9. Not recommended $$\frac{a}{a+d+2(b+c)}$$	10. Rogers–Tanimoto $$\frac{a+d}{a+d+2(b+c)}$$
	Excluded	11. Unnamed $$\frac{a}{a+2(b+c)}$$	12. Nonsense $$\frac{a+d}{a+2(b+c)}$$
Matched pairs excluded from denominator	—	13. Kulczynski $$\frac{a}{b+c}$$	14. Unnamed $$\frac{a+d}{b+c}$$

6. Sokal and Sneath (1963) include this measure in their list without attribution. It may be viewed as an extension of measure number 2 such that matched pairs are given double weight. The double weighting seems to exclude any possibility for a probabilistic interpretation.

7. This measure excludes 0–0 matches entirely while double weighting 1–1 matches. It may be viewed as an extension of measure number 3, though the probabilistic interpretation is lost. Hall (1969, p. 322) offers an alternative interpretation:

> However, for 0, 1 mismatches the zero is just as trivial as in the 0, 0 case. Mismatches should then lie about midway along the scale of significance between the 0,0 and 1,1 cases respectively. The number of mismatches in the coefficient should by this reasoning be multiplied by $\frac{1}{2}$.

Clearing the $\frac{1}{2}$ fraction then results in double weight for the 1–1 matches.

10. In the context of association among variables, this coefficient is best viewed as an extension of measure number 2 based on double weighting of unmatched pairs. In the context of association among data units it has an interesting interpretation given in Section 5.3.1.

13. This measure is the ratio of matches to mismatches with 0–0 matches excluded.

14. This measure is the ratio of matches to mismatches including 0–0 matches.

Measures 3, 7, and 11 are all monotonic to each other. To illustrate the method of proof, suppose there are two tables denoted by 1 and 2 and that measure number 7 gives the result

$$\frac{2a_1}{2a_1 + b_1 + c_1} \geq \frac{2a_2}{2a_2 + b_2 + c_2}.$$

Since the table entries are all nonnegative, the fractions may be cleared to give

$$4a_1 a_2 + 2a_1(b_2 + c_2) \geq 4a_1 a_2 + 2a_2(b_1 + c_1).$$

Subtracting $2a_1 a_2$ from both sides and dividing by 2 gives

$$a_1 a_2 + a_1(b_2 + c_2) \geq a_1 a_2 + a_2(b_1 + c_1),$$

which implies

$$\frac{a_1}{a_1 + b_1 + c_1} \geq \frac{a_2}{a_2 + b_2 + c_2}$$

or monotonicity with measure number 3. The same method may be used to prove that measures 2, 6, and 10 are also monotonic to each other. This result is important because when using monotonically invariant clustering techniques, such as single linkage and complete linkage, measures 3, 7, and 11 are equivalent to each other; measures 2, 6, and 10 are likewise equivalent to each other when using such techniques.

4.3.3 PROBABILITY BASED MEASURES

Among the matching measures, numbers 1, 2, and 3 possess reasonably useful probabilistic interpretations. There are several additional measures with probabilistic foundations.

The quantity $a/(a + b)$ is the conditional probability that a randomly chosen data unit scores 1 on variable B given that it scored 1 on variable A. Likewise the quantity $a/(a + c)$ is the conditional probability of scoring a 1 on variable A given that a 1 was scored on variable B. In Section 4.2.2, while developing the lambda measure, it was found useful to adopt the view that variable A is estimated half the time and variable B the other half; this same device then gives the symmetric measure

$$\frac{1}{2}\left[\frac{a}{a+b} + \frac{a}{a+c}\right], \tag{4.19}$$

which is the conditional probability of scoring a 1 on one variable given a score of 1 on the other. Sokal and Sneath (1963, p. 130) attribute this measure to Kulczynski.

It may happen that it is not clear which half of each dichotomy should receive the score of 1; by taking the arithmetic mean of expression (4.19) computed with the scores of Table 4.2 and the same expression computed with the reverse scoring the following measure is obtained:

$$\frac{1}{4}\left[\frac{a}{a+b} + \frac{a}{a+c} + \frac{d}{b+d} + \frac{d}{c+d}\right]. \tag{4.20}$$

Note that expressions (4.12) and (4.13) are very similar to these last two expressions, the difference being geometric averaging versus arithmetic averaging of the conditional probabilities.

The final 2×2 measure is

$$\frac{(a+d)-(b+c)}{a+b+c+d}, \tag{4.21}$$

which is the probability that a randomly chosen data unit will score the same on both variables minus the probability it will score differently on the two variables. Since $b + c = n - (a + d)$ this measure may also be written as

$$\frac{2(a+d)}{a+b+c+d} - 1,$$

which is related monotonically to measures 2, 6, and 10 of the matching coefficients. Sokal and Sneath (1963) credit this measure to Hamann.

4.3.4   VARIOUS ASPECTS OF USING BINARY VARIABLES

It may be advantageous to formulate a clustering problem in terms of binary variables because of certain attractive practical consequences. First, a dichotomy is often a quite reliable distinction in that it usually is possible to find a way to divide a set of data so that the two subsets have strongly different behavior in terms of the variable of interest. The emphasis in this statement is on the differences between groups rather than any homogeneity *within* groups; it may be that one-half of a dichotomy consists of several distinct subgroups. As compared to multiclass groupings, a dichotomy is less likely to rest on finely divided differences.

Second, a larger variety of association measures apply to binary variables than any other variable type. The analyst has a greater opportunity to highlight the sense of association most relevant to the problem at hand. By the same token the actual choice of a measure may be more difficult because of the larger variety. Time and resources permitting, a possibly useful device is to try several measures in an effort to investigate different senses of association among the variables.

Third, the use of binary variables provides an opportunity to achieve massive data compression in computer storage and a substantial reduction in computational effort. In FORTRAN IV the unit of storage is the computer word consisting of 32 (IBM 360) to 60 (CDC 6000-series) bits. Usually an entire word is used to store the score for one data unit on one variable. In the case of binary variables this same information can be stored in a single bit. To illustrate the magnitude of the advantage, a data set of 300 variables and 5000 data units would require 1,500,000 words of storage. But if the variables are binary, then the same data set can be stored at the bit level in 25,000 words on a machine with a 60-bit word length. Using word level storage the problem is impossible to manage within core storage; but with bit level storage, it will fit nicely in several existing machines. Appendix D includes a discussion of the principal technical considerations and listings of computer programs for utilizing bit level storage.

4.4   Strategies for Mixed Variable Data Sets

Up to this point attention has been focused on the measurement of association between variables of the same type. But real data sets often involve a mixture of nominal, ordinal, and interval variables. The previous discussion includes no provision for measuring association between variables of different types, much less the more difficult problem of obtaining a consistent measure across all pairwise combinations of variables in a mixed data set. A variety of difficulties always will surround this problem, but there seem to be at least three workable strategies for dealing with mixed variable types.

4.4.1 PARTITIONING OF VARIABLES

Perhaps the most obvious approach is to partition the variables into types and confine the analysis to the dominant type. The question of which type is "dominant" is a matter for informed judgement and may depend on factors such as the number of variables of each type, the variables thought to be most important to the analysis, relevant theory, and such other considerations. In one way or another many statistical analyses are restricted to avoid the problems of heterogeneous data sets. Often the problem is formulated at the outset in terms of only one variable type, perhaps at the risk of overlooking important factors. If the data set precedes the decision to perform an analysis, then concentrating attention on only one type involves ignoring or discarding data which obviously may be relevant but of the wrong type, a wasteful and unpleasing practice.

A logical extension of this approach is to partition the variables into types and perform separate independent analyses for each type; at least all the variables are considered. But then, how are the results of several parallel analyses to be integrated into one composite result? In many kinds of analysis only a qualitative and intuitive approach is available.† Cluster analysis may be enough of an aid to the analyst's intuition that mixed groups can be identified "by inspection" of the results. As a hypothetical example, suppose a data set involves twenty variables, ten nominal (denoted by N) and ten interval (denoted by I). Further suppose that there are three (unknown) "objective" clusters of variables as follows:

Cluster	Variables
1	I1, I2, N1, N2, N3, N4
2	I3, I4, I5, I6, I7, N5
3	I8, I9, I10, N6, N7, N8, N9, N10

If the chosen measures of association and clustering criterion reflect the structural characteristics of the "objective" classification, then clustering

† A formal method for integrating separate analyses in the context of discriminant analysis has been investigated by Kossack and his students at the University of Georgia. The most accessible description is provided by Henschke (1969). The method is in need of further development and does not seem presently suitable for adaptation to cluster analysis problems.

each type of variable separately should tend to give the following groups:

Interval clusters	Nominal clusters
I1, I2	N1, N2, N3, N4
I3, I4, I5, I6, I7	N5
I8, I9, I10	N6, N7, N8, N9, N10

Now these six clusters are all subsets of the "objective" clusters. They represent at least a partial solution though there are too many clusters relative to the "objective" classification. However, even without a formal analysis cutting across variable types, the aggregation of twenty variables into six clusters may have condensed the problem to the point that the analyst can identify intuitively the proper pairings to form the "objective" clusters. And even if only one of the three pairings can be established, the result certainly will contain more information than would be available had the problem been restricted to a single variable type. It is even possible that the six cluster solution itself will be adequate for the analyst's purpose; it might be seen as a little more detailed than necessary, but satisfactory in that each of the six groups contain variables from only one "objective" cluster.

4.4.2   CONVERSION OF VARIABLES

The major portion of Chapter 3 was devoted to methods of converting a variable from one type to another while retaining as much of the original information as possible. All of this material now may be brought to bear in the effort to transform a set of mixed variables into a new set of variables, all of a single type. The primary question to be faced is which variable type should be chosen as the single type for the analysis. From a practical point of view, this choice probably will be determined by which variable type is most numerous in the data set and the relative effort required for each kind of conversion. Consider each of the possibilities in turn.

Conversion to interval variables permits use of the product-moment correlation coefficient which is both a powerful measure of association and a familiar statistical quantity. Ordinal variables may be converted by using their ranks as class scores, a technique discussed in Section 3.2.5. Conversion of nominal variables requires the use of a reference variable in order to establish an ordering of the classes; the class scores then may be set either to ranks or to optimal scores which maximize the correlation with the reference variable. Section 3.2.6 includes the relevant details. When converting several nominal variables there is the option of associating each one with its own unique reference variable versus using some reference variables repeatedly. It might be useful to employ just one reference variable for the entire set of conversions in which case the reference variable would have substantial influence, even if it was not explicitly part of the actual data set. Section 9.4.2 includes discussion of a strategy for wholesale conversions to maximize the influence of a reference variable and thereby permit clustering with respect to an external criterion. Finally, as discussed in Section 3.1.4, binary variables do not require conversion when using the product moment correlation coefficient.

Conversion to nominal variables permits use of either of the optimal prediction measures (Section 4.2.2) or the canonical correlation method (Section 4.2.3). Converting interval variables is essentially a one-dimensional grouping problem for which many methods are discussed in Section 3.2.1. Converting ordinal variables is achieved simply by forgetting the order information and perhaps combining some classes (see Section 3.2.3). And, of course, binary variables are just nominal variables with two categories.

Conversion to binary variables permits use of a wide array of association measures, many of which have probabilistic interpretations. Also, the use of binary variables opens the way to substantial compression of storage and increased computational efficiency. Of course the problem is how to dichotomize all the variables that are not already in binary form. For interval variables the problem is a special case of interval to nominal conversion. For ordinal and nominal variables it is a problem of combining groups into dichotomies (see Section 3.2.7 for some useful techniques).

Without actually considering any alternatives the reader probably has assumed that the foregoing discussion necessarily implies one-to-one conversions; that is, one original variable is converted to exactly one new variable of the desired type. But why should the analysis be restricted in this manner? Some conversions are difficult because there are so many alternatives available; several reference variables may be of potential or actual interest; there may be many useful groupings in conversions to nominal variables. Rather than force a choice, there is much to be gained by allowing the old variable to be represented by several new variables of the desired type. To illustrate, consider a nominal variable with four categories: A, B, C, and D. If the analysis is to be carried out in terms of binary variables, there are seven alternative dichotomies:

1. (A) (B, C, D)	5. (A, B) (C, D)
2. (B) (A, C, D)	6. (A, C) (B, D)
3. (C) (A, B, D)	7. (A, D) (B, C)
4. (D) (A, B, C)	

Suppose prior knowledge indicates that B and D should go together but it is not clear what should be done with A and C. Dichotomies 1, 3, and 6 are still

candidates and it is difficult to choose among them. Consider what would
happen if all three binary variables corresponding to these alternatives were
used in a cluster analysis of variables. If the three are somehow equivalent
they should all turn up in the same cluster. If they are different, they should
fall into separate clusters. How do you choose which dichotomy is best?
Actually it is probably best not to force a choice but to see how many of the
three have some meaningful characteristics based on their cluster memberships. Each dichotomy is a different representation of the original variable;
if three different representations fall into three separate clusters it may happen
that they tap three distinct aspects of the variable. In effect, a single variable is
given a multidimensional representation.

Using multiple representations for the transformed variable is not limited
to dichotomization problems as in the example. The idea can be applied with
any of the conversion schemes described in Chapter 3. For example, several
different sets of class scores might be used to represent a nominal variable in
an analysis based on interval variables. Multiple representations may be
beneficial in many statistical problems. Indeed the use of polynomial and
cross-product terms in the general linear model of statistics might be viewed
as the utilization of multiple representations.

In cluster analysis the use of several new variables to represent one original
variable may cause considerable growth in the size of the problem. In most
cases the matter should not cause serious concern because the hierarchical
methods of Chapter 6 will accommodate up to 300 variables in a single problem
on the CDC 6600. Outside of some problems in zoology, information retrieval,
and signal detection, such a number of variables would be considered quite
large, even with the augmentation of multiple representations. If the problem
does grow out of bounds, the large problem strategies of Chapter 9 might lead
to a satisfactory solution.

4.4.3   COMPATIBLE MEASURES

The third strategy for dealing with mixed variable data sets is to use a set
of measures which are compatible with each other and collectively cover every
pairwise combination of variable types. This strategy would provide an ideal
solution if all the measures were fully compatible. As will be seen, the following
proposal falls short of achieving perfection but comes close enough to be
worthy of serious consideration.

The proposed set of measures for each combination of variable types is as
follows:

1. Interval–interval: product moment correlation coefficient of Section
4.1.2.

2. Nominal–interval: product moment correlation coefficient computed with optimal scores assigned to each nominal class. See Appendix A, Section A.1, and Section 3.2.6.2 for two different but equivalent developments of the optimal class scores.

3. Nominal–nominal: canonical correlation which is the product moment correlation coefficient computed using optimal scores for both sets of nominal classes. See Appendix A for development of the technique and Section 4.2.3 for additional discussion.

Ordinal variables are scored with their ranks and thereafter treated as interval variables in accordance with the discussion of Section 3.2.5. Binary variables may be treated as nominal variables with two classes or they may be treated as interval variables as discussed in Section 3.1.4.

The key element in this proposal is the assignment of scores to classes. Once these scores are available, every pairwise association is computed using the product moment correlation. The scores for interval variables are the observed values while scores for ordinal variables are ranks and binary variables are scored with any two distinct numbers. Thus for interval, ordinal, and binary variables a single set of scores is used throughout all comparisons. However, the optimal scores for nominal variables are determined anew for each pairwise comparison. If a single reference variable could be chosen for a nominal variable, then a single set of scores could be used throughout all comparisons. In the absence of a reference variable, the scores are determined so as to maximize the correlation computed from each individual pairwise comparison. Since these correlations are the maximum attainable with any set of scores, there is at least a suspicion that correlations involving nominal variables might tend to be biased upward as compared to correlations between interval variables. A proposed modification is then to adopt the ordering of the optimal scores but replace these scores by their ranks. The nominal classes then are treated as being equally spaced just like the ordinal variables; however the ordering is allowed to be determined separately for each pairwise comparison. It seems reasonable to allow a changeable order for otherwise the nominal variable would become an ordinal variable. The details behind both the optimal scores and rank scores are discussed in Appendix A. This approach is implemented with computer programs in Appendix C.

5

MEASURES OF ASSOCIATION AMONG DATA UNITS

Given a data set of n variables and m data units, a common device for displaying the measured values is the data matrix of n rows and m columns. The ith row of the matrix contains all scores pertaining to the ith variable and the jth column contains all scores for the jth data unit. Within this setting the methods of Chapter 4 deal with association among the rows of the data matrix while those of this chapter pertain to association among columns.

Shifting attention from rows to columns in the data matrix involves a major change in the character of the score vectors to be studied. A row vector of scores is the collective response of all data units to a single variable and consequently all scores are comparable to each other. On the other hand, a column vector of scores for a data unit cuts across all the variables. There may be quite a variety of measurement units and variable types. This heterogeneity makes it especially difficult to define meaningful measures of association between data units within the context of a given set of variables.

In a simple problem with only two variables the analyst can plot the data units in two dimensions as in Fig. 5.1. The distances between points can be assessed visually and the clusters identified by inspection. In problems of many variables selected two-dimensional plots may be beneficial but rarely informative enough to suggest a complete set of clusters; the problem is beyond the grasp of unaided intuition and needs the help of a systematic cluster analysis algorithm. Visual assessment of distances is impossible in spaces of more than three dimensions and must give way to computational methods.

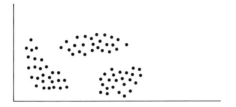

Fig. 5.1. Two-dimensional clustering.

This chapter gives a rather comprehensive discussion of the problems of combining scores on disparate variables to compute distances between data units. Following the same general format as in Chapter 4, the various distance measures are classified by the variable types for which they are defined; methods for mixed variable data sets are presented last. Continuing attention is given to the problem of finding suitable bases for equating the effects of variables such that relative weights can be assigned with some confidence. As will be seen, the problem of assigning weights to the variables requires the active participation of the analyst as a source of judgements and subjective appraisals; the blind application of prescribed rules and "standard formulas" cannot be expected to suffice.

5.1 Metric Measures for Interval Variables

The most mathematically sophisticated of the distance functions are those called metrics. This class of function is of general mathematical interest and consequently has received considerable study. This discussion will highlight only certain results most directly applicable in cluster analysis.

5.1.1 FORMAL PROPERTIES OF A METRIC

Let E be a symbolic representation for a measurement space and let X, Y, and Z be any three points in E. Then a distance function D is a metric if and only if it satisfies the following conditions:

1. $D(X, Y) = 0$ if and only if $X = Y$,
2. $D(X, Y) \geqslant 0$ for all X and Y in E,
3. $D(X, Y) = D(Y, X)$ for all X and Y in E,
4. $D(X, Y) \leqslant D(X,Z) + D(Y,Z)$ for all X, Y, and Z in E.

The first property implies that X is zero distance from itself and that any two points zero distance apart must be identical. The second property prohibits negative distances. The third property imposes symmetry by requiring the distance from X to Y to be the same as the distance from Y to X. The fourth property is known as the triangle inequality and it requires that the length of one side of a triangle be no longer than the sum of the lengths of the other two sides. These properties are in accordance with intuitive notions because the popular conception of a distance is the Euclidean distance of elementary geometry, itself a metric.

It may be verified quite easily that the sum of two metrics is also a metric. However, the product of two metrics (in particular the square of a metric) does not necessarily satisfy the triangle inequality and so may not be a metric. Any positive multiple of a metric is a metric. If D is a metric and w is any positive number, then

$$D' = \frac{D}{w + D}$$

is also a metric. To show this assertion to be true, let r, s, and t correspond to the distances in the triangle inequality such that $r + s \geq t$; then it must be shown that

$$\frac{r}{w + r} + \frac{s}{w + s} \geq \frac{t}{w + t}.$$

Now consider three cases: (1) if $t = 0$ the relation is satisfied trivially; (2) if $s \geq t$, then

$$\frac{1}{s} \leq \frac{1}{t}, \quad \frac{w}{s} \leq \frac{w}{t}, \quad \frac{w + s}{s} \leq \frac{w + t}{t}, \quad \frac{s}{w + s} \geq \frac{t}{w + t},$$

and similarly if $r \geq t$; (3) if $s < t$ and $r < t$ then

$$s < t, \quad \frac{w + s}{s} < \frac{w + t}{s}, \quad \frac{s}{w + s} > \frac{t}{w + t},$$

and similarly for r; then

$$\frac{r}{w + r} + \frac{s}{w + s} > \frac{r}{w + t} + \frac{s}{w + t} > \frac{t}{w + t}.$$

A function which satisfies the first three conditions of a metric but not the triangle inequality is known as a semimetric. A metric which additionally satisfies

$$D(X, Y) \leq \max\{D(X, Z), D(Y, Z)\} \qquad \text{for all} \quad X, Y, Z \text{ in } E,$$

is called an ultrametric (Johnson, 1967). This latter property is considerably stronger than the triangle inequality.

5.1.2   THE MINKOWSKI METRIC AND SPECIAL CASES

Let x_{ij} be the score achieved by the jth data unit on the ith variable and let the vector of scores for the jth data unit be $X_j^{\mathrm{T}} = (x_{1j}, \ldots, x_{nj})$. Then the Minkowski metric between data units j and k is

$$D_p(X_j, X_k) = \left[\sum_{i=1}^{i=n} |x_{ij} - x_{ik}|^p \right]^{1/p}, \tag{5.1}$$

where $p \geqslant 1$. By choosing various values of p many different metric distance functions can be obtained. The so-called "city block," "taxicab," or L_1 metric is obtained by taking $p = 1$:

$$D_1(X_j, X_k) = \sum_{i=1}^{i=n} |x_{ij} - x_{ik}|. \tag{5.2}$$

The familiar Euclidean distance or L_2 metric is obtained by taking $p = 2$:

$$D_2(x_j, X_k) = \left[\sum_{i=1}^{i=n} (x_{ij} - x_{ik})^2 \right]^{1/2}. \tag{5.3}$$

The Chebychev metric is obtained as the limit of $D_p(X_j, X_k)$ as p increases without bound and so sometimes is called the L_∞ (L-infinity) metric:

$$D_\infty(X_j, X_k) = \max_{i=1,\ldots,n} |x_{ij} - x_{ik}|. \tag{5.4}$$

The set of points for which $D_p(X_j, X_k) = 1$ is called the unit ball or unit sphere of the metric. For two dimensions, the unit balls of the L_1, L_2, and L_∞ metrics can be plotted as in Fig. 5.2. The unit balls of L_p metrics for $1 < p < 2$ are convex curves lying between the unit balls for L_1 and L_2 metrics. For the case $2 < p < \infty$, the unit ball for a given p is a convex curve lying between the unit balls for the L_2 and L_∞ metrics. The unit ball shows the shape of the surface formed by the collection of points which are at a unit distance from another point.

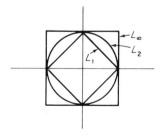

Fig. 5.2. Unit balls for common metrics.

Of all possible metrics, most attention is given to the Euclidean or L_2 metric. The L_1 metric occasionally is encountered and metrics based on other values of p hardly ever are of more than theoretical interest.

5.1.3   TRANSFORMATION OF VARIABLES

All varieties of the Minkowski metric have three characteristics that should be kept in mind:

1. Variables and their manner of representation are taken as given. If variable 1 is expressed in feet and variable 2 in pounds, then the metric involves the sum of the pth power of a difference in feet and the pth power of a difference in pounds.

2. Each variable is treated in a linear manner; that is, each variable appears as itself; more complex functional forms such as polynomials are not included explicitly.

3. Each variable is treated independently of all the others. The contribution of each variable is the pth power of the difference in score for two data units and this quantity depends in no way whatever on the scores achieved on other variables.

If taken quite literally, these three characteristics would be very restrictive. However, the analyst has the opportunity to generate dummy variables by revising measurement scales or creating new composites as functions of the original variables. This section discusses various aspects of transforming the original variables and provides a conceptual approach which highlights the points where the analyst needs to inject his judgement to shape the outcome.

Before proceeding with the problems of comparing two data units across an array of variables, it is wise to make sure that the data units are comparable in the desired sense on each individual variable. For example, suppose the purpose of the analysis is to investigate the public health characteristics of all the counties in a state. Without giving the matter much thought it might seem appropriate to describe each county in terms of the number of reported cases of various diseases; but this kind of data will reflect the differences in population much more than the differences in the state of health for each county; per capita figures would be more revealing. In many instances it may be necessary to remove the effect of some pervasive variable (like population) before taking other steps in the analysis.

5.1.3.1   *Assignment of Weights for Variables*

In a typical data set the variables are measured in a variety of units. Using the variables in this raw state has two serious drawbacks:

1. The units of measurement determine the magnitude of the scores. The choice between feet and yards on one variable and between pounds and ounces on another involves an implicit weighting of the variables.

2. The units for the different variables are combined to achieve a single measure of distance which implies a composite of these units. How is the sum of feet and ounces to be interpreted? How may the influence of each variable be assessed in the presence of mixed units of measurement?

With these problems in mind the first order of business is to equalize the variables in some appropriate manner. The principal idea in equalization is to remove the artifact of the measurement unit and anchor each variable to some common numerical property. Disposing of the measurement unit involves dividing all the scores for a variable by a suitable equalizing factor expressed in the same units. The sense in which the variables are equalized depends on which kind of equalizing factor is chosen. Three of the most appealing alternatives for equalizing factors are discussed below.

1. *Mean score on the variable.* The expected score on a variable equalized by the mean is one and the expected difference on the equalized scores for two randomly chosen data units is zero. Strictly speaking the mean should be used as an equalizing factor only for ratio variables, that is, variables with meaningful zero points. The mean for strictly interval variables is an arbitrary quantity which may be changed by uniform addition of any desired constant. For example, take the six numbers $(-10, -5, 0, 5, 10, 30)$ and divide by the mean of 5 to obtain $(-2, -1, 0, 1, 2, 6)$. If 5 is added to all the original scores, the new scores $(-5, 0, 5, 10, 15, 35)$ preserve the differences between data units; but dividing by the new mean of 10 gives $(-0.5, 0, 0.5, 1, 1.5, 3.5)$, quite a different set of scores than that obtained when the mean was 5.

2. *Range of the variable.* If x_{ij} is the score on the ith equalized variable for the jth data unit, then the quantity $|x_{ij} - x_{ik}|$ lies between 0 and 1. The range of the variable is the largest observed difference between any two data units so that the difference in equalized scores may be viewed as the fractional disagreement (relative to the maximum possible) between two data units. Since the observed range of a variable is the difference between its two most extreme scores, the analyst should examine such scores and their associated data units for errors of observation or indications that the data units really do not belong with the data set. Special scrutiny is indicated when an extreme score is grossly different from the next most extreme.

3. *Standard deviation of the variable.* The equalized variables have unit variance and the quantity $|x_{ij} - x_{ik}|$ is in terms of standard deviations. Using this equalizing factor gives the same results as the often recommended procedure of standardizing the data to zero mean and unit variance; since the quantity of interest is the score difference between data units, it does not matter whether or not the mean is subtracted out first.

Many authors would consider these equalizing transformations as complete weighting methods and devote no further attention to the topic. Perhaps a more rational view is that the scores are ready for weighting only after the variables have been equated to some common basis such as unit mean, unit range, or unit variance. After applying an equalizing transformation the analyst can utilize his judgement and his understanding of the problem to assign weights so that each variable contributes to the mean, range, or variance of the composite in a manner consistent with his objectives and interests in the analysis. From this point of view, it is an abdication of responsibility for the analyst merely to standardize the data and let it go at that; rigid adherence to such a practice is tantamount to saying that an increment in standard deviation is equally important for all variables regardless of the purpose of analysis. The choice of weights is not a matter for neat mechanical devices; it is one of the principal means through which the analyst can shape the analysis to his objectives.

There is an alternative to the method of equalizing variables and applying subjective weights as just described. If the desired distances between data units can be estimated or are known to be related closely to a single variable, then a criterion variable can be established and the weights can be determined by simple linear regression. In particular, let y_j be the criterion value for the jth data unit; then the weights t_i are found as the solution to the regression problem

$$|y_j - y_k|^p = \sum_{i=1}^{i=n} |t_i(x_{ij} - x_{ik})|^p + e_{jk}. \tag{5.5}$$

That is, the t_i weights are chosen so as to minimize the squared error e_{jk}^2 between the predicted and criterion distances. The ith term in the summation may be written as

$$|t_i(x_{ij} - x_{ik})|^p = |t_i|^p |x_{ij} - x_{ik}|^p = b_i |x_{ij} - x_{ik}|^p.$$

If p is an odd integer, then b_i may have any sign and the problem is one of ordinary linear regression; otherwise it is necessary to constrain b_i to be nonnegative because a negative b_i would give a t_i which is either imaginary or undefined.

If there existed a measurable criterion variable such that the score difference on this variable really was the desired distance, then the problem would be reduced to a one-dimensional form and there would be no need to deal with the x_{ij} scores or t_i weights. But lacking such a perfect criterion, there are at least three instances in which it would be beneficial to use this regression approach for determining the t_i weights.

1. If the criterion score y_j or the criterion distance $|y_j - y_k|$ is estimated subjectively, then the t_i weights can be determined to "capture" or simulate the analyst's policy for assigning scores or distances.

2. If the criterion is difficult or costly to determine, then the t_i weights might be estimated from a subset of cases and then applied to the determination of distances for the entire data set.

3. Even though a perfect criterion is not available, it may be desired to make the computed distances maximally related to some criterion variable; the other variables exert their influence to the extent that the criterion is predicted imperfectly.

The primary problem with this regression approach is that of finding a suitable criterion. As has been mentioned, subjective estimation might provide such a criterion; if not, some measured variable is needed. The basic idea is much the same as that of using a reference variable, a device frequently encountered in Chapter 3 as a means of determining suitable scale conversions.

There is also a question of what to estimate. If the criterion scores y_j can themselves be estimated, then for m data units only m scores are required. But, if the distances between data units, $|y_j - y_k|$, are estimated, then distances for all $\frac{1}{2}m(m-1)$ distinct pairs of data units are required to resolve all ambiguities.

Regardless of the method used for obtaining the equalizing factor and weight for each variable, their combined effect can be expressed as a single multiplier t_i, which transforms the original score x_{ij} into the weighted score $\hat{x}_{ij} = t_i x_{ij}$. This scaling of the variables can be conveniently represented in terms of a diagonal matrix $\mathbf{T}$ in which the ith diagonal element is t_i and all off-diagonal elements are zero. Then the vector of transformed scores for the jth data unit is†

$$\hat{\mathbf{X}}_j = \mathbf{T}^T \mathbf{X}_j.$$

Using Eq. (5.3), the squared Euclidean distance between two data units (as measured on the transformed variables) may be written in various forms

$$D_2{}^2(\hat{\mathbf{X}}_j, \hat{\mathbf{X}}_k) = D_2{}^2(\mathbf{T}^T \mathbf{X}_j, \mathbf{T}^T \mathbf{X}_k) = \sum_{i=1}^{i=n} [t_i(x_{ij} - x_{ik})]^2$$
$$= [\mathbf{T}^T(\mathbf{X}_j - \mathbf{X}_k)]^T \, [\mathbf{T}^T(\mathbf{X}_j - \mathbf{X}_k)]$$
$$= (\mathbf{X}_j - \mathbf{X}_k)^T \, \mathbf{T}\mathbf{T}^T(\mathbf{X}_j - \mathbf{X}_k).$$

5.1.3.2   Representation Spaces

As the analyst selects, equalizes, and weights variables he effectively builds a representation space which is a transformation of the original space of measured values. The representation space need not be restricted to simple

---

† Since $\mathbf{T}$ is a symmetric matrix, $\mathbf{T} = \mathbf{T}^T$, and it makes no difference whether the transpose is used or not. The transpose is used here for consistency with later notation.

multiples of the original variables; dummy variables defined as powers, polynomials or more complex functions of one or several of the original variables also may be constructed to represent nonlinearities and interactions. The representation space may be of either larger or smaller dimension than the original space depending on whether the dummy variables are used in place of or in addition to some equalized and weighted original variables.

The use of such constructed dummy variables is a familiar device in regression analysis. However, in the regression setting the contribution of a particular dummy variable can be assessed in terms of the marginal reduction in the error sum of squares. In cluster analysis no such criterion is available; the analyst can rely only on his knowledge of the problem to choose relevant transformations and on his judgement to assess the results. However, the effectiveness of such judgements may be enhanced by an appreciation of the geometric implications of the dummy variables.

The unit ball for the Euclidean metric in the representation space is a hypersphere. To understand the effect of various transformations on the original data, it is informative to map the unit ball from the representation space back to the original space and observe the resulting shape. The effects of various

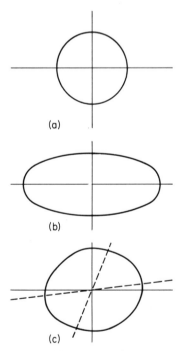

Fig. 5.3. Image of unit ball in original space. (a) Identity: $\hat{x}_{ij} = x_{ij}$. (b) Multiplicative scaling: $\hat{x}_{ij} = t_i x_{ij}$. (c) Linear combination: $\hat{x}_{ij} = \sum_{k=1}^{k=n} t_{kj} x_{ik}$.

linear transformations are illustrated in Fig. 5.3. Let $\hat{x}_{ij}$ be the score of the jth data unit on the ith variable in the representation space and x_{ij} be the corresponding score in the original space. The identity transformation makes no change so the image of the unit ball is itself a sphere. Multiplicative scaling of each variable as in the preceding section causes some axes to stretch and others to shrink so the image is an ellipsoid.

By defining the new variables as linear combinations of the old variables, the image becomes a pseudoellipsoidal form with nonorthogonal principal axes; the dotted lines in Fig. 5.3 show the images of the representation variables in the space of the original variables. Note that each axis may be rotated to a different extent than the other axes. A special linear transformation called principal components analysis is discussed further in Section 5.1.3.3. Addition of a scalar constant to any of these transformations causes the image to be displaced from the origin.

When squares and cross products are also introduced the images may assume the shape of a torus (doughnut), paraboloid, or hyperboloid. The drawback with using such transformations is that it is difficult to know which shapes to anticipate in each pair of dimensions. If it is expected that clusters should have a paraboloid shape, then the projection on some planes will be parabolic while on others the projection will be elliptical. Using such transformations to recover particular cluster shapes requires a great deal of prior knowledge about the dimensionality and orientation of the clusters. Rohlf (1970) discusses the use of squares and cross products further and gives some interesting two-dimensional examples.

5.1.3.3   Principal Components Analysis

Given a data set of n variables, some or all of which may be intercorrelated, it is possible to construct a set of n or fewer orthogonal composite variables which are linear combinations of the original variables and which account for the variance in the original data. Put another way, the axes representing the original variables may be rotated individually to be orthogonal with each other and in the process it may be found that fewer than n orthogonal axes will span the space. A particular technique for finding such a set of orthogonal variables is the method of principal components.

Let $\mathbf{X} = \{x_{ij}\}$ be the data matrix whose typical element x_{ij} is the score on the ith variable (either a constructed composite or a measured variable) for the jth data unit. The mean for the ith variable is $\bar{x}_i = \sum_{j=1}^{j=m} x_{ij}/m$. The matrix of centered or mean-deviated scores is $\hat{\mathbf{X}} = \{x_{ij} - \bar{x}_i\}$. The sample variance–covariance matrix for the data set is then

$$\frac{\hat{\mathbf{X}}\hat{\mathbf{X}}^T}{m} = \left\{ \sum_{j=1}^{j=m} (x_{ij} - \bar{x}_i)(x_{kj} - \bar{x}_k)/m \right\} = \{s_{ik}\} = \mathbf{S}.$$

The matrix S is $n \times n$, real, symmetric, and positive semidefinite (some number of the n eigenvalues, say r, are positive and the remainder are zero). The ordinary algebraic eigenproblem for S is to find the set of nonzero eigenvalues $(\lambda_1, \ldots, \lambda_r)$ and the corresponding set of eigenvectors $(a_1, \ldots, a_r)$ which are the r nontrivial solutions to the equation

$$[S - \lambda I]\, a = 0,$$

where 0 is the n-component vector of zeroes and I is the $n \times n$ identity matrix. The eigenvectors corresponding to the r nonzero eigenvalues may be chosen orthonormal to each other, that is,

$$a_i^T\, a_i = 1, \qquad i = 1, \ldots, r,$$
$$a_i^T\, a_j = 0, \qquad i \neq j.$$

With these preliminaries about the eigenvalues and eigenvectors of S in hand, the principal components of S may be described. Let Y be a new variable defined as a linear combination of the X_i variables

$$Y = \sum_{i=1}^{i=n} b_i\, X_i,$$

where b_i is the ith element in a vector of coefficients b. If the data units are evaluated on this new variable, the (row) vector of scores is

$$Y = b^T X,$$

where X is the original data matrix. The vector of coefficients b could be chosen in infinitely many ways; but for the principal components solution b is chosen to maximize the variance of Y, or equivalently, to find the line in the space of X on which the projections of the data units are dispersed maximally. The (row) vector of centered scores for Y is $\hat{Y} = b^T \hat{X}$ so that

$$\mathrm{var}\,(Y) = \frac{\hat{Y}\hat{Y}^T}{m} = \frac{b^T \hat{X}\hat{X}^T b}{m} = b^T S b.$$

The variance of Y may be made arbitrarily large by a suitable choice of the elements of b; therefore it is conventional to norm b to unit length, $b^T b = 1$. The problem now may be stated formally as

$$\text{maximize} \quad b^T S b, \qquad \text{subject to} \quad b^T b = 1.$$

Joining the constraint to the objective function with a Lagrange multiplier μ and differentiating, the solution b must satisfy

$$\frac{\partial}{\partial b}[b^T S b + \mu(1 - b^T b)] = 2Sb - 2\mu b = 0,$$

or

$$[S - \mu I] \, b = 0,$$

which is recognized to be an ordinary eigenproblem. Thus the vector of coefficients b is one of the eigenvectors a_i of S and μ is the associated eigenvalue λ_i. Since there are r nonzero eigenvalues of S it still remains to choose the appropriate solution. As found above $Sb = \mu b$; multiplying from the left by b^T then gives

$$b^T S b = \mu b^T b = \mu = \text{var}(Y),$$

since $b^T b = 1$. Therefore, the vector of coefficients for Y is the eigenvector of S corresponding to the *largest eigenvalue* and this eigenvalue is itself equal to the variance of Y. The linear composite Y is called the first principal component and the linear composite formed by using the eigenvector corresponding to the ith largest eigenvalue is called the ith principal component. For a matrix with r nonzero eigenvalues there are r principal components.

The principal components are of great interest not only because their coefficients define a set of orthogonal vectors, but also because of their maximum variance properties. The first principal component has the largest variance of any linear combination of the variables represented in the data matrix; the second principal component has the largest variance of any linear combination orthogonal to the first principal component; the third has the largest variance of any linear combination orthogonal to the first two, and so forth. Rao (1964) presents several useful interpretations of the principal components, foremost of which is that the best least-squares fit of the original space in a subspace of k dimensions is achieved by using the first k principal components. A measure of the degree of fit obtained with the first k principal components is

$$\frac{\lambda_1 + \ldots + \lambda_k}{\lambda_1 + \ldots + \lambda_r},$$

which may be interpreted as the fraction of the total variance "explained." In many situations only a few principal components are needed to make this ratio quite high, say 0.8 to 0.9; in such cases the space of the data matrix may be compressed greatly. In particular, if $A(k)$ is an $n \times k$ matrix whose columns are the k eigenvectors of S corresponding to the k largest eigenvalues, then the $k \times m$ matrix of principal component scores is $Y(k) = A(k)^T X$. These scores may be used in Eqs. (5.1)–(5.4) to compute metric distances in the principal components space.

The principal components actually obtained in this analysis are scale dependent (Kendall, 1968, p. 17). If the variables are standardized to unit variance, then the S matrix is a correlation matrix with ones in the diagonal.

Equalizing the variables to unit range or unit mean will give yet other forms for the **S** matrix and different sets of principal components. Most authors seem to ignore this scale dependency, or deal with it by standardizing blindly to unit variance when using principal components analysis as a prelude to cluster analysis.

The computation of the eigenvectors of a real symmetric matrix such as **S** may be accomplished quickly and accurately using subroutine EIGEN in Appendix C. Many factor analysis programs also give principal components. Additional discussion of the fundamental elements of principal components analysis may be found in Morrison (1967, pp. 82–85, 221–258). Some rather advanced aspects of this subject are discussed by Gower (1966).

5.2   Nonmetric Measures for Interval Variables

When the data units are described by interval variables most attention is given to the metric measures just discussed. However, some investigators have devised nonmetric measures which may be applied usefully in some problems. This section includes a review of two such measures and also some consideration of using correlation as a measure of association between data units.

5.2.1   CALHOUN DISTANCE

Bartels *et al.* (1970) have proposed a distance measure which utilizes only the ordering of data units along each dimension of the space. The basic concept for measuring the distance between two data units is to imagine them as opposite vertices of a hypervolume whose sides are parallel to the axes of the space. The distance is basically the fraction of all data units which fall into this hypervolume and its extensions. The idea is grasped most easily in pictorial form as in Fig. 5.4. The distance between points 1 and 2 is determined by the number of points in the shaded volume, that is, points numbered 3, 4, 6, and 7; these four data units have a value on at least one variable which falls between the values scored by data units 1 and 2. Bartels *et al.* (1970) have found it necessary to employ a weighting scheme to handle tied values. It is convenient to use the following defined counts:

N_i is the number of points which fall between the two points of interest on at least one variable, that is, points in the interior of the hypervolume and its extensions;

N_b is the number of points which do not fall between the two points of interest on any dimension but tie in value on one or more variables with one or

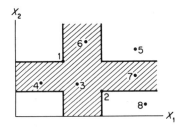

Fig. 5.4. Illustration of Calhoun distance.

the other of the data units, that is, points on the boundary of the hypervolume and its extensions;

N_z is the number of points which tie in value on at least one variable simultaneously with both points of interest but do not otherwise lie in the
interior or on the boundary of the hypervolume; that is, the two data
units of interest tie on one or more variables so that the associated portion
of the extended hypervolume is of zero thickness; points in this count
intersect the hypervolume only in this zero thickness portion.

Then the raw Calhoun distance is defined as

$$D_c = 6N_i + 3N_b + 2N_z. \tag{5.6}$$

Boundary points receive only half and "zero" points one-third the weight of
interior points. This choice of weighting factors is intuitive and offered only
as a suggestion which Bartels *et al.* found to be satisfactory for their purposes.
If the total number of data units is N, then the maximum possible value of D_c
is $6(N - 2)$, in which case the normalized Calhoun distance is

$$D_{nc} = (6N_i + 3N_b + 2N_z)/6(N - 2). \tag{5.7}$$

The Calhoun distance fails as a metric in two regards. First, two points
need not be identical to be zero distance apart as may be seen in Fig. 5.5.

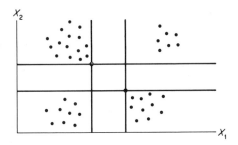

Fig. 5.5. Example of zero Calhoun distance.

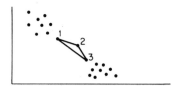

Fig. 5.6. Triangle inequality illustration.

Thus two points which probably would be considered as belonging to different clusters are zero distance apart. However, this anomaly can occur only if there is separation between the clusters in every dimension, a condition which probably does not occur often and which can be detected easily by most clustering methods using other measures. This measure can be expected to remain useful in the more usual case where clusters overlap on one or more dimensions as in Fig. 5.1. The second way in which the Calhoun distance fails as a metric is in the triangle inequality. Figure 5.6 illustrates the point. Using Eq. (5.6), the distances are $D(1,2) = D(2,3) = 0$, $D(1,3) = 6$ and consequently $D(1,3) > D(1,2) + D(2,3)$.

The Calhoun distance is invariant to transformations which preserve the order of the data units along the measurement axes. One kind of transformation to which the Calhoun distance is not invariant is an orthogonal rotation. Referring to Fig. 5.5, simultaneously rotating both axes a few degrees either direction brings a few points into the extended hypervolume. Deleting a variable can also have a substantial effect; referring to Fig. 5.4, if variable X_2 is deleted, then all points are projected onto the X_1 axis so that points 4 and 7 no longer lie "between" points 1 and 2. The ideas behind this measure are quite interesting and probably deserve more study.

5.2.2   LANCE AND WILLIAMS' NONMETRIC MEASURE

Lance and Williams (1966) suggest

$$D_{\mathrm{LW}} = \frac{\sum_{i=1}^{i=n} |x_{ij} - x_{ik}|}{\sum_{i=1}^{i=n} (x_{ij} + x_{ik})} \tag{5.8}$$

as a distance measure between data units j and k. The numerator is simply the L_1 metric of Eq. (5.2); the denominator may be viewed as a measure of the gross magnitude of the two data units. While Lance and Williams do not insist on it, it would appear wisest to use this measure only with nonnegative scores for otherwise substantial cancellations may occur in the denominator and possibly give negative distances. Because of this feature it would not be desirable to center the data by subtracting out the means. Since a different denominator is used for each pair of data units, the measure does not always satisfy the triangle inequality.

Apparently this measure was developed as a generalization of the Czekanow-ski or Dice coefficient, measure number 7 in Table 4.5. In the case of binary variables, D_{LW} reduces to

$$\frac{\sum\limits_{i=1}^{i=n} |x_{ij} - x_{ik}|}{\sum\limits_{i=1}^{i=n} (x_{ij} + x_{ik})} = \frac{b+c}{(a+b)+(a+c)} = 1 - \frac{2a}{2a+b+c}$$

or the complement of the Dice coefficient.

5.2.3   CORRELATION

In a purely formal sense, the product moment correlation coefficient can be computed as a measure of association between the score vectors of the two data units. Correlation is used in just this way in Q-mode factor analysis, a technique encountered with some frequency in psychology; Stephenson (1935, 1938, 1953) originated the technique as a means of identifying types of individuals through use of the methods of factor analysis which are well known among psychologists. The idea has been the subject of some contro-versy in psychology and has been severely criticized in other fields (Baker, 1965; Eades, 1965; Majone and Sanday, 1968). The central problem is that a vector of scores for a data unit involves many different units of measurement so that the mean and variance of such scores are rather meaningless. Yet these two statistics along with the covariance are needed in the computation of the "correlation" with another data unit.

Cronbach and Gleser (1953) show that the correlation coefficient has a limited metric character. In particular, let x_{ij} be the score for the jth data unit on the ith variable; then define the mean and scatter for the jth data unit as

$$\bar{x}_j = \sum_{i=1}^{i=n} x_{ij}/n, \qquad S_j = \left[\sum_{i=1}^{i=n} (x_{ij} - \bar{x}_j)^2\right]^{1/2}.$$

Now use these statistics to transform the scores for each data unit to zero mean and unit scatter. The Euclidean distance between the data units as measured on the transformed scores $\hat{x}_{ij} = (x_{ij} - \bar{x}_j)/S_j$ is then

$$D^2(\hat{X}_j, \hat{X}_k) = \sum_{i=1}^{i=n} \left[\frac{x_{ij} - \bar{x}_j}{S_j} - \frac{x_{ik} - \bar{x}_k}{S_k}\right]^2$$

$$= \frac{1}{S_j^2 S_k^2} \left[S_j^2 S_k^2 - 2S_j S_k \sum_{i=1}^{i=n} (x_{ij} - \bar{x}_j)(x_{ik} - \bar{x}_k) + S_j^2 S_k^2\right]$$

$$= 2\left[1 - \frac{\sum\limits_{i=1}^{i=n} (x_{ij} - \bar{x}_j)(x_{ik} - \bar{x}_k)}{S_j S_k}\right] = 2(1 - r_{jk}),$$

where r_{jk} is the product moment correlation between the data unit score vectors. Notice carefully the operations employed. The data units were subjected to linear transformations; the two scores on the ith variable, x_{ij} and x_{ik}, were subjected to different transformations so that the transformed scores are not comparable to each other. As a result of this demonstration it may be said that the complement of the product moment correlation coefficient (i.e., $1 - r$) is a metric; but it is a metric only in the space in which data units have been transformed individually to a zero mean and unit scatter, a most unappealing and counterintuitive representation space. It also should be noted that changing the unit of measurement for one variable changes one component in each of the data unit score vectors, which in turn changes the mean and variance of each of these vectors and causes the correlation to change. Hence, correlations between data units are very much dependent on the units of measurement which implicitly give relative weights to the variables.

Transforming the scores for each data unit as

$$\hat{x}_{ij} = x_{ij} \bigg/ \left[\sum_{i=1}^{i=n} x_{ij}^2 \right]^{1/2}$$

and following the same procedures as above, it can be shown that

$$D^2(\hat{X}_j, \hat{X}_k) = 2(1 - \cos \alpha_{jk}),$$

where α_{jk} is the angle between vectors discussed in Section 4.1.1. Again, the transformation within a data unit destroys the comparability of scores between data units.

5.3 Measures Using Binary Variables

With several exceptions, the measures of association between binary variables also may be applied to measuring association between data units. Table 4.3 is altered slightly by changing the labels from "Variable" to "Data Unit" as in Table 5.1. Whereas in Table 4.3 the entry in the "a" cell was the

TABLE 5.1
2 × 2 CONTINGENCY TABLE

Data unit j \ Data unit k	1	0	Totals
1	a	b	$a+b$
0	c	d	$c+d$
Totals	$a+c$	$b+d$	n

number of cases in which data units scored 1 on both variables A and B, the "a" cell entry in Table 5.1 is the number of cases in which data units j and k simultaneously score one on a variable. Similar contrasts of interpretation may be stated for each of the other three cells.

5.3.1 BINARY MEASURES OF SECTION 4.3 REVISITED

The cosine of the angle between vectors and the product moment correlation coefficient seem ill suited to measuring association among data units in the 2×2 case. The shortcomings of the correlation coefficient were examined in Section 5.2.3. The cosine of the angle between vectors has similar problems, though it could be of some interest if the geometric mean of the conditional probabilities for a 1–1 match could be interpreted usefully.

Somewhat more useful are the lambda and D measures of Section 4.2.2. The lambda measure gives the relative decrease in error probability for predicting the score achieved by one data unit on a given variable as a result of the added knowledge of the score achieved by the other data unit on the same variable. The D measure is similar except that it is the actual as opposed to relative decrease in error probability. The special 2×2 case for fixed marginal distributions developed in Section 4.3.1.3 is not applicable in this setting; transforming the table to have uniform marginal distributions involves a change in the definitions (and perhaps the number) of variables to achieve a balance in the numbers of ones and zeroes achieved by each data unit.

The matching coefficients of Section 4.3.2 also seem suitable for measuring associations between both variables and data units. Using the numbering scheme of Table 4.5, the measures are reviewed as follows for their interpretations in the setting of associations among data units:

1. This measure is the probability that the two data units both score 1 on a randomly selected variable. Note that 0–0 matches are not "relevant matches" but are counted in the denominator for the determination of the number of possible matches.

2. This measure is the probability that the two data units have the same score (either 0 or 1) on a randomly chosen variable. The complement of this measure is discussed further in Section 5.3.2.

3. This measure is the conditional probability that both data units score 1 on a randomly chosen variable given that all 0–0 matches are discarded.

4, 8, and 12. These measures are nonsense when treating association between variables and there seems to be no good reason for considering them otherwise in measuring association between data units.

5 and 9. As in Section 4.3.2, these measures seem to admit of no useful interpretation.

6. This measure gives matched pairs double the weight of unmatched pairs and gives 0–0 matches full credit in both numerator and denominator. A possible interpretation for this measure is discussed in Section 5.4.2.

7. This measure is analogous to 6 except that 0–0 matches are uniformly ignored; it is the complement of the Lance and Williams' nonmetric coefficient as already noted in Section 5.2.2.

10. This measure gives unmatched pairs double weight and 0–0 matches full credit. When measuring association between data units, this measure has an interesting interpretation. Each variable involves two classes and therefore the numerator is the number of classes which are shared by the two data units; then the denominator is the number of classes represented in the pair of data units. In Section 5.3.2 the complement of this measure is shown to be a metric.

11. This measure is quite analogous to 10 except that 0–0 matches are excluded in both the numerator and denominator. It may be interpreted as the conditional probability that the two data units both score one on a class randomly chosen from all those classes on which either of the data units scores one. This measure treats as irrelevant any classes which are not represented positively in at least one of the data units. Of course these same classes may be relevant in the comparison of other data units in the same sample.

13. This measure is the ratio of the number of variables on which the data units exhibit a 1–1 match to the number of mismatches.

14. This measure is the ratio of the number of matches (either 1–1 or 0–0) to the number of mismatches between the data units.

The probability based measures of Section 4.3.3 may also prove useful for measuring association between the data units. Expression (4.19) is the conditional probability that one data unit will score 1 on a randomly chosen variable given that the other data unit scored 1 on the same variable. An additional qualification to expression (4.19) is that the direction of prediction is randomized so that each data unit is the predictor half the time and the predictand the other half. Expression (4.20) may be interpreted the same as (4.19) except for the additional condition that the direction of scoring also is randomized. Expression (4.21) is the probability that the data units match on a randomly chosen variable less the probability that they differ.

All of these 2×2 measures may be computed as measures of association between data units by using the computer programs of Appendix D. These latter programs are explained in terms of association between variables but may be used for association between data units merely by reversing the roles of variables and data units; that is, input will consist of a vector of scores across all variables for one data unit rather than a vector of scores across all data units for one variable.

5.3.2 METRIC DISTANCES IN BINARY FORM

In the case of binary variables, the quantity $|x_{ij} - x_{ik}|$ in Eq. (5.1) is zero if data units j and k match on variable i and is one otherwise. Therefore, Eq. (5.1) may be rewritten as

$$\frac{D_p^{\,p}}{n} = \frac{1}{n} \sum_{i=1}^{i=n} |x_{ij} - x_{ik}|^p = \frac{b+c}{n} = 1 - \frac{a+d}{n}, \tag{5.9}$$

for all values of $p \geqslant 1$. Division by n does not affect the metric character of the distance as can be verified by examining the metric properties in Section 5.1.1. In the case $p = 1$, Eq. (5.9) is known widely as the mean character difference or MCD. The quantity $(b+c)/n$ is proportional to both the L_1 metric and the square of the Euclidean metric; the complement of $(b+c)/n$ is the simple matching coefficient, measure number 2 in Table 4.5.

It is suggested sometimes that the divisor in Eq. (5.9) need not be fixed as n but should be allowed to vary to suit the particular pairwise comparison of data units. Occasionally the data set is not complete and so some data units may not be scored on some variables. In such cases each pairwise comparison can be limited to the variables on which both data units have scores and the divisor is adjusted to the number of variables actually used in the comparison. Because the divisor depends on the particular pairwise comparison, the triangle inequality may be violated.

As shown in Section 5.1.1., if D is a metric then $D/(w + D)$ is also a metric. Letting D be $(b+c)/n$ and w be 1, then

$$2\frac{(b+c)/n}{1+(b+c)/n} = \frac{2(b+c)}{(a+d)+2(b+c)} = 1 - \frac{a+d}{(a+d)+2(b+c)}.$$

Hence the complement of the Rogers and Tanimoto coefficient (measure number 10, Table 4.5) is a metric. Majone and Sanday (1968) claim that the complement of the Jaccard coefficient

$$1 - \frac{a}{a+b+c} = \frac{b+c}{a+b+c}$$

is also a metric; if so, then the complement of measure number 11 in Table 4.5 may be shown to be a metric by taking D to be $(b+c)/(a+b+c)$ and constructing $2D/(1+D)$ in the same manner as above.

5.3.3 CODING CATEGORICAL VARIABLES

A number of different schemes have been proposed for coding nominal and ordinal variables in terms of binary variables so that the distinctions of multiple classes can be retained while enjoying the advantages of binary measures. These schemes are rather plausible sounding and frequently are offered without qualification. This section includes a review of two of the most promising suggestions; it is shown that their uses are quite limited and necessarily require certain precautions to retain any degree of rationality in the results. The reader should be impressed with the necessity for examining carefully the consequences of any coding scheme; no similar analysis of the effects on 2×2 tables given in the following two sections has been found in the literature; yet, these effects should be central considerations in the decision to use a particular method of coding.

5.3.3.1   *Simple Coding of Nominal Variables*

The simplest method of coding a nominal variable of k categories is to define one binary variable for each category; a data unit falling into the ith category scores one on the ith binary variable and zero on the other $k - 1$ variables. This kind of system is often adopted in regression studies as an expedient means of handling categorical variables.

When a nominal variable is coded this way, the comparison between two data units on the k associated binary variables contributes to the 2×2 table as follows:

match		mismatch	
1	0	0	1
0	$k-1$	1	$k-2$

Obviously 0–0 matches are irrelevant because the number of such matches merely reflects the number of classes in the original nominal variable. Interest then centers on measures 3, 7, and 11 in Table 4.5, all of which are monotonic to each other. Notice that a mismatch contributes 2 to the sum of cells b and c, whereas a match contributes only 1 to cell a. Therefore using this coding scheme with measure number 3 of Table 4.5 (the Jaccard coefficient) is equivalent to using measure number 11 on a 2×2 table of ordinary binary variables. Likewise, using this coding scheme with measure number 7 (the Dice coefficient) is equivalent to using measure number 3 with ordinary binary variables. Perhaps often overlooked when using this scheme is the additional necessity for coding an original binary variable with two dummy binary variables; otherwise, mismatches on original binary variables receive only half the weight of mismatches on nominal variables coded with dummy binary variables. This entire scheme may be seen as an elaborate attempt to count matches and mismatches which may be accomplished more directly with the method of Section 5.5.3. However, by using this binary coding scheme the data may be stored using the bit level storage method discussed in Section D.1

5.3.3.2   *Additive Coding for Ordinal Variables*

Sokal and Sneath (1963, pp. 76–77) suggest coding an ordinal variable of k classes with $k - 1$ dummy binary variables such that if the data unit is in the ith class then $i - 1$ of the dummy variables score one and the remainder are zero. For a variable with five classes, the four dummy variables would be scored as in Table 5.2. The contribution to a 2×2 table depends on the particular pairwise combination of classes for the two data units. Table 5.3

TABLE 5.2
EXAMPLE OF ADDITIVE CODING

Ordinal class \ Dummy binary variables	1	2	3	4
1	0	0	0	0
2	1	0	0	0
3	1	1	0	0
4	1	1	1	0
5	1	1	1	1

TABLE 5.3
CONTRIBUTION TO A 2×2 TABLE FOR EACH POSSIBLE
COMPARISON OF FIVE ORDINAL CLASSES

Class for data unit I \ Class for data unit J	1		2		3		4		5	
1	0	0	0	0	0	0	0	0	0	0
	0	4	1	3	2	2	3	1	4	0
2	0	1	1	0	1	0	1	0	1	0
	0	3	0	3	1	2	2	1	3	0
3	0	2	1	1	2	0	2	0	2	0
	0	2	0	2	0	2	1	1	2	0
4	0	3	1	2	2	1	3	0	3	0
	0	1	0	1	0	1	0	1	1	0
5	0	4	1	3	2	2	3	1	4	0
	0	0	0	0	0	0	0	0	0	0

shows all the possible contributions to the 2×2 table made by a variable of
five classes; note that none of the 25 possibilities coincide. However the
number of matches and mismatches between the dummy variables depends
only on the rank difference between the data units as measured on the ordinal
variable. That is, if i is the rank difference between the data units and the
variable has k ordinal classes, then the contributions are

$$\Delta(a + d) = k - i - 1, \qquad \Delta(b + c) = i.$$

To illustrate, suppose the variable has five classes, data unit I falls in class 3
and data unit J falls in class 2; then $i = 1$, $\Delta(a + d) = 3$, and $\Delta(b + c) = 1$, as
may be verified from Table 5.3. It is apparent that this coding scheme makes
sense only if 0–0 matches are given full weight as in measures 2, 6, 10, and 14
of Table 4.5. Thus, this coding scheme and that of Section 5.3.3.1 may not be
used together because they require opposing treatments for 0–0 matches.

Unfortunately, the size of the contribution to the 2×2 table depends on
the number of classes for the variable. To remove this artifact it is desirable
to norm the contributions by the number of dummy variables $k - 1$ to give

$$\Delta(a + d) = \frac{k - i - 1}{k - 1} = 1 - \frac{i}{k - 1}, \qquad \Delta(b + c) = \frac{i}{k - 1}.$$

Using this method, $\Delta(a + d)$ ranges from zero (one data unit falls in class 1
and the other falls in class k) to one (both data units in the same class). This
revised scheme is consistent with ordinary binary variables. However, the
necessity for weighting the contribution of each block of dummy variables in
inverse proportion to the number of such variables in the block largely negates
any advantages to be gained by binary coding. In particular the bit level
storage method may not be used because each block of dummy binary variables
requires special treatment.

5.3.4   WEIGHTING CONSIDERATIONS

Up to this point all the binary variables have been taken as equally important. While the units of measurement do not cloud the issue as in the case of
interval variables, there is still room for differential weighting to reflect the
greater importance of one variable over another.

From the purely pragmatic view of computational simplicity, it would be
preferable to use unit weights for most variables and small integer weights for
the most important; such a weighting scheme can be implemented very easily
by duplicating the weighted variables. For example, in a problem of 25
variables where it is desired to give variable 6 double weight and variable 17
triple weight, the weighting can be put into practice by treating the problem
as one of 28 variables with the three repeated scores (x_6, x_{17}, x_{17}) appended to

the end of each of the original score vectors. This scheme is compatible with the bit level storage technique and the other features of the computer programs in Appendix D. Of course a decision to use a different weight for every variable will result in quite a large number of repetitions and perhaps an unacceptable growth of the problem size. This approach is most suitable for use when the weights are determined by judgement; then the analyst can choose weights convenient for computation but with minimal distortion from the "ideal" weights he would choose in the absence of computational considerations.

When integer weights are not suitable or the repetition of variables is too extensive, then a general approach is to assign a weight of w_i to each variable and simply accumulate the table entries as usual except for one step: as the data units are compared on the ith variable the quantity w_i is added to the appropriate cell in the table rather than 1. The bit level storage method of Appendix D may be used with these fully variable weights but it is necessary to check individual bits through masking operations and look up the appropriate weights in a table rather than use the simpler method of just counting the number of bits set to 1. The technique is straightforward and is illustrated by the last subprogram in Appendix D; the subprogram is written very generally and can accommodate any desired association measure simply by inserting a single statement to calculate that measure from a given 2 × 2 table.

Now that the computational considerations of actually using weights are in hand, attention can be given to the problems of equalizing and weighting the variables. The equalizing methods of Section 5.1.3.1 are specialized to the case of binary variables as follows:

1. If m_1 out of m data units score one on a binary variable, then the *mean* is m_1/m. The mean is also the probability that a randomly chosen data unit scores one on the variable. Equalizing binary variables to have unit mean gives results which are quite dependent on the choice of which half of each dichotomy is to be scored zero. If the more numerous half is scored zero, then the equalized score or weight is at least two and may be quite large; if the rare half is scored zero, then the equalized score lies between one and two. This method is more likely to obscure than clarify in the case of binary variables.

2. Binary variables have unit range and therefore are equalized in one sense just as they stand.

3. The standard deviation of a binary variable does not depend on which half of the dichotomy is scored zero since the variance of a binary variable B_j is

$$\text{var}(B_j) = E(B_j^2) - [E(B_j)]^2 = \frac{m_1}{m} - \left[\frac{m_1}{m}\right]^2 = \frac{m_1(m - m_1)}{m^2}.$$

The variance is equal to half the probability that two randomly chosen data units do not match on the variable. The equalizing factor for variable B_j is then

$$e_j = m/[m_1(m - m_1)]^{1/2}.$$

Table 5.4 shows values of m_1/m needed to achieve certain values of e_j. The most notable aspect of this table is that there must be quite a substantial variation in m_1/m to achieve even a moderate variation in e_j. Accordingly, it probably is not worth the trouble to equalize by the standard deviation.

TABLE 5.4
STANDARD DEVIATION
EQUALIZATION FACTORS

e_j	m_1/m	e_j	m_1/m
2	0.5000	6	0.0286
3	0.1250	7	0.0208
4	0.0666	8	0.0159
5	0.0417	9	0.0125

As in the case of interval variables weighting is primarily a matter of subjective appraisal and judgement. However, Williams *et al.* (1964) propose as a squared weighted distance between data units j and k

$$D_W{}^2 = \sum_{i=1}^{i=n} \left\{ (x_{ij} - x_{ik})^2 \sum_{\substack{g \ne i \\ g=1}}^{g=n} \chi_{ig}^2 \right\}$$

where χ_{ig}^2 is the sample chi-square statistic computed between binary variables i and g. As mentioned in Section 4.3.1.2, the chi-square statistic divided by m is the squared correlation when computed in a 2×2 table; hence this scheme may be viewed as weighting each variable by the sum of its squared correlations with the other variables. Variables which are the most redundant receive the highest weight, while variables which are nearly orthogonal to the rest (and therefore have a unique contribution to make) receive nearly zero weight. Intuitively, the opposite effect should be desired; perhaps weighting each variable inversely to the sum of its association measures with the other variables would be more rational.

5.4   Measures Using Nominal Variables

The computation of association measures between data units described by nominal variables has received very little attention. Except for the Rogers and Tanimoto coefficient (measure number 10, Table 4.5), apparently it is not widely known that the matching coefficients of Table 4.5 may be extended for use with nominal variables; this extension is discussed along with a novel method of weighting.

5.4.1   EXTENSION OF BINARY MEASURES

When nominal variables are employed, the comparison of one data unit with another can only be in terms of whether the data units score the same or different on the variables. It then becomes interesting to investigate matching measures analogous to those in Table 4.5. Just as in the case of binary variables, some variables may include a "not applicable" category; in case two data units match in such a category it probably would be desirable to exclude the associated variable from the comparison. With this view in mind, define the following counts for the comparison between two data units:

n_{a+d} is the number of variables on which the data units match;
n_d is the number of variables on which the data units match in a "not applicable" category;
n_{b+c} is the number of variables on which the data units do not match.

The subscripts on these counts are intended to convey a strong analogy with the notation of Table 5.1. If the variables are all taken as being of equal importance, then the measures of Table 4.5 (excluding the five "nonsense" and "not recommended" measures) may be redefined for the present context as in Table 5.5. The interpretations of these measures is much the same as given in Section 5.3.1 for their binary counterparts and are not repeated here. It also should be noted that the monotonicity relations within the trio of measures 3, 7, 11 and the trio of measures 2, 6, 10 carry over to their analogs

TABLE 5.5
MATCHING COEFFICIENTS FOR NOMINAL
VARIABLES[a]

1. $\dfrac{n_{a+d} - n_d}{n_{a+d} + n_{b+c}}$	2. $\dfrac{n_{a+d}}{n_{a+d} + n_{b+c}}$
3. $\dfrac{n_{a+d} - n_d}{n_{a+d} - n_d + n_{b+c}}$	
7. $\dfrac{2(n_{a+d} - n_d)}{2(n_{a+d} - n_d) + n_{b+c}}$	6. $\dfrac{2n_{a+d}}{2n_{a+d} + n_{b+c}}$
11. $\dfrac{n_{a+d} - n_d}{n_{a+d} - n_d + 2n_{b+c}}$	10. $\dfrac{n_{a+d}}{n_{a+d} + 2n_{b+c}}$
13. $\dfrac{n_{a+d} - n_d}{n_{b+c}}$	14. $\dfrac{n_{a+d}}{n_{b+c}}$

[a] Numbered as in Table 4.5.

in Table 5.5. The measure of expression (4.21) also may be defined for nominal variables as

$$\frac{n_{a+d} - n_{b+c}}{n_{a+d} + n_{b+c}}. \tag{5.10}$$

Expression (5.10) is monotonic to measures 2, 6, and 10 in Table 5.5.

5.4.2 WEIGHTING CLASSES

In at least some cases it is of interest to score an agreement or disagreement on a particular variable according to which class or classes might be involved, especially if some classes are rare. To illustrate, suppose eye color is a variable and the recorded classes are brown, blue, hazel, and pink (for albinoes). The first three colors are fairly common and it would probably suffice to consider matches and mismatches among these three as being of approximately equal significance. However, if both persons were recorded as having pink eyes the rare nature of the match is indicative of a strong association and probably should be given an extraordinary weight. Likewise if only one person of a pair has pink eyes, the distinctive mismatch should be of special significance in distinguishing between the two persons.

Sokal and Sneath (1963, pp. 135–139) describe a complex weighting scheme originally proposed by Smirnov (in a Russian journal). Their description is relatively accessible so that the comments here will be confined to few observations. Smirnov's method assigns weights according to which classes appear in the two data units; however, it has some curious features which inhibit an operational interpretation. When two data units are compared on a variable, every class enters into the comparison. For example, if one person has blue eyes and the other brown, then the pair receive positive credit for matching on their joint failure to have pink or hazel eyes and negative credit for their mismatches in the categories of brown and blue; if the two persons have the same color eyes then the pair receives positive credit for every class. Positive credits are based on frequency counts and different values are given depending on whether the match is in regard to joint presence or joint absence in the class; negative credits for mismatches are taken universally as −1 so that particularly significant mismatches are not given special weight. The attendant computational effort and the lack of a suitable rationale are strong deterrents to the use of this method.

However, the idea of weighting according to which classes are represented in the data is not so easily set aside. The desire to give rare classes extra weight appears frequently in the biological literature though systematic methods suitable for assigning such weights are not offered. Perhaps the method described below will be interesting and appealing.

Since the idea is to give status to rare classes, it seems only natural to assign each class a weight which is a function of the frequency with which that class occurs in the data set (or in the population if earlier studies can provide such information). Let m be the number of data units in the data set and let m_i be the number in the ith class for the variable of interest. The probability that a randomly chosen data unit falls in class i is m_i/m. Then for two randomly chosen data units, the following event probabilities are of interest:

$$P_{ii} = \left(\frac{m_i}{m}\right)^2 \quad \text{Probability that both data units fall in class } i,$$

$$P_{ij} = \frac{2m_i m_j}{m^2} \quad \begin{array}{l} \text{Probability that one data unit falls in class } i \\ \text{while the other falls in class } j. \end{array}$$

The factor of 2 in the latter expression reflects the fact that the first data unit may fall in either of the two classes so that there are two ways for the event to occur. With these results, the probability of occurrence for any observed pair of classes may be computed and taken as an estimate of how commonly that combination occurs.

Since rare events have low probabilities, the probability of an event is not a suitable weight; however any inverse function of the probability is potentially interesting. The simplest such functions are $1 - P_{ij}$ and $1/P_{ij}$. The first function is not particularly discriminating as may be seen from an example. Let the variable have four classes which occur with frequencies 0.05, 0.20, 0.25, and 0.50, respectively. Then the values of P_{ij} and $1 - P_{ij}$ for each possible combination are as in Table 5.6. Examining the values of $1 - P_{ij}$, it is readily apparent that such weights give little discrimination among the events. Something like $(1 - P_{ij})^{10}$ would give a more appealing separation of events. However, such a procedure would be inconvenient computationally, and the results would lack a useful operational interpretation. Looking back at the array of P_{ij} values in Table 5.6, there is some intuitive satisfaction in saying

TABLE 5.6
EXAMPLE PROBLEM FOR PROBABILITY WEIGHTS

	Values of P_{ij}				Values of $1 - P_{ij}$			
	1	2	3	4	1	2	3	4
1	0.0025				0.9975			
2	0.0200	0.0400			0.9800	0.9600		
3	0.0250	0.1000	0.0625		0.9750	0.9000	0.9375	
4	0.0500	0.2000	0.2500	0.2500	0.9500	0.8000	0.7500	0.7500

that the occurrence of an event with a 0.10 probability should receive twice
the weight of an event with a 0.20 probability. The weights obtained according
to $W_{ij} = 1/P_{ij}$ have this property as may be verified readily in Table 5.7 where
such weights are given for the same four class example used in Table 5.6.

TABLE 5.7
$1/P_{ij}$ WEIGHTS FOR EXAMPLE

	1	2	3	4
1	400			
2	50	25		
3	40	10	16	
4	20	5	4	4

The magnitude of the weights obtained with this latter scheme depends on
the number of classes in the variable. The probability of using the weight
$1/P_{ij}$ is P_{ij}, so that the expected or average weight for a variable of k classes is

$$E(W) = \sum_{i=1}^{i=k} \sum_{j=1}^{j=i} P_{ij}(1/P_{ij}) = 1 + 2 + 3 + \ldots + k,$$

$$E(W) = \tfrac{1}{2}k(k + 1). \tag{5.11}$$

The expected weight for a variable of four classes is 10, while that for a variable
of five classes is 15. Dividing the class weights by the appropriate expected
weight equalizes all variables to unit mean. Equalizing the variables to unit
range seems inappropriate since the range depends on the relative frequencies
of the classes, the information which is being utilized to generate class weights.
Equalizing to unit variance similarly depends on the class frequencies since

$$\mathrm{var}\,(W) = E(W^2) - [E(W)]^2 = \sum_{i=1}^{i=k} \sum_{j=1}^{j=i} \frac{1}{P_{ij}} - \left[\frac{k(k + 1)}{2}\right]^2.$$

Thus Eq. (5.11) seems to provide the most appropriate equalization factor for
each variable. The analyst may then apply whatever subjective or judgemental
weights he desires to the equalized variables.

Class weights may be used directly with the matching measures of Table 5.5.
Instead of accumulating n_{a+d}, n_d, and n_{b+c} as the *numbers* of matches and
mismatches, accumulate sums of equalized class weights; that is:

1. If the two data units both fall in class i add W_{ii} to n_{a+d}.
2. If the two data units both fall in class i and the class is "not applicable"
add W_{ii} to n_d.
3. If the data units fall in classes i and j, respectively, add W_{ij} to n_{b+c}.

Matching measures 1, 2, and 3 of Table 5.5 along with expression (5.10) probably are the most suitable measures for use with this weighting scheme.

It is interesting to note that if all classes for a variable are equally likely ($m_i/m = 1/k$ for all i) then the equalized class weights are

$$W_{ii} = 2k/(1 + k), \qquad W_{ij} = k/(1 + k).$$

Matches are given double the weight of mismatches and therefore measures 6 and 7 in Table 5.5 may be interpreted as being based on class weighting with a prior hypothesis of equally probable classes.

5.5   Mixed Variable Strategies

Mixed variable data sets can be even more troublesome when clustering data units than when clustering variables. Not only is there the problem of measuring association between data units while using different types of variables, but there is also the problem of weighting the contributions of the different variables. Binary variables present no difficulties since they may be treated directly within the framework of either nominal or interval variables. Likewise ordinal variables are not too troublesome since they may be treated as nominal variables by forgetting their order properties or as interval variables by using ranks as scores. The difficult problems arise in the simultaneous use of interval and nominal variables. This section offers three strategies for dealing with such problems.

5.5.1   Partitioning of Variables

Even though assignment of weights for variables of a single type is not a clear-cut process, it is simpler than trying to assign weights to variables of different types. Accordingly, if one variable type clearly dominates the analysis, it might be worthwhile to separate the nonconforming variables and use them in only a qualitative supplementary fashion. It is undeniable that information may be lost by ignoring some of the variables when computing association measures, but the analysis will be simplified and the contributions of the variables actually used will be easier to assess.

If restricting attention to a dominant variable type is unsatisfactory, the next step is to consider performing two parallel but separate analyses, one based on interval variables and the other on nominal variables. Data units which are close together or far apart in both analyses likely would bear the same relation in any composite dealing with all the variables together. Somewhat more complex to resolve is the case of data units which are close in one analysis but distant in another. Not only would relative weighting of the variable types be of importance, but also perhaps the joint or interactive effects between variables of different types. A formal, systematic, and meaningful method of integrating such separate analyses is not available at present. However, parallel analyses of the partitioned data set will condense the problem, reveal some groups through their simultaneous occurrence in both analyses, and highlight the differences between the two sets of variables in regard to their ability to discriminate among data units. Perhaps this degree of processing will be sufficient to permit the analyst to exercise his powers of insight and bring to bear additional nonquantifiable information such that important results can be discovered "by inspection."

5.5.2 CONVERSION OF VARIABLES

Chapter 3 included a wide variety of techniques for converting variables from one type to another. To maintain visibility of effects it is desirable to limit the number of conversions to a minimum; accordingly the choice of how to homogenize the set of variables should be influenced strongly by which variable type is the most numerous. If it is decided to make conversions such that all variables are expressed on interval scales, then the measures of Sections 5.1 and 5.2 are available. Alternatively, a decision to convert to nominal scales requires the use of the matching measures described in Section 5.4.1.

In Section 4.4.2 it was suggested that variable conversions might be treated as a one-to-many transformation giving a converted variable multiple representations. The same notion can be employed in the context of measuring association among data units. However, the weights assigned to the converted variables should reflect the presence of several representatives so that one of the original variables is not overweighted inadvertently while trying to describe adequately its various aspects.

5.5.3 DISAGREEMENT INDICES

As a prerequisite to assigning weights it is necessary to equalize the variables in some appropriate sense. This section presents a novel proposal for equalizing the variable-by-variable disagreements between data units.

When two data units have identical responses on a variable there is zero difference or disagreement between them. And within a finite data set there is a maximum level of observed disagreement on any variable. If the maximum disagreement is scored as one, then all disagreements on a variable may be represented by a disagreement index ranging from zero to one. Disagreements on variables of every type may be expressed in this manner.

1. *Binary variables.* All one can say is that the data units either agree or disagree. Therefore a match is scored zero and mismatch is scored one.

2. *Nominal variables.* As with binary variables, one may be able to do no better than say the data units either do or do not match. On the other hand it may be useful to consider some mismatches as showing less disagreement than others. Recalling the eye color example of Section 5.4.2, it might be desirable to adopt the following array of disagreement indices. The numbers in Table 5.8 are not meant to be taken as objectively meaningful; they are merely a numerical statement of some subjective viewpoint.

TABLE 5.8
DISAGREEMENT INDICES FOR EYE COLOR
EXAMPLE

	Brown	Blue	Hazel	Pink
Brown	0			
Blue	0.8	0		
Hazel	0.8	0.6	0	
Pink	1	1	1	0

3. *Ordinal variables.* If an ordinal variable involves k distinct ranks then the maximum rank difference between any two data units is $k-1$. Taking the ranks of data units i and j as R_i and R_j, a simple linear disagreement index between them is

$$\frac{|R_i - R_j|}{k-1}. \tag{5.12}$$

The square root of expression (5.12) gives larger scores and the square of the linear index gives smaller scores. For a variable with five ranks, the values of these three forms of disagreement index are shown in Table 5.9. The square root index strongly penalizes any disagreement, whereas the squared index treats small disagreements little differently than full agreement. The analyst

TABLE 5.9
COMPARISON OF ORDINAL DISAGREEMENT INDICES

Form of index	Rank difference			
	1	2	3	4
Square Root	0.5000	0.7071	0.8660	1.000
Linear	0.2500	0.5000	0.7500	1.000
Squared	0.0625	0.2500	0.5625	1.000

may also wish to assign indices without regard to any particular formula. For one variable of five classes the disagreement indices (0.1, 0.3, 0.4, 1.0) might be appropriate while for another variable in the same data set the indices (0.90, 0.95, 0.99, 1.0) could be suitable.

   4.   *Interval variables.*   Within a finite data set the maximum disagreement on an interval variable is the range of the variable. Then a simple linear disagreement index for variable i between data units j and k is

$$\frac{|x_{ij} - x_{ij}|}{\text{range } x_i}. \tag{5.13}$$

The square root, square, or other function of expression (5.13) might be an appropriate alternative for some variables.

   With all variables equalized in terms of disagreement indices the analyst can apply his judgement and subjectively appraise the relative importance of each variable in order to assign weights. A weighted sum of disagreement indices probably will suffice unless prior knowledge suggests that the product of two or more variables might be important. In the latter case the analyst may choose between forming the product in terms of the original variables and using a disagreement index for this product or in terms of disagreement indices computed for each variable; these two alternatives can be expected to behave in distinctively different ways.

CHAPTER

6

HIERARCHICAL CLUSTERING METHODS

The association measures discussed in Chapters 4 and 5 may be used to construct a similarity matrix describing the strength of all pairwise relationships among the entities (variables or data units) in the data set. The methods of hierarchical cluster analysis operate on this similarity matrix to construct a tree depicting specified relationships among the entities as in Fig. 6.1 (see also Figs. 3.8 and 3.9). The branches on the left each represent one entity while the root represents the entire collection of entities. Moving down the tree from the branches toward the root depicts increasing aggregation of the entities into clusters. Hierarchical clustering methods which build a tree from branches to root often are called agglomerative methods. Less common are the divisive hierarchical methods which begin at the root and work toward the branches.

Once a tree is constructed for n entities, the analyst may choose from as many as n sets of clusters. Taking the example of Fig. 6.1, the possible sets of clusters are as shown in Table 6.1. Notice that these clusters are nested: from the agglomerative view, when two entities are merged they are joined together permanently and become a building block for later merges; from the divisive view, when a group of entities is split into two parts, the parts are separated permanently and may be treated independently for the remainder of the

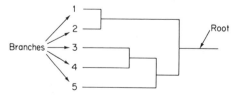

Fig. 6.1. Tree for hierarchical clustering.

131

TABLE 6.1
CLUSTERS FROM FIG. 6.1

Number of clusters	Clusters
5	(1), (2), (3), (4), (5)
4	(1, 2), (3), (4), (5)
3	(1, 2), (3, 4), (5)
2	(1, 2), (3, 4, 5)
1	(1, 2, 3, 4, 5)

analysis. Herein lie both the strength and weakness of hierarchical methods: by taking early decisions as permanent the number of possibilities that need be examined is reduced greatly as compared with complete enumeration; but this same convention precludes discovering early mistakes or capitalizing on later opportunities.

While there may appear to be an overwhelming abundance of hierarchical clustering methods treated in the literature, all the methods seem to be alternative formulations or minor variations of three major clustering concepts:

1. Linkage methods (Sections 6.2.1, 6.2.2, 6.2.3, 6.2.4, 6.4.1, and 6.4.2),
2. Centroid methods (Sections 6.2.5 and 6.3.4),
3. Error sum of squares or variance methods (Sections 6.2.6, 6.3.1, 6.3.2, 6.3.3).

All of these methods are suitable for clustering data units. However, the linkage methods are the only techniques in this book which may be used for clustering variables. These various methods have individual computational characteristics so that several different approaches to the problem of computation are of interest. This chapter begins with a presentation of a general procedure for agglomerative hierarchical clustering and then explores the details of three computational approaches for implementation. Particular variations of the three major hierarchical concepts are described in conjunction with the applicable computational approaches. All these methods are implemented through computer programs given in Appendix E. Finally, in Section 6.5 some alternative computational approaches used by other authors are discussed and a brief guide to divisive hierarchical methods is offered.

6.1 The Central Agglomerative Procedure

The agglomerative hierarchical methods described in this chapter are treated as variations on a single major technique. Most agglomerative methods can be implemented efficiently within this framework.

$$S = \begin{vmatrix} s_{21} & & & & & \\ s_{31} & s_{32} & & & & \\ s_{41} & s_{42} & s_{43} & & & \\ \vdots & \vdots & \vdots & \ddots & & \\ s_{n1} & s_{n2} & s_{n3} & \cdots & s_{n(n-1)} \end{vmatrix}$$

Fig. 6.2. Lower triangular similarity matrix.

Let s_{ij} be the similarity between entities i and j as defined by one of the association measures in Chapter 4 or 5. Assuming that the similarity is symmetric (i.e., $s_{ij} = s_{ji}$), the complete schedule of similarities for all $\binom{n}{2} = \frac{1}{2}n(n-1)$ possible pairwise combinations of entities may be arrayed in a lower triangular similarity matrix as in Fig. 6.2. All the methods discussed in this chapter assume that the s_{ij} entries are nonnegative. This limitation is of consequence only for correlation and the cosine of the angle between vectors; the distinction between positive and negative association cannot be utilized in these clustering methods. A simple remedy is to use the absolute value or the square of the measure if it can assume negative values. The primary purpose for studying variables and association measures in Chapters 3, 4, and 5 was to provide the means for creating such a similarity matrix. Once the matrix is defined, the process of clustering entities is almost trivially simple. The general procedure for agglomerative clustering on a data matrix is as follows:

1. Begin with n clusters each consisting of exactly one entity. Let the clusters be labeled with the numbers 1 through n.
2. Search the similarity matrix for the most similar pair of clusters. Let the chosen clusters be labeled p and q and let their associated similarity be $s_{pq}, p > q$.
3. Reduce the number of clusters by 1 through merger of clusters p and q. Label the product of the merger q and update the similarity matrix entries in order to reflect the revised similarities between cluster q and all other existing clusters. Delete the row and column of S pertaining to cluster p.
4. Perform steps 2 and 3 a total of $n - 1$ times (at which point all entities will be in one cluster). At each stage record the identity of the clusters which are merged and the value of similarity between them in order to have a complete record of the results.

Different agglomerative methods are implemented by varying the procedures used for defining the most similar pair at step 2 and for updating the revised similarity matrix at step 3.

Hierarchical clustering on a similarity matrix may be carried out using at least three different computational approaches; each approach has its own unique advantages and limitations. No single approach is to be preferred in all circumstances; each has its own realm of application. All three approaches are discussed in the following sections with particular emphasis on their computational characteristics. The clustering techniques suited for each approach are described in detail. Appendix E provides computer programs which implement all these approaches.

6.2   The Stored Matrix Approach

In this variation the similarity matrix is stored in its entirety in the computer's central memory so that the similarity values may be accessed directly in any sequence. This approach has the following important characteristics.

1. The similarity matrix is a given array of numbers. The numerical execution of the clustering procedure is completely independent of how the similarity values were generated or whether the entities to be clustered are variables or data units. However, it is necessary to make a direct distinction between distance-like measures (the smallest values correspond to the most similar pairs) and correlation-like measures (the largest values correspond to the most similar pairs); the essential difference is whether the search for the most similar pair involves seeking the minimum or maximum entry in the similarity matrix.

2. The use of in-core storage for the entire similarity matrix severely limits the number of entities which may be clustered. Table 6.2 illustrates how the storage requirement for the similarity matrix grows with the number of entities. Small and moderate problems of up to 150 entities can be handled on many computers; larger problems rapidly consume additional space so that even the largest of today's computers can handle only relatively modest problems. Further, the figures in Table 6.2 allow for the similarity matrix

TABLE 6.2
STORAGE REQUIREMENTS FOR SIMILARITY
MATRICES

Number of entities	Storage required	Number of entities	Storage required
50	1225	300	44,850
100	4950	350	61,075
150	11,175	400	79,800
200	19,900	450	101,025
250	31,125	500	124,750

alone and do not include storage for the program or various bookkeeping arrays. Chapter 9 includes some strategies for circumventing this limitation.

3. A clustering problem involves both variables and data units but the storage limits just cited apply only to the entities which are being clustered. When clustering variables there is no limit to the number of data units that may be involved; the data units are never "seen" by the clustering algorithm because the entire data set is summarized in the similarity matrix whose size depends only on the number of variables. Likewise, clustering data units can involve an unlimited number of variables, at least in regard to the computational aspects of the cluster analysis.

The search for the most similar pair and the subsequent updating of the similarity matrix is virtually the entire computational burden for a hierarchical cluster analysis based on a given similarity matrix. It is then desirable to minimize the effort expended in these two steps. The following algorithm is the one used in the computer programs in Appendix E; it appears to be at least as efficient as any other reported algorithm based on storing the entire similarity matrix in the central memory. The algorithm is stated in terms of distance-like similarities; necessary changes for correlation-like similarities are given in parentheses. Frequent reference to the lower triangular similarity matrix in Fig. 6.2 should help to clarify the algorithm.

1. For each row of the initial similarity matrix find the minimum (maximum) entry and record the column in which it occurs. The ith row contains $i - 1$ entries so that the search in this row involves $i - 2$ comparisons; the entire search over rows 2 through n involves $\sum_{i=2}^{i=n} (i - 2) = \frac{1}{2}(n - 1)(n - 2)$ comparisons.

2. Prior to searching for the pair of clusters to be used in the kth merge, there are $n + 1 - k$ clusters remaining, and they are described by $n - k$ active rows of the similarity matrix. To find the most similar pair, search for the minimum (maximum) of the row minima (maxima) which were found initially in step 1 and repeatedly updated in step 3. At the kth stage the search of $n - k$ rows involves $n - k - 1$ comparisons. The total number of comparisons over $n - 1$ stages is then $\sum_{k=1}^{k=n-1} (n - k - 1) = \frac{1}{2}n(n - 3) + 1$.

3. After the most similar pair is found at the kth stage, it is necessary to update the $n - k$ similarity matrix entries corresponding to the cluster resulting from the merger. Additionally, it is necessary to find the new row minimum corresponding to the new cluster and check the remaining updated values against their respective row minima (maxima). For $n - 1$ stages the number of updates and the number of comparisons are each $\sum_{k=1}^{k=n-1} (n - k) = \frac{1}{2}n(n - 1)$. Finally it is necessary to search for a new row minimum (maximum) in any row where the cluster or column corresponding to the minimum (maximum) is involved in the merger; that is, the most similar pair in such rows includes

one of the clusters involved in the merger. The computational burden for this latter aspect of updating is highly contingent on the pattern of numbers in the similarity matrix and the particular clustering method employed; however, in a typical case it would be expected that one row would be updated for this reason at each stage and the expected length of the row at the kth stage is $(n - k)/2$. Therefore, the expected number of comparisons at each stage is $\frac{1}{2}(n - k) - 1$ and the total over $n - 1$ stages is $\sum_{k=1}^{k=n-1} [\frac{1}{2}(n - k) - 1] = \frac{1}{4}n(n-5) + 1$.

4. To avoid testing each row of the similarity matrix as to whether it represents an active cluster, a list of active clusters is maintained; at each stage one element of the list is removed and the remaining elements are "pushed down." It is necessary to search this list for the location of the cluster to be deleted; the expected number of comparisons at each stage is half the number of clusters and for $n - 1$ stages the total number of comparisons is $\sum_{k=1}^{k=n-1} \frac{1}{2}(n + 1 - k) = \frac{1}{4}(n + 2)(n - 1)$.

Adding up the comparisons required for all steps gives a total of $2n^2 - 9n/2$. The computational effort increases at a rate slightly less than $2n^2$. Table 6.3

TABLE 6.3
NUMBER OF COMPARISONS
FOR HIERARCHICAL CLUSTERING

n	$2n^2 - 9n/2$
100	19,550
200	79,100
300	178,650
400	318,200
500	497,750

illustrates how the number of comparisons grows with the number of entities. Even these large numbers of comparisons can be accomplished in modest amounts of computer time on modern machines.

The following sections describe various clustering methods which can be implemented with a stored data matrix. The first four methods may be applied with any association measure and may involve either data units or variables as the entities. The last two methods are developed around squared Euclidean distance for data units. Of course these methods may be applied to *any* representation space because the effects of desired transformations are embedded in the similarity matrix. This stored matrix approach is implemented through subroutines CNTRL, CLSTR, and METHOD in Appendix E.

6.2.1   SINGLE LINKAGE

The method of single-linkage cluster analysis is the simplest of all hierarchical techniques. At each stage, after clusters p and q have been merged, the similarity between the new cluster (labeled t) and some other cluster r is determined as follows:

1. If s_{ij} is a distance-like measure

$$s_{tr} = \min(s_{pr}, s_{qr}).$$

The quantity s_{tr} is the distance between the two closest members of clusters t and r. If clusters t and r were to be merged, then for any entity in the resulting cluster the distance to its nearest neighbor would be at most s_{tr}.

2. If s_{ij} is a correlation-like measure

$$s_{tr} = \max(s_{pr}, s_{qr}).$$

The quantity s_{tr} is the similarity between the two most similar entities in clusters t and r. If clusters t and r were to be merged, then for any entity in the resulting cluster there would be at least one other entity in the same cluster such that the pair would have a similarity at least as large as s_{tr}.

The method is known as single linkage because clusters are joined at each stage by the single shortest or strongest link between them. For example, Fig. 6.3a shows two clusters with their mutually closest members linked. Moving these clusters slightly closer together would make this pair the most similar in the entire array. This example illustrates the fact that single linkage clustering is incapable of delineating poorly separated clusters. On the other hand, if the two clusters are moved farther apart then single-linkage clustering will distinguish between them quite well. For any cluster of two or more entities produced by the single-linkage method, every member is more similar to some other member of the same cluster than to any other entity not in the cluster. The single-linkage method is one of the very few clustering techniques

(a)   (b)

Fig. 6.3. Single-linkage clustering examples.

which can outline nonellipsoidal clusters (Fig. 6.3b shows a U-shaped cluster as delineated by the single-linkage method). The tendency to give long serpentine clusters is sometimes called "chaining"; this property is often criticized because entities at opposite ends of a cluster may be markedly dissimilar.

Since the updating process involves choosing only the minimum or maximum, single-linkage clustering is invariant to any transformation which leaves the ordering of the similarities unchanged, that is, any monotonic transformation. Johnson (1967) and Jardine *et al.* (1967) make a variety of theoretical observations on this property. From a practical point of view, this property means that some measures of association are equivalent to each other (such as measures 2, 6, and 10 in Table 4.5) when used with this method.

This method of clustering appears in the literature in many different forms. Zahn (1971) gives a most enlightening discussion of the intuitive aspects of single-linkage clustering and provides a profusion of two-dimensional examples. Johnson (1967) describes a single-linkage clustering algorithm which inspired the design of the stored matrix hierarchical algorithm given here. The method also has interesting connections with the problem of finding the minimum spanning tree of a graph; this latter topic is explored further in Section 6.4.2.

6.2.2   COMPLETE LINKAGE

The complete-linkage method is related closely to the single-linkage method and is no more difficult to execute. At each stage, after clusters p and q have been merged, the similarity between the new cluster (labeled t) and some other cluster r is determined as follows:

1. If s_{ij} is a distance-like measure

$$s_{tr} = \max(s_{pr}, s_{qr}).$$

The quantity s_{tr} is the distance between the most distant members of clusters t and r. If clusters t and r were merged, then every entity in the resulting cluster would be no farther than s_{tr} from every other entity in the cluster. The value of s_{tr} is the diameter of the smallest sphere which can enclose the cluster resulting from the merger of clusters t and r.

2. If s_{ij} is a correlation-like measure

$$s_{tr} = \min(s_{pr}, s_{qr}).$$

The quantity s_{tr} is the similarity between the two most dissimilar entities in clusters t and r. If clusters t and r were to be merged, then every entity in the resulting cluster would have a similarity of at least s_{tr} with every other entity in the cluster.

The method is called complete linkage because all entities in a cluster are linked to each other at some maximum distance or minimum similarity. Such a cluster is called a "maximally connected subgraph" in graph theory. In contrast to the single-linkage method, the interpretation of the clusters can be made only in terms of the relationships within individual clusters; there is no particularly useful interpretation involving the differences between clusters. Like the single-linkage method, complete-linkage cluster analysis is invariant to monotonic transformations of the similarity measure. Johnson (1967) discusses this property and relates it to the ultrametric inequality.

6.2.3   AVERAGE LINKAGE WITHIN THE NEW GROUP

In the single-linkage method each cluster is characterized by the longest link needed to connect any member of the cluster to some other member of the cluster; in the complete-linkage method each cluster is characterized by the longest link needed to connect every member of a cluster to every other member. Instead of relying on extreme values as in these two cases, it may be of interest to characterize a cluster by the average of all links within it.

The s_{ij} entries in the initial similarity matrix may be built up as the sum of similarities associated with all pairwise combinations formed by taking one entity from cluster i and the other from cluster j. Before any merges have occurred each cluster consists of just one such entity and there is only one such pair of entities for each pair of clusters. When clusters p and q are merged, the sum of pairwise similarities between the new cluster (labeled t) and some other cluster r is updated as

$$s_{tr} = s_{pr} + s_{qr}$$

to maintain this form for the similarity matrix.

Also, let SUM_i be the sum of all pairwise similarities among entities within cluster i and let N_i be the number of entities in cluster i. Before the first merge $SUM_i = 0$ and $N_i = 1$ for all clusters. When clusters p and q are merged, the new cluster is labeled t and the description for the cluster is updated as

$$SUM_t = SUM_p + SUM_q + s_{pq}$$
$$N_t = N_p + N_q.$$

Then, when searching for the most similar pair, the average within group similarity for the cluster formed by merging the candidate pair of clusters i and j is

$$\frac{SUM_i + SUM_j + s_{ij}}{(N_i + N_j)(N_i + N_j - 1)/2}.$$

The denominator in the last expression is the number of distinct pairwise combinations that may be formed from the entities in the merger of i and j.

This method of clustering is not dependent on extreme values for defining clusters and accordingly it is not possible to make any statements about the minimum or maximum similarity within a cluster. However, as a practical matter this method frequently gives results that are little different from those obtained with the complete linkage method.

6.2.4   AVERAGE LINKAGE BETWEEN MERGED GROUPS

An alternative method based on average linkage consists of evaluating the potential merger of clusters i and j in terms of the average similarity for links between the two clusters. If the similarity matrix is treated as in Section 6.2.3, then s_{ij} is the sum of pairwise similarities between clusters i and j. The number of such between group pairwise similarities is the product of N_i and N_j, where N_i is the number of entities in cluster i. Then the average between group similarity for clusters i and j is

$$\frac{s_{ij}}{N_i N_j}.$$

The principal difference between this method and that of Section 6.2.3 is that the sums of within group pairwise similarities, SUM_i and SUM_j, are ignored. Lance and Williams (1967a) label this method as "group average" and provide a different computing expression. As might be expected, this method gives results which are not radically different from those obtained with the method of Section 6.2.3.

6.2.5   CENTROID METHODS

In statistical analysis the mean is often the basic summary statistic for a set of data. The familiar t-test and the analysis of variance technique both are used to identify differences between groups by testing for differences between their means. It then should be appealing to cluster hierarchically by merging at each stage those two clusters with the most similar mean vectors or centroids. The method is easy to implement with a stored similarity matrix only for squared Euclidean distance. Lance and Williams (1967a) give the update equation for this case as

$$s_{tr} = \frac{N_p}{N_p + N_p} s_{pr} + \frac{N_q}{N_p + N_q} s_{qr} - \frac{N_p N_q}{(N_p + N_q)^2} s_{pq},$$

where p and q are the labels for the clusters just merged, t is the label adopted for the new cluster, and r is any other existing cluster. This equation could be used with any similarity measure for either variables or data units; however, the results would lack a useful interpretation and could be quite strange unless s_{ij} is the squared Euclidean distance between the centroids of cluster i and j. Section 6.3.4 also treats the centroid method but in a context amenable to a wider variety of distance functions.

   This method of clustering has enjoyed some considerable popularity in the community of biologists. Sokal and Sneath (1963, pp. 183–185) describe the so-called "pair group" methods which are variations on the centroid approach. The essential difference is that the centroids of the merged group are weighted to compute the centroid of the new group and these weights are not necessarily proportional to the number of entities in each group. Another variation on the same approach is the "median method" first proposed by Gower (1967, p. 626). The general idea is that the centroids are weighted equally regardless of how many entities are in the respective clusters. When s_{ij} is a distance-like function, the update equation for the median method is

$$s_{tr} = \tfrac{1}{2}(s_{pr} + s_{qr}) - \tfrac{1}{4}s_{pq}.$$

When s_{ij} is a correlation-like function, the update equation is

$$s_{tr} = \tfrac{1}{2}(s_{pr} + s_{qr}) + \tfrac{1}{4}(1 - s_{pq}).$$

Lance and Williams (1967a, p. 375) note that the line between the centroid of some cluster r and the centroid of the cluster formed by the merger of clusters p and q lies along the median of the triangle formed by p, q, and r, hence the name for the method.

   A characteristic of the centroid method and its variants is that the similarity value associated with the merger of the most similar clusters may rise and fall from stage to stage. For example, when the similarity measure is a distance, the distance between the centroids of some pair may be less than that between another pair merged at an earlier stage; the following sequence of distances between cluster centroids could occur in three successive mergers: $s_{21} = 8$, $s_{34} = 9$, $s_{31} = 7$. The resulting tree is shown in Fig. 6.4a; the last merger occurs at a lower level than either of the preceding two so that depicting the tree requires crossovers. With only a few reversals the tree begins to look like a wiring diagram for a color television set. Reversals occur because cluster centroids can migrate as mergers take place. Referring to Fig. 6.4b, if clusters with centroids p and q are merged, the centroid of the resulting cluster t will be closer to point r than were either p or q. This problem is not shared by any of the other hierarchical methods in this chapter.

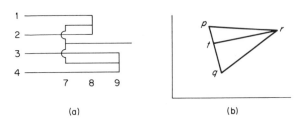

Fig. 6.4. Centroid reversal phenomenon.

6.2.6   THE WARD METHOD

Ward (1963) and Ward and Hook (1963) describe a very general hierarchical clustering method in which the merges at each stage are chosen so as to maximize an objective function which reflects the investigator's purpose in the particular problem. This general formulation encompasses all of the hierarchical clustering methods in the present chapter. However, Ward illustrated his method with an error sum of squares objective function, and the example came to be better known than the more general procedure. This particular method originally was implemented in the form described in Section 6.3.1. However, Wishart (1969a) showed how the Ward algorithm could be implemented through updating a stored matrix of squared Euclidean distances between cluster centroids.

Define the following quantities:

x_{ijk} = score on the ith of n variables for the jth of m_k data units in the kth of h clusters,

$$\bar{x}_{ik} = \sum_{j=1}^{j=m_k} x_{ijk}/m_k,$$

= mean on the ith variable for data units in the kth cluster,

$$E_k = \sum_{i=1}^{i=n} \sum_{j=1}^{j=m_k} (x_{ijk} - \bar{x}_{ik})^2 = \sum_{i=1}^{i=n} \sum_{j=1}^{j=m_k} x_{ijk}^2 - m_k \sum_{i=1}^{i=n} \bar{x}_{ik}^2$$

= error sum of squares for cluster k; sum of Euclidean distances from each data point in cluster k to the mean vector of cluster k;

within group squared deviations about the mean for cluster k,

$$E = \sum_{k=1}^{k=h} E_k$$

= total within group error sum of squares for the collection of clusters.

To begin there are m data units, each a cluster unto itself; consequently the membership and the mean of each cluster coincide so that $E_k = 0$ for all clusters. The Ward objective is to find at each stage those two clusters whose merger gives the *minimum increase* in the total within group error sum of squares E. Suppose clusters p and q are chosen to be merged and the resulting cluster is denoted as t. Then the increase in E is

$$\Delta E_{pq} = E_t - E_p - E_q$$

$$= \left[\sum_{i=1}^{i=n} \sum_{j=1}^{j=m_t} x_{ijt}^2 - m_t \sum_{i=1}^{i=n} \bar{x}_{it}^2 \right] - \left[\sum_{i=1}^{i=n} \sum_{j=1}^{j=m_p} x_{ijp}^2 - m_p \sum_{i=1}^{i=n} \bar{x}_{ip}^2 \right]$$

$$- \left[\sum_{i=1}^{i=n} \sum_{j=1}^{j=m_q} x_{ijq}^2 - m_q \sum_{i=1}^{i=n} \bar{x}_{iq}^2 \right]$$

$$= m_p \sum_{i=1}^{i=n} \bar{x}_{ip}^2 + m_q \sum_{i=1}^{i=n} \bar{x}_{iq}^2 - m_t \sum_{i=1}^{i=n} \bar{x}_{it}^2, \tag{6.1}$$

since the terms involving x_{ijk}^2 all cancel. The mean on the ith variable for the new cluster is found from the relation $m_t \bar{x}_{it} = m_p \bar{x}_{ip} + m_q \bar{x}_{iq}$; squaring both sides of this equation gives

$$m_t^2 \bar{x}_{it}^2 = m_p^2 \bar{x}_{ip}^2 + m_q^2 \bar{x}_{iq}^2 + 2 m_p m_q \bar{x}_{ip} \bar{x}_{iq}.$$

The product of the means can be rewritten as

$$2 \bar{x}_{ip} \bar{x}_{iq} = \bar{x}_{ip}^2 + \bar{x}_{iq}^2 - (\bar{x}_{ip} - \bar{x}_{iq})^2,$$

so that

$$m_t^2 \bar{x}_{it}^2 = m_p(m_p + m_q) \bar{x}_{ip}^2 + m_q(m_p + m_q) \bar{x}_{iq}^2 - m_p m_q (\bar{x}_{ip} - \bar{x}_{iq})^2.$$

Noting that $m_t = m_p + m_q$ and dividing both sides of the last equation by m_t^2,

$$\bar{x}_{it}^2 = \frac{m_p}{m_t} \bar{x}_{ip}^2 + \frac{m_q}{m_t} \bar{x}_{iq}^2 - \frac{m_p m_q}{m_t^2} (\bar{x}_{ip} - \bar{x}_{iq})^2. \tag{6.2}$$

Substituting Eq. (6.2) into Eq. (6.1) then gives

$$\Delta E_{pq} = \frac{m_p m_q}{m_p + m_q} \sum_{i=1}^{i=n} (\bar{x}_{ip} - \bar{x}_{iq})^2. \tag{6.3}$$

Thus, the minimum increase in the error sum of squares is proportional to the squared Euclidean distance between the centroids of the merged clusters. This result is different from the centroid clustering methods in that it involves weighting of the distance between centroids when computing the distances. The error sum of squares function E is nondecreasing and the method is not subject to reversals.

Equation (6.3) is sufficient to define the algorithm when using the stored data approach as in Section 6.3.1. However, for the stored similarity matrix approach an update equation still needs to be developed. Letting t denote the result of merging clusters p and q, and letting r denote any other cluster, the increase in E that would result from the potential merger of clusters r and t is obtained from Eq. (6.3) as

$$\Delta E_{rt} = \frac{m_r m_t}{m_r + m_t} \sum_{i=1}^{i=n} (\bar{x}_{ir} - \bar{x}_{it})^2. \tag{6.4}$$

By substituting $\bar{x}_{it} = (m_p \bar{x}_{ip} + m_q \bar{x}_{iq})/m_t$, $m_t = m_p + m_q$, and using Eq. (6.2), the squared terms in the summation of Eq. (6.4) may be written

$$(\bar{x}_{ir} - \bar{x}_{it})^2 = \frac{m_p}{m_t}(\bar{x}_{ir} - \bar{x}_{ip})^2 + \frac{m_q}{m_t}(\bar{x}_{ir} - \bar{x}_{iq})^2 - \frac{m_p m_q}{m_t^2}(\bar{x}_{ip} - \bar{x}_{iq})^2.$$

Substituting in Eq. (6.4) and gathering terms

$$\Delta E_{rt} = \frac{1}{m_r + m_t} \sum_{i=1}^{i=n} [m_r m_p (\bar{x}_{ir} - \bar{x}_{ip})^2 + m_r m_q (\bar{x}_{ir} - \bar{x}_{iq})^2$$
$$- \frac{m_r m_p m_q}{m_p + m_q} (\bar{x}_{ip} - \bar{x}_{iq})^2],$$

and using Eq. (6.3) again

$$\Delta E_{rt} = \frac{1}{m_r + m_t} [(m_r + m_p) \Delta E_{rp} + (m_r + m_q) \Delta E_{rq} - m_r \Delta E_{pq}]. \tag{6.5}$$

It would be simplest to construct the similarity matrix such that $s_{ij} = E_{ij}$ and then use Eq. (6.5) as an update equation. Since the initial clusters each consist of one data unit, Eq. (6.3) shows that the initial entries in such a similarity matrix are $E_{ij} = \frac{1}{2} d_{ij}^2$, half the squared Euclidean distance between data units i and j. However, it is not necessary to actually divide the entries in the initial matrix of squared Euclidean distances by 2; the updates of Eq. (6.5) may be carried out using the squared distances and when the increment to the error sum of squares function is added, merely use half the computed value. The entire algorithm then involves the following steps:

1. Store the lower triangular similarity matrix of squared distances. Treat s_{ij} as E_{ij}.
2. At each stage merge the two clusters which as a pair give the smallest increment to the error sum of squares. Let these clusters be labeled p and q with the resulting cluster labeled t.

3. Update entries in the similarity matrix according to Eq. (6.5). Increment the error sum of squares function by using the replacement relation

$$E = E + \tfrac{1}{2}\Delta E_{pq}.$$

4. Repeat steps 2 and 3 for $m - 1$ stages where m is the number of data units.

As with the centroid methods, the algorithm operates directly on the similarity matrix which is just an array of numbers; hence, the entries in this matrix could be computed using any association measure for either variables or data units. However, it is not known what properties the resulting clusters would have unless the similarity is the squared Euclidean distance.

While the similarity matrix should contain squared Euclidean distance, these distances may be computed in any desired representation space such as one involving principal components, weighted variables, or nonlinear composites of variables. By keeping this fact in mind, the Ward method can be quite a versatile technique for cluster analysis even though nominally limited to Euclidean distances.

As mentioned in the introduction to this chapter, hierarchical methods do not guarantee an optimal solution in terms of the clustering criterion; a set of h clusters produced by the Ward method may or may not give the minimum possible error sum of squares over all possible sets of h clusters formed from the m data units. However, the Ward solution is usually very good even if it is not optimal on this criterion.

6.3   The Stored Data Approach

While the stored matrix approach is very versatile, it is limited to fairly modest problems because of the size of the similarity matrix; Table 6.2 showed how the demands for storage space increase with problem size. By storing the similarity matrix in core, the similarity between any two clusters can be retrieved for comparison or update without a search; much the same effect can be achieved if the similarity between clusters can be computed on demand from stored information about each cluster. This latter notion becomes very interesting when one realizes that the similarity matrix may be larger than the original data set. For example, in a problem of 500 data units and 50 variables, a similarity matrix among the data units contains 124,750 elements while the entire data set consists of only 25,000 elements.

Storing all of the original data in the central memory of the computer makes available the entire body of numerical information about the problem; then, in principle it is possible to use any clustering method with any similarity measure because all original information is available on all the data units in a cluster. However, computing association measures between clusters by references to the original data tends to become a combinatorial problem of infeasible size. To keep the problem manageable, it is necessary to store either the similarity values themselves or a small number of summary statistics for each cluster from which the desired similarity values may be computed. The latter approach is utilized for the four methods described in the following sections. These methods apply only to the clustering of data units described by interval variables. It may be possible to extend this approach to other kinds of variables and perhaps even to the clustering of variables; however, the means for realizing these extensions are not presently known and must await future research.

The conceptual approach is much the same as that employed with the stored matrix approach. In fact the same algorithm as described in Section 6.2 is utilized; the only difference is that similarity values are computed when needed rather than retrieved from storage. In particular, the steps are as follows for a problem of n entities:

1. For each row of the similarity matrix compute each similarity value, save the minimum (maximum) entry in the row and record the column in which this extreme value occurs. This step requires $\frac{1}{2}(n-1)(n-2)$ comparisons and computation of $\frac{1}{2}n(n-1)$ similarity values.

2. Search the row minima (maxima) for the most similar pair. This step requires $\frac{1}{2}n(n-3)+1$ comparisons when repeated over $n-1$ stages.

3. Update the representation for the cluster resulting from the merger of the most similar pair. For each row in which the cluster or column corresponding to the row minimum (maximum) is involved in the merger, it is necessary to recompute the similarity values for the row and find a new row minimum (maximum). One such row is that corresponding to the cluster resulting from the merger; assuming an average of one additional row per stage and $\frac{1}{2}(n-k)$ elements per row at the kth stage, the total expected number of recomputed similarity values over $n-1$ stages is $2\sum_{k=1}^{i=n-1}\frac{1}{2}(n-k) = \frac{1}{2}n(n-1)$ and the total expected number of comparisons for updating the row minima (maxima) is $2\sum_{k=1}^{k=n-1}[\frac{1}{2}(n-k)-1] = \frac{1}{2}n(n-5)+2$.

4. And as with the stored matrix approach, a pushdown list of active clusters is maintained at a cost of $\frac{1}{4}(n+2)(n-1)$ comparisons.

In sum, the algorithm requires computation of $n(n-1)$ similarity values (twice that of the stored matrix approach) and $\frac{1}{2}(3n^2-11n)$ comparisons (about $\frac{3}{4}$ that of the stored matrix approach). This stored data approach is implemented through the subroutines MANAGE, GROUP, and PROC in Appendix E.

6.3.1   THE WARD METHOD

The Ward method was reviewed thoroughly in Section 6.2.6. Equation (6.3) gives the increase in the error sum of squares due to the merger of clusters p and q; this latter equation requires only the storage of the mean vector and the number of data units in each cluster. Since the clustering begins with each data unit as a cluster of one member, the storage requirement is equal to the size of the data set plus a vector containing the number of data units in each cluster. However, the constant recomputation of cluster means gives the opportunity for accumulation of round-off error which could be important in large problems. Accordingly a slightly altered method of computation is suggested. In addition to the notation of Section 6.2.6, define the following quantities:

$$T_{ik} = \sum_{j=1}^{j=m_k} x_{ijk} = m_k \bar{x}_{ik}$$

= total of scores on ith variable for data units in the kth cluster,

$$S_k = \sum_{i=1}^{i=n} \sum_{j=1}^{j=m_k} x_{ijk}^2$$

= sum of squared scores on all variables for all
     data units in the kth cluster.

Then, the error sum of squares for cluster k may be written as

$$E_k = S_k - \sum_{i=1}^{i=n} T_{ik}^2/m_k.$$

The increase in the total error sum of squares due to the merger of clusters p and q to form the new cluster t is

$$\Delta E_{pq} = E_t - E_p - E_q$$
$$= S_t - \sum_{i=1}^{i=n} T_{it}^2/m_t - E_p - E_q$$
$$= S_p + S_q - \sum_{i=1}^{i=n} (T_{ip} + T_{iq})^2/(m_p + m_q) - E_p - E_q, \tag{6.6}$$

since $m_t = m_p + m_q$, $S_t = S_p + S_q$, and $T_{it} = T_{ip} + T_{iq}$. This expression could easily be reduced further or put in many other forms; however, accumulating S_k, E_k, m_k, and $\{T_{ik}: i = 1,\ldots,n\}$ for each cluster primarily involves simple addition and avoids the repeated multiplication and division required when using cluster means. This formulation also permits an easy implementation of the other methods given in the following sections. In particular, the other methods can be utilized to control the clustering but at the same time the Ward criterion can be computed easily to give a common base of comparison for results from the different methods.

6.3.2   MINIMUM TOTAL WITHIN GROUP SUM OF SQUARES IN THE NEW CLUSTER

In this variation the merger at each stage is chosen such that the error sum of squares in the new cluster

$$E_t = E_p + E_q + \Delta E_{pq}$$

is a minimum. In contrast, the Ward criterion is to choose the merger such ΔE_{pq} is a minimum. This approach tends to produce clusters with approximately equal error sums of squares. By placing attention on E_t rather than ΔE_{pq}, the growth of relatively larger clusters is retarded and isolated points are likely to merge with some cluster at an earlier stage in the clustering.

6.3.3   MINIMUM AVERAGE WITHIN GROUP SUM OF SQUARES IN THE NEW CLUSTER

In this variation the merger at each stage is chosen to minimize E_t/m_t which is the average contribution to the error sum of squares for each member of the cluster; this quantity is also the variance in the new cluster. This method has a strong tendency to produce clusters of approximately equal variance; consequently, if clusters are all of approximately equal density, then there will be a tendency for large natural groups to appear as several smaller clusters or for small natural groups to merge into larger clusters. Thus, it may be necessary to consider some branches of the tree in greater detail because the "true" clusters are not really homogeneous in variance.

6.3.4   CENTROID METHODS

Since T_{ik} and m_k are elements of the preceding methods, it is convenient to calculate the squared Euclidean distance between cluster centroids as

$$d_{pq}^2 = \sum_{i=1}^{i=n} \left[\frac{T_{ip}}{m_p} - \frac{T_{iq}}{m_q} \right]^2 .$$

Of course reversals are just as likely using the stored data approach as they are when using the stored matrix approach as described in Section 6.2.5. However, it is a simple matter to compute the total within group error sum of squares and plot the tree against this criterion rather than the squared distance between centroids. Also, the analyst could easily use any other desired distance function between cluster centroids.

6.4 The Sorted Matrix Approach

This section is devoted to a relatively unexploited approach to hierarchical clustering; it is useful especially in cases where the similarity matrix is too large to fit into the central memory of the computer. Like the stored matrix approach of Section 6.2, the similarity matrix is a given element and the approach is not affected by any aspect of the chosen measure of association other than whether it is distance-like or correlation-like. Hence, this approach is suitable for clustering either variables or data units.

6.4.1 SINGLE LINKAGE

When the entities to be clustered are data units and the similarity matrix is computed from a distance function, the single linkage method of Section 6.2.1 begins by finding the link between the two closest data units; the succeeding stages consist of finding the shortest remaining link which directly joins together data units not already joined indirectly through a chain of links. In effect this algorithm goes through the similarity matrix seeking the shortest link that can contribute to further agglomeration of the current set of clusters. It is apparent that the following algorithm will achieve the same result.

1. Compute the lower triangular similarity matrix and write the result on magnetic tape as the triples (s_{ij}, i, j).

2. Sort the collection of triples into ascending sequence on s_{ij}.

3. Take for the first merger the pair of data units corresponding to the first similarity value in the sorted list, that is, the smallest similarity value in the entire similarity matrix.

4. At each succeeding stage choose for the next link or merger that pair of data units corresponding to the next largest similarity value, provided the two data units are not already in the same cluster (connected to each other through a chain of links). If the data units are already in the same cluster bypass the link and move on to the next triple.

5. Continue step 4 until a total of $m - 1$ merges have been made, m being the number of data units in the data set.

The advantage of this algorithm is that it does not require storage for either the original data or for the similarity matrix; the sorted similarity matrix may be stored on tape and simply read one triple at a time in the sorted sequence. The search phase of the earlier approaches is replaced by the sort phase in step 2 above. Most sizeable computation facilities include as part of their standard program library an efficient, high capacity sort/merge package which can perform this sort phase with little effort on the part of the analyst. Even though a problem of 1000 data units involves a similarity matrix of 499,500 entries, sorting such an array is not an overwhelming task with these sort/merge packages. The initial computation of the similarity matrix is likely to be a more onerous task than the sort. However, this algorithm permits the problem to be decomposed so that computing the similarity matrix, sorting, and clustering may be performed in three separate phases; each phase may even be performed in several subphases.

The discussion to this point has been in terms of data units and distances. However, single linkage clustering may be performed in this same manner for variables and correlation-like measures simply by sorting the similarity matrix in descending sequence. Subroutines ALLIN1 and PREP in Appendix E implement this algorithm given the sorted similarity matrix as input.

6.4.2   GRAPH THEORY CONNECTIONS

The single-linkage method has a close connection with certain aspects of graph theory. A graph is a set of nodes or points together with a set of edges or arcs joining together some or all of the nodes. Figure 6.5 shows an example

Fig. 6.5. Example of a graph.

of a graph with 9 nodes and 11 edges. A graph in which there is a path from any node to every other node is *connected*; if there is an edge connecting each pair of nodes directly, then the graph is *fully connected*. A *spanning tree* of a connected graph of n nodes is any set of $n - 1$ edges which provide one and only one path between any pair of nodes; that is, a spanning tree links together all the nodes but with a minimum number of edges. If the edges are taken to have lengths associated with them, then the *minimal spanning tree* (MST) is the shortest of all spanning trees that may be constructed from the given set of edges.

Given a similarity matrix in terms of distances between data units, the data units may be taken as the nodes of a fully connected graph with the entries in the similarity matrix specifying the lengths of the edges. Since the single linkage and complete linkage methods choose $n - 1$ links connecting together all of the clustered entities, these two methods find spanning trees of the graph defined by the similarity matrix. The single-linkage method finds the minimal spanning tree; the complete-linkage method begins with the shortest link and terminates with the longest link to construct a spanning tree which seems to have no special interpretation in graph theory terms. If the similarity matrix contains correlation-like measures between variables, then the variables may be taken as nodes and the links found by the single-linkage method form the maximal spanning tree of the graph defined by the similarity matrix.

Gower and Ross (1969) and Zahn (1971) also note the correspondence between single linkage cluster analysis and the minimal spanning tree of a graph. One source of their interest is the fact that very efficient algorithms exist for computing the minimal spanning tree; indeed, the method discussed in Section 6.4.1 is exactly the same as Kruskal's (1956) algorithm for finding the MST. An alternative is the Prim or "greedy" algorithm which begins at any node and connects that node to its nearest neighbor; at each succeeding stage an additional node is added to the set of connected nodes by choosing the shortest remaining edge connecting a node in the tree and another node not yet in the tree. Both these algorithms are simple and efficient; normally one or the other is superior to any alternative method of finding the minimal spanning tree.

Another reason for possible interest in the minimal spanning tree is that many different groupings could be achieved by finding the MST of a graph with some of the edges removed, i.e., a restricted graph. Indeed, any spanning tree of a graph can be found as the MST of a restricted version of that graph; at the very least, the graph could be restricted to just those edges in the desired spanning tree. If the similarity matrix is based on a distance-like measure, eliminated edges may be represented by very large distances, whereas in a matrix based on a correlation-like measure eliminated edges may be represented by zeroes. Removing edges would also give the opportunity to incorporate constraints in the clustering. This idea apparently has not been exploited; particular methods to nominate edges for removal must await further research. However, one interesting possibility is to utilize the analyst's judgement in an interactive computer program where he could compute the MST, examine the result to identify edges to be deleted, find the MST in the restricted graph, identify more edges to be deleted and so forth until a satisfactory solution is found. The algorithm of Section 6.4.1 using the sorted similarity matrix would probably be preferred as the basis for such an interactive method. When an edge in the current MST is identified for elimination, all the edges found prior to the rejected edge will be in the new MST; therefore, the search for the rest of the edges can begin at that point on the sorted tape where the rejected edge was found originally.

6.5   Other Approaches

The clustering methods described in the preceding sections are all implemented in the computer programs of Appendix E. Alternative computational
approaches to agglomerative hierarchical clustering have been utilized in three
other computer programs described in the following sections. In addition, a
review of divisive hierarchical methods is provided to acquaint the reader
with the full range of hierarchical techniques; these divisive methods are
treated only briefly because they do not seem to be as widely applicable or as
flexible as the agglomerative methods.

6.5.1   Parks' Clustering Program

Parks (1969, 1970) describes a computer program for hierarchical clustering
of data units. The 1970 publication includes a listing of the program dimensioned to handle problems of up to 200 variables and 1000 data units; this
version can be accommodated on a CDC 6400 computer with a 65k memory.
The program implements the following algorithm:

1. All variables are transformed to have range 0 to 1.
2. A similarity matrix is computed among the transformed variables.
Two options are available:
      a. product moment correlation coefficient,
      b. the complement of the mean square difference between variables
         i and j computed over all data units

$$1 - \left[\sum_{k=1}^{k=m} (x_{ik} - x_{jk})^2/m \right]^{1/2}.$$

3. The principal components of the similarity matrix are computed and
various criteria are applied to choose the number of components to use in
subsequent computations.
4. Data units are scored on the chosen principal components.
5. A new similarity matrix of mean square difference between data units
(proportional to Euclidean distance) is computed in the space of principal
components.
6. The data units are clustered by the ordinary centroid clustering
method (the centroids of the groups merged at each stage are weighted according to the number of data units in each cluster).

To make this algorithm computationally feasible, all storage of original
data, similarity matrices, and principal component scores is on disk or scratch

tapes. The actual clustering of data units does not use the entire similarity matrix; instead, for m data units only the m smallest distances are saved and the first $m/2$ merges are performed by scanning and updating this shortened list of distances; after these first $m/2$ merges, the distance matrix among centroids is recomputed and sorted for the smallest values; this recomputation and sort is performed a third time after 80% of the merges are completed.

The program has some interesting features such as sorting the similarity matrix and using only the smallest distances for the first 50% of the merges. On the other hand, some features seem objectionable, especially the following:

1. After the variables are equalized to unit range there is no opportunity for weighting.
2. The principal components of a matrix of complemented mean-square differences between variables are uninterpretable.
3. The centroid method is subject to annoying reversals.

The principal virtue of the program is that it can handle fairly large problems. However, this capacity is purchased at the expense of many input/output transactions with scratch tapes; this aspect of the program puts a premium on the efficiency of the input/output routines in the local computer system. Further, reliance on peripheral storage devices rather than the central memory will have the effect of increasing computation time substantially for modest problems.

6.5.2   WISHART'S HIERAR PROGRAM

Wishart (1969b) describes a package of computer programs titled CLUSTAN I and written in FORTRAN II for the IBM 1620 computer. Subsequently, the package was expanded, rewritten in FORTRAN IV for the IBM 360/44 and offered under the title CLUSTAN IA. The entire CLUSTAN IA package consists of about 18,000 card images.†

Program HIERAR is a part of the CLUSTAN package (both versions); this program performs hierarchical clustering on an input similarity matrix essentially as described in Section 6.2. However, the program has the capacity to cluster data sets of up to 1000 data units because the similarity matrix is stored on disk. When the similarity matrix is small enough to fit in the central memory, the input/output transactions with the disk are simulated in core resulting in a substantial saving in peripheral processing time.

† The CLUSTAN IA package is available through The Manager, Computing Laboratory, Mathematical Institute, University of St. Andrews, North Haugh, St. Andrews, Fife, Scotland.

6.5.3   WOLFE'S APPROACH

Wolfe (1970) describes a clustering method based on the decomposition of a mixture of distributions. Wolfe (1971) provides complete documentation and listings for an IBM 360/65 version of the program. The method requires an initial set of clusters for the data units in order to start the main clustering method. These initial clusters are obtained by an interesting combination of hierarchical clustering methods.

If the entire data set will fit in the central memory, then the clustering technique is the Ward method described in Section 6.3.1. If the data set will not fit in the central memory, then the following algorithm is used:

1. Store as many data units as possible in the central memory and store the remainder on disk.

2. At each stage, use centroid clustering to merge two groups and read a new data unit from the disk into the space left by the merger.

3. After the disk file is emptied, represent the clusters by their centroids and finish clustering with the Ward method.

This method essentially restricts the number of different possible mergers at early stages of clustering; that is, some of the initial mergers may be less desirable than those which could be achieved if every data unit could be considered for merger with each of the other data units. It seems unlikely that these restrictions at early stages will have more than minor marginal effects on the membership of the resulting clusters. This innovation could be a profitable addition to the programs in Appendix E implementing the stored data method described in Section 6.3.

6.5.4   A BRIEF REVIEW OF DIVISIVE METHODS

This chapter has concentrated on agglomerative hierarchical methods. There exists a variety of other hierarchical techniques based on dividing the entire data set into subgroups. These methods are focused on finding the groups which are the best separated from each other or most distinctive as opposed to the agglomerative notion of pulling together the entities which are most alike.

The conceptually simplest methods are known as "monothetic division." Data units are described by binary variables and the objective is to split the data set on one of these variables so as to minimize the value of some appropriate measure of similarity between the two groups. Each of these groups may then be divided on any of the remaining variables and so on until some satisfactory configuration is obtained. Williams and Lambert (1959) and

Lance and Williams (1965) discuss the basic approach under the label of "association analysis." Crawford and Wishart (1967, 1968) present a variant on monothetic division built around an interaction measure. Wishart's CLUSTAN IA package includes a program titled DIVIDE which performs both the ordinary association analysis and the Crawford-Wishart variation. The AID technique of Sonquist and Morgan (1964) is a monothetic division technique where each split of the data set is chosen so as to minimize the within group error sum of squares for a criterion or reference variable.

A different approach is suggested by Edwards and Cavalli-Sforza (1965). Their method is to choose that division into two groups which minimizes the total error sum of squares for the two groups, the same criterion as used with the Ward method of Sections 6.2.6 and 6.3.1. As originally proposed, the method proceeds by considering all $2^{m-1} - 1$ partitions of m data units into two groups. Recently, Scott and Symons (1971b) showed that the number of relevant partitions is somewhat less than $2^{m-1} - 1$, but they lack a method of enumerating these partitions and so are not yet able to capitalize on the discovery. The method is limited to small problems because of the large number of possible partitions. Harding (1971) has studied the sampling properties of trees generated by this method and his results may be of some value in the study of other hierarchical methods.

A third approach is based on the use of discriminant analysis. The essential idea is to begin with an initial partition, compute a linear discriminant function, and then iteratively reassign points and recompute discriminant functions so as to find the most strongly separable groups. Casetti (1964) and Hung and Dubes (1970) both provide FORTRAN programs for this method. Mayer (1971a) discusses a variant on this approach which uses a single dominant variable to specify an initial partition; this variable is also used to impose an ordering on the data units. The Mayer method may be implemented with most discriminant analysis programs.

7

NONHIERARCHICAL CLUSTERING METHODS

For a data set of m entities the hierarchical methods of Chapter 6 give m nested classifications ranging from m clusters of one member each to one cluster of m members. The methods of this chapter are designed to cluster data units (these methods are not suitable for variables) into a single classification of k clusters, where k either is specified *a priori* or is determined as part of the clustering method.

The central idea in most of these methods is to choose some initial partition of the data units and then alter cluster memberships so as to obtain a better partition. The various algorithms which have been proposed differ as to what constitutes a "better partition" and what methods may be used for achieving improvements. The broad concept for these methods is very similar to that underlying the steepest descent algorithms used for unconstrained optimization in nonlinear programming. Such algorithms begin with an initial point and then generate a sequence of moves from one point to another, each giving an improved value of the objective function, until a local optimum is found. Apparantly this analogy has not been explored to any great extent; it would be a substantial contribution if the accumulated research and experience on unconstrained optimization problems could be used to devise more effective clustering methods. See Himmelblau (197) for a discussion of these optimization techniques, especially the so-called variable metric methods.

The methods of this chapter typically may be used with much larger problems than the hierarchical methods because it is not necessary to calculate and store the similarity matrix; it is not even necessary to store the data set.

156

In general, the data units are processed serially and can be read from tape or disk as needed. This characteristic makes it possible, at least in principle, to cluster arbitrarily large collections of data units.

7.1   Initial Configurations

The methods discussed in this chapter begin with an initial partition of the data units into groups or with a set of seed points around which clusters may be formed. In most cases a technique for establishing an initial partition is given as part of the original published clustering algorithm; however, these techniques usually are provided as a convenience to the user rather than as an integral part of the clustering algorithm. This section reviews a variety of such techniques which are more or less interchangeable with each other.

7.1.1   SEED POINTS

A set of k seed points can be used as cluster nuclei around which the set of m data units can be grouped. The following methods are representative examples of how such seed points can be generated.

1. Choose the first k data units in the data set (MacQueen, 1967). If the initial configuration does not influence the ultimate outcome in any important way, then this method is the cheapest and simplest.

2. Label the data units from 1 to m and choose those labeled m/k, $2m/k$, ..., $(k-1)m/k$, and m. This method is almost as simple as method 1 but tries to compensate for a natural tendency to arrange the data units in the order of collection or some other nonrandom sequence.

3. Subjectively choose any k data units from the data set.

4. Label the data units from 1 to m and choose the data units corresponding to k different random numbers in the range 1 to m (McRae, 1971).

5. Generate k synthetic points as vectors of coordinates where each coordinate is a random number from the range of the associated variable. Unless the data set "fills" the space, some of these seed points may be quite distant from any of the data units.

6. Take any desired partition of the data units into k mutually exclusive groups and compute the group centroids as seed points (Forgy, 1965). Methods of generating such partitions are considered in Section 7.1.2.

7. An intuitively appealing goal is to choose seed points which span the data set, that is, most data units are relatively close to a seed point but the seed points are well separated from each other. Astrahan (1970) strives for this goal by using the following procedure:

   a. Compute the "density" for each data unit as the number of other data units within some specified distance, say d_1;

b. Order the data units by "density" and choose the one with the highest "density" as the first seed point;

c. Choose subsequent seed points in order of decreasing "density," subject to the stipulation that each new seed point be at least a minimum distance, say d_2, from all other previously chosen seed points. Continue choosing seed points until all remaining data units have zero "density," that is, they are at least a distance of d_1 from every other data unit;

d. Assuming that an excess of seed points are generated by this method, hierarchically group the seed points until there are just k such points. The centroid clustering method of Sections 6.2.5 and 6.3.4 probably would be most suitable for this last step.

The choice of the d_1 and d_2 parameters is likely to require good judgement or several guesses; if d_1 is chosen too small there may be many isolated data units with zero density whereas if d_1 is too large a few seed points will cover the entire data set. In general d_2 should be larger than d_1; unless d_2 is at least twice d_1 some data units may contribute to the density value of more than one of the chosen seed points. The elaborate nature of the method makes it more expensive than the alternatives; however, Astrahan used it with a data set of 3231 data units at acceptable cost.

8. Ball and Hall (1967, pp. 72–74) suggest a somewhat simpler approach than that used in method 7 above. Take the overall mean vector of the data set as the first seed point; select subsequent seed points by examining the data units in their input sequence and accept any data unit which is at least some specified distance, say d, from all previously chosen seed points; continue choosing points until k seed points are accumulated or the data set is exhausted. This method is sufficiently simple that two or three values of the threshold distance d could be tried if the first value gave too few seed points or examined too little of the data set.

This list of methods certainly is not exhaustive, but it does provide a setting for a few observations. First, methods 1, 2, 3, 4, and 8 all share the property that every seed point is itself a data unit, and therefore any cluster built around such a point will have at least one member; the seed points from method 5 easily could result in one or more empty clusters, whereas methods 6 and 7 are relatively immune to such oddities. Second is the topic of randomness: methods 1, 2, 4, and 5 have elements of randomness about them, either through an implicit assumption of random ordering of data units within the data set or through explicit random selection. Doyle (1966, p. 45) astutely observes that the analyst's primary interest is not randomness per se but indifference; that is, the goal is an initial configuration free of overt biases. Ultimately, indifference probably is effected through random selection; but the set of possibilities from which random selections are made is likely to be shaped by the implications of indifference in each particular problem. Third, indifference may be cast aside in preference to a deliberate effort to span the data set with seed points as in methods 7 and 8; such methods seem less prone to giving distorted or badly balanced configurations than are methods involving random selection. Methods like 7 and 8 are also computationally more expensive, but some economy may be achieved by employing a random subset of the full data set. For example, Astrahan (1970) was interested in clustering about 16,000 data units, but used only a little over 3000 data units to establish the initial set of seed points.

7.1.2   INITIAL PARTITIONS

In some clustering methods the emphasis is on generating an initial partition of the data units into k mutually exclusive clusters rather than finding a set of seed points. Some methods of generating such partitions are considered below; note how a set of seed points is used in several cases.

1.   For a given set of seed points, assign each data unit to the cluster built around the nearest seed point (Forgy, 1965). The seed points remain stationary throughout the assignment of the full data set; consequently, the resulting set of clusters is independent of the sequence in which data units are assigned. Incidentally, the clusters are separated by piecewise linear boundaries; recall from elementary geometry that the locus of points equidistant from two given points is a straight line perpendicular to the line joining the given points. For three seed points, repeated application of the principle gives cluster boundaries shown as solid lines in Fig. 7.1. In higher dimensional spaces the boundaries are segments of hyperplanes.

Fig. 7.1.   Cluster boundaries equidistant from seed points.

2.   Given a set of seed points, let each seed point initially be a cluster of one member; then assign data units one at a time to the cluster with the nearest centroid; after a data unit is assigned to a cluster, update the centroid so that it is the true mean vector for all the data units currently in that cluster (MacQueen, 1967). This method bears a strong resemblance to the hierarchical centroid methods of Sections 6.2.5 and 6.3.4. As in the centroid methods, the cluster centroids migrate so the distance between a given data unit and the centroid of a particular cluster may vary widely during the assignment process;

accordingly, the resulting set of initial clusters is dependent on the order in which data units are assigned. MacQueen's suggestion of using the first k data units as seed points permits the assignment process to begin with the data unit numbered $k + 1$; therefore it is unnecessary to be concerned with the possibility of using a data unit twice, once as a seed point and once in the assignment process.

3. In most cases a hierarchical clustering method can produce an excellent initial partition. Wolfe (1970) uses the Ward hierarchical clustering method to provide an initial set of clusters for his algorithm. However, a complete hierarchical clustering of the entire data set may require more effort than the rest of the analysis and certainly tends to limit the size of the problems that may be considered. Lance and Williams (1967b) suggest using hierarchical methods on one or more subsets of convenient size and then use the resulting groups as nuclei for assignment of the remaining clusters. Subsets of 150 to 250 data units are quite manageable; hierarchical grouping ordinarily will give group centroids relatively distinct from each other but with reasonable constituencies of data units.

4. Various random allocation schemes could be used. For example, assign a data unit to one of the k clusters by generating a random number between 1 and k. The difficulty with all such random schemes is that the resulting groups are spread more or less uniformly over the entire data set and their centroids are k different estimates of the data set mean vector. Such groups have no properties of internal homogeneity and are not clusters at all. In general, random allocation to groups is not an attractive alternative.

5. The analyst could use his judgement to sort data units into an initial partition. It might be of interest in this regard to sort the data units on a single concept or variable and thereby deliberately bias the clustering in favor of a particular aspect of the problem.

7.2   Nearest Centroid Sorting—Fixed Number of Clusters

A set of seed points can be computed as the centroids of a set of clusters, and a set of clusters can be constructed by assigning each data unit to the cluster with the nearest seed point. The simplest iterative clustering methods merely consist of alternating these two processes until they converge to a stable configuration. This section presents several such methods suited to the basic problem of sorting the data units into a fixed number of clusters such that every data unit belongs to one and only one cluster. These simple methods are implemented through subroutines EXEC and KMEAN in Appendix F. Some more elaborate variations on this approach are treated in later sections of this chapter.

7.2.1   Forgy's Method and Jancey's Variant

Forgy (1965) suggests a very simple algorithm consisting of the following sequence of steps:

1.   Begin with any desired initial configuration. Go to step 2 if beginning with a set of seed points; go to step 3 if beginning with a partition of the data units.

2.   Allocate each data unit to the cluster with the nearest seed point. The seed points remain fixed for a full cycle through the entire data set.

3.   Compute new seed points as the centroids of the clusters of data units.

4.   Alternate steps 2 and 3 until the process converges; that is, continue until no data units change their cluster membership at step 2.

It is not possible to say how many repetitions of steps 2 and 3 will be required to achieve convergence in any particular problem; however, empirical evidence indicates that five repetitions or less ordinarily will suffice; only infrequently will more than ten repetitions be needed. The subject of convergence is given further treatment in Section 7.2.4.

At each repetition the assignment of m data units to k clusters requires mk distance computations and $m(k-1)$ comparisons of distances. Since k is ordinarily much smaller than m and the number of repetitions to convergence is small, the analyst can often examine sets of clusters associated with several different values of k at less cost than for a full hierarchical analysis using the methods of Chapter 6.

Jancey (1966) independently suggested the same method except for a modification at step 3. The first set of cluster seed points is either given or computed as the centroids of clusters in the initial partition; at all succeeding stages each new seed point is found by reflecting the old seed point through the new centroid for the cluster. Figure 7.2 illustrates the process. The line from point 1 to point 2 may be viewed as an approximation to the local gradient, the direction in which the seed point should move for greatest improvement in the partition. However, since data units were assigned to the

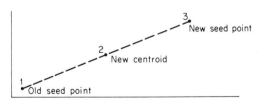

Fig. 7.2. Jancey's seed point reflection method.

cluster on the basis of their proximity to point 1 rather than point 2, it might be inferred that the movement of the centroid was retarded, and therefore the new seed point should overshoot the computed centroid. Jancey suggests that this technique will accelerate convergence and possibly lead to a better overall solution through bypassing inferior local minima.

Both Jancey's and Forgy's methods implicitly minimize a within cluster error function as is discussed in Section 7.2.4. As noted in Section 7.1.2, the cluster boundaries are piecewise linear for both these methods because the boundaries are equidistant from the nearest centroids. Since the seed points are recomputed only after the full data set has been reallocated, the results of these two methods are not affected by the sequence of the data units within the data set.

7.2.2   MacQueen's k-Means Method and a Variant

MacQueen (1967) uses the term "k-means" to denote the process of assigning each data unit to that cluster (of k clusters) with the nearest centroid (mean). The key implication in this process is that the cluster centroid is computed on the basis of the cluster's current membership rather than its membership at the end of the last reallocation cycle as with the Forgy and Jancey methods. MacQueen's algorithm for sorting m data units into k clusters is composed of the following steps:

1. Take the first k data units in the data set as clusters of one member each.

2. Assign each of the remaining $m - k$ data units to the cluster with the nearest centroid. After each assignment, recompute the centroid of the gaining cluster.

3. After all data units have been assigned in step 2, take the existing cluster centroids as fixed seed points and make one more pass through the data set assigning each data unit to the nearest seed point.

The last step is the same as the Forgy method except that the reallocation phase is performed just once rather than being continued until convergence is achieved. Like the Forgy method, the resulting clusters have piecewise linear boundaries.

By using the first k data units as seed points and relying on only one reallocation pass, this method achieves the distinction of being the least expensive of all clustering methods discussed in this book. The total effort from the initial configuration through to the final clusters involves only $k(2m - k)$ distance computations, $(k - 1)(2m - k)$ distance comparisons, and $m - k$ centroid updates. This computational workload is only a small fraction of that involved in a hierarchical cluster analysis because k is usually much smaller than m.

However, blindly using the first k data units may be less than satisfactory unless the analyst can arrange to place his choices for the initial centroids at the front of the data set.

The set of clusters constructed in step 2 of the algorithm depends on the sequence in which the data units are processed. MacQueen (1967, p. 290) reports some preliminary investigations into this effect. His experience indicates that the ordering of the data units has only a marginal effect when the clusters are well separated; differences from one ordering to the next are due largely to ambiguities arising from data units which fall between clusters. MacQueen also reports that he tried three different orderings when grouping a set of 250 data units into 18 clusters; the within group error sum of squares differed by at most 7% among the three sets of clusters. These results are encouraging but deserve further exploration.

The economy of effort inherent in this method stems from acceptance of the first reallocation of data units as opposed to continued processing until convergence is achieved. Apparently the method gives useful results because most major changes in cluster membership occur with the first reallocation; subsequent reallocations usually result in relatively few reassignments.

A *convergent* clustering method using the k-means process can be implemented through the following sequence of steps.

1. Begin with an initial partition of the data units into clusters. If desired, the partition could be constructed by using steps 1 and 2 of the ordinary MacQueen method, though any of the methods given in Section 7.1.2 could also be used.

2. Take each data unit in sequence and compute the distances to all cluster centroids; if the nearest centroid is not that of the data unit's parent cluster, then reassign the data unit and update the centroids of the losing and gaining clusters.

3. Repeat step 2 until convergence is achieved; that is, continue until a full cycle through the data set fails to cause any changes in cluster membership.

This convergent k-means process is a basic element of Wishart's RELOC and McRae's MIKCA computer programs which are discussed in Sections 7.3.2 and 7.4, respectively. This method also is included among the options in the computer programs of Appendix F.

7.2.3 A RESEARCH OPPORTUNITY

The hierarchical methods of Chapter 6 are distinguished fairly well from each other by the properties they impose on the hierarchical classifications they produce. The four iterative methods introduced so far in this chapter are not so well distinguished from each other. The Forgy, Jancey, and convergent

k-means methods all use variations on one central process, and they exhibit hardly any differences in computational workload. These three methods also converge in the same way, so a final set of clusters produced by one method would satisfy the convergence criteria of the other methods. Thus the three methods could converge to the same partition and if they did so regularly, then the analyst would choose the method that is most convenient to use or easiest to explain. If the methods tended to give different partitions, it would be of interest to characterize these differences so they could be used to guide the selection of a method appropriate to the data and the analyst's interests. The MacQueen k-means method is distinctive for its economy of effort but only because it does not carry to completion the same basic cluster-seeking process used by the other three methods.

These four methods of cluster analysis are simple, economical and fairly popular. Yet, it is not clear whether they differ systematically in regard to the substantive properties of the clusters they produce. It would be valuable to have the results of a comprehensive comparative study of these four methods. Particular questions which could be studied profitably include the following.

1. Does the initial configuration have any substantial effect on the final clusters? If so, do the four clustering methods differ in their sensitivity to this factor? Is there a clearly superior method of generating an initial configuration? Is it worth the effort to generate a set of seed points which span the data set as in methods 7 and 8 of Section 7.1.1?

2. Jancey's method differs from Forgy's only in regard to the generation of successive sets of seed points. Does this difference have any beneficial effects such as faster convergence, or reduced sensitivity to the initial configuration? Jancey suggested that reflecting the seed point about the centroid would tend to give partitions with smaller within group error sum of squares; is this hope fulfilled?

3. Does the sequence of data units within the data set make a substantial difference in the final clusters produced by the k-means methods? Do the extra iterations in the convergent k-means method tend to diminish sensitivity to the initial partition?

4. Are there any systematic distinctions in the final partitions that may be attributed to updating the centroids after every reallocation of a data unit (as in the k-means methods) versus updating only after a full cycle through the data set? Does this update policy affect the speed of convergence?

5. Is it worthwhile to continue reallocating data units until the partition converges? Does MacQueen's k-means method reveal most of what is to be learned through cluster analysis? Would the three convergent methods give substantially the same final partition, but at reduced cost, by stopping when a complete cycle through the data set results in fewer than some small number of reallocations, for example, 1 % of the number of data units?

6. Does the number of variables or number of data units affect the number of iterations needed to attain convergence?

These questions merely touch on some of the more obvious topics which might be studied. The answers to these and other questions would be of great value not only in choosing among these four particular methods but also in identifying pertinent criteria for the evaluation of clustering methods in general.

7.2.4   CONVERGENCE PROPERTIES

The discussion up to this point has included repeated references to the convergence of these methods but without any indication of why convergence should be expected. The broad outline of the convergence proof is not difficult to grasp, but a detailed and rigorous proof is sufficiently long and tedious as to obscure rather than illuminate the central issues. The reader who wants a complete proof can obtain one by filling in the background of the following steps.

1. For a given cluster of data units, the sum of squared deviations about a reference point is a unique minimum when the reference point is chosen as the mean vector (centroid) of the cluster. The sum of squared deviations about the centroid for the kth cluster is written as

$$E_k = \sum_{j=1}^{j=m_k} \sum_{i=1}^{i=n} (x_{ijk} - \bar{x}_{ik})^2$$

(using the notation of Section 6.2.6) and is denoted as the error sum of squares for cluster k. For a given partition of a data set into h clusters, the total within group error sum of squares is

$$E = \sum_{k=1}^{k=h} E_k,$$

and E has a characteristic value for the partition. Note that $\sum_{i=1}^{i=n} (x_{ijk} - \bar{x}_{ik})^2$ is the squared Euclidean distance between the centroid of cluster k and the jth data unit in that cluster.

2. The number of different ways a data set of m data units may be partitioned into h clusters is a Stirling number of the second kind

$$\mathfrak{S}_m^{(h)} = \frac{1}{h!} \sum_{i=0}^{i=h} (-1)^{h-i} \binom{h}{i} i^m,$$

which is a finite number if m is finite. Therefore any method which generates each partition at most once is finitely convergent because there are only finitely many different partitions. Consider methods in which the current partition is altered only if the change gives a new partition with a smaller total within group error sum of squares E. Since each partition has a characteristic value of E, such methods cannot regenerate a partition which was abandoned at an earlier stage and therefore such methods are convergent. Thus a method is convergent if the successive partitions which it generates exhibit a strictly decreasing sequence of values for E.

3. The "convergent k-means" method and Forgy's method can now be shown to be convergent. Let the most recently computed set of cluster centroids be denoted as the seed points. In both methods, a data unit is reallocated only if it is nearer to the seed point of the gaining cluster than to the seed point of the losing cluster; if the distance function is chosen as Euclidean distance (or any power thereof), then the sum of squared deviations about the seed point decreases more for the losing cluster than it increases for the gaining cluster, thereby giving an overall decrease in sum of squared deviations about the seed points for the partition as a whole. In addition, this sum of squared deviations is decreased even further if it is computed about the new centroids of the clusters rather than the old seed points. Thus, each new partition has a lower value of E than does the partition from which the seed points were computed and therefore these methods are convergent.

Strictly speaking, convergence has been shown only for the case in which the objective is to find partitions with minimum E; these methods explicitly minimize E when Euclidean distance is used and they may or may not converge with regard to some other objective function such as those in Section 7.4. If the analyst is interested in weighting the variables to minimize some weighted sum of squares, then weighted Euclidean distance can be used or more simply the variables can be redefined to absorb the weights. If one chooses to use the L_1 or "city block" metric as the measure of divergence between a data unit and a seed point, then the unique minimum of

$$E_k = \sum_{j=1}^{j=m_k} \sum_{i=1}^{i=n} |x_{ijk} - c_{ik}|$$

is approached when c_{ik} is chosen as the median value of the ith variable in the kth cluster. Revising the methods to compute seed points as the cluster medians and using the L_1 metric then minimizes the total within group sum of absolute errors. This latter scheme and weighted versions thereof can also be shown to converge by the same argument as presented above.

The Jancey method is somewhat problematic because neither a proof of convergence nor an example of nonconvergence is available. The method of proof applied to the convergent k-means method and Forgy's method cannot be used because successive partitions may have either larger or smaller values of E than their immediate predecessors. The possibility of nonconvergence arises because there is no apparent barrier to repeating some partitions and cycling through them again and again. The E function has many local minima which may be represented as valleys and potholes on a smooth hypersurface. When an iterative method converges it simply has become trapped in the bottom of one of these depressions. Apparently Jancey hoped that the local trend of the hypersurface (as sampled by successive partitions) would be representative of the global trend so that further movement in that direction ultimately would lead to convergence at a smaller value of E. Along the way it might be necessary to climb over local hills and ridges to reach a deeper depression; therefore, occasional increases in the value of E are to be expected. Of course such increases may lead to entrapment at a higher rather than a lower value of E.

The criterion chosen for deciding convergence of these clustering methods is stability of cluster membership; an alternative criterion is stability of the cluster seed points. These two criteria are equivalent for the Forgy and convergent k-means methods because the seed points are the cluster centroids which are dependent only on the cluster memberships. However, the alternative criterion is not suitable for the Jancey method. Refer back to Fig. 7.2 where the reflection of seed points is illustrated. Suppose a cluster has its centroid at point 2 and that all other data units are sufficiently far away so that grouping about point 1 or point 3 would give the same membership. Then taking point 1 as the initial seed point would give point 3 as the second seed point, which in turn would reflect back to point 1 and so forth for an indefinite number of cycles. Thus, once the membership of a cluster stabilizes the seed point oscillates between two fixed positions, unless it coincides with the centroid.

7.3 Nearest Centroid Sorting—Variable Number of Clusters

This section includes a discussion of three more elaborate clustering methods which employ heuristic devices for adjusting the number of clusters to conform to the apparent "natural structure" of the data set. Some of the most prominent motivations for allowing the number of clusters to vary can be illustrated with the aid of Fig. 7.3, which shows a data set consisting of five "natural clusters" and one outlier not belonging to any of the five clusters.

1. Sorting this data set into three or more clusters would almost surely result in one cluster consisting of the outlier alone. By allowing the number of clusters to increase the outlier could be structured as a cluster unto itself and the remainder of the data set classified as though the outlier was not there.

2. If only one seed point was located in the vicinity of clusters 1 and 2 these two clusters probably would appear as one but with some relatively

Fig. 7.3. Variable number of clusters example.

large discrepancies among their members. By allowing the number of clusters to increase, this "unnatural" cluster could be split into two more compact pieces.

   3. On the other hand, if there were several seed points in the vicinity of cluster 2, then several poorly differentiated clusters would result. By allowing the number of clusters to decrease these fragments could be combined into a distinctive whole.

   4. There may be more than one meaningful level of aggregation within the data set. In the case of Fig. 7.3, clusters 3, 4, and 5 could be merged to form one big cluster and give an additional classification of three clusters. On the other hand, an effort to force creation of four or six clusters (the outlier not included) probably would give some poorly defined clusters.

   While these difficulties are easy to recognize and appreciate in two dimensions, it may be rather difficult to recognize them in a problem of 40 variables and hundreds of data units. The three clustering methods discussed below demonstrate some techniques devised to tackle such problems. It appears that there is room for much more work in this area.

7.3.1   MacQueen's *k*-Mean Method with Coarsening and Refining Parameters

   MacQueen (1967) proposed a variation on his basic *k*-means method which permits the number of clusters to vary during the initial assignment of the data units to clusters. The steps of the method are as follows:

   1. Choose values for three parameters *k*, *C*, and *R*, which are used in the subsequent steps of the method.
   2. Take the first *k* data units as initial clusters of one member each.
   3. Compute all pairwise distances among these first *k* data units. If the smallest distance is less than the "coarsening parameter" *C*, then merge the two associated clusters and compute the distances between the centroid of the new cluster and all remaining clusters. Continue merging nearest clusters as necessary until all centroids are separated by a distance at least as large as *C*.

4. Assign the remaining $m - k$ data units one at a time to the cluster with the nearest centroid. After each such assignment update the centroid of the gaining cluster and compute the distance to the centroids of the other clusters; merge the new cluster with the cluster having the nearest centroid if the distance between centroids is less than C and continue merging as necessary until all centroids are at least C distance apart. If the distance to the nearest centroid is greater than a "refining parameter" R, then take the data unit as a new cluster of one member rather than assigning it to an existing cluster.

5. After data units $k + 1$ through m have been processed through step 4, take the existing cluster centroids (however many there may be) as fixed seed points and reallocate each data unit to its nearest seed point.

As with MacQueen's basic k-means method, the process ends after the first reallocation rather than continuing to convergence. Consequently, this method ranks as the least expensive of the alternatives in this section.

By allowing clusters with nearby centroids to merge the method avoids creating fine distinctions which artificially divide natural clusters. By creating new clusters when a data unit is distant from all existing centroids, the final set of centroids tends to span the data set and outliers tend to be set off by themselves rather than to be forced into a cluster. On the other hand, the number of clusters which actually come out of the process is a surprise and their centroids are not necessarily separated by distances greater than C; the analyst may find it necessary to merge nearby clusters by hand to obtain a satisfactory partition.

It appears that some experience is necessary to make wise choices for the C and R parameters; quite obviously one should choose $R > C$ but other useful guidelines do not seem available. Perhaps fractions of the variance computed over all the variables (weighted in the same way as in the chosen distance function) would be suitable choices for R and C; by taking R quite large and C fairly small this method might be a useful device for identifying outliers as a preliminary step before using other analysis methods.

7.3.2   Wishart's Variant on k-Means

Wishart's CLUSTAN IA package includes a program called RELOC which combines the convergent k-means method with systematic reductions in the number of clusters. The steps in this method are as follows:

1. Choose values for the control parameters: THRESH, MINSIZ, MAXIT, and MINC. The roles of these parameters will be clear in the subsequent steps of the algorithm.

2. Begin with an initial partition of the data set into clusters and compute the centroids.

3. Let k be the current number of clusters. Consider each data unit in sequence and compute the distance to all k centroids.

    a. If the smallest distance exceeds THRESH, then assign the data unit to a residue of unclassified data units and update the centroid of the losing cluster.

    b. If the nearest centroid is not that of the data unit's parent cluster and the distance does not exceed THRESH, then reassign the data unit and update the centroids of the losing and gaining clusters.

    c. If the data unit is currently in the residue and the distance to the nearest centroid does not exceed THRESH, reassign the data unit and update the centroid of the gaining cluster.

4. After all data units have been reallocated as appropriate at step 3, assign clusters with fewer than MINSIZ members to the residue; note that such assignments cause a reduction in the number of clusters.

5. Repeat steps 3 and 4 until the partition converges (no data units change their memberships at step 3) or until the number of cycles through these steps reaches MAXIT.

6. Compute the pairwise similarities between clusters and merge the two most similar clusters. Repeat steps 2–5 to obtain another partition. Continue in this manner until the number of clusters is reduced to MINC.

This method produces several partitions ranging from the initial partition to the final one of MINC clusters. The method will not necessarily give partitions for all numbers of clusters between the initial number and MINC, because step 4 may cause the process to step over some possibilities; for example, beginning with 15 clusters the method might produce partitions only for 15, 9, and 4 clusters. The method might be thought of as constructing partitions only for the most likely levels of aggregation. Note that the clusters in one partition need not be related hierarchically to those in another partition because the data units are reallocated iteratively at each cycle.

True outliers seem likely to remain in the residue once they are allocated to it. However, ambivalent data units lying between clusters may make frequent moves in and out of the residue; this possibility prompts the need for the MAXIT parameter because cycling data units can impede convergence. Step 3 of the method is recognized to be the convergent k-means method except for the movements to and from the residue; perhaps the convergence properties of this latter method could be regained by leaving data units in the residue until the number of clusters changes.

7.3.3 THE ISODATA METHOD

The most elaborate of the nearest centroid clustering methods is called ISODATA and was developed at Stanford Research Institute over a period of several years. Ball and Hall (1965) present a complete step-by-step description of the original method and illustrate it in great detail with a two-dimensional example. Ball (1966, pp. 39–40) describes subsequent improvements inspired by his wide-ranging survey of clustering methods in general. Ball and Hall (1967, pp. 70–82, 102–112) and Hall *et al.* (1969, pp. 41–46) describe a version of the method incorporated into the PROMENADE system, an on-line data analysis package utilizing interactive graphics. Wolf (1968) provides complete program listings for the PROMENADE system. Sammon (1968) also has included the ISODATA method in his on-line interactive system, OLPARS. Dubes (1970, pp. 139–151, 180–196) presents program listings in FORTRAN IV for a batch-oriented version of ISODATA. This latter program is a modified version of programs obtained from Stanford Research Institute; consequently it uses the same mnemonics for variable and subroutine names as are encountered throughout the Ball and Hall reports.

The ISODATA method has been the subject of extended research, and it is not surprising that several versions exist. The version presented here is representative of the major features of the method but may differ in some details from other versions. The method consists of the following steps:

1. Choose values for seven control parameters: NPARTS, NRWDSD, NCLST, THETAN, THETAE, THETAC, and ITERMAX. The role of each parameter will be clear in the subsequent steps.

2. If a set of seed points is not provided as part of the input, then generate a set using method number 8 of Section 7.1.1.

3. Assign each data unit to the cluster with the nearest seed point. The seed points remain fixed for a complete cycle through the data set. Recompute the seed points as cluster centroids. Repeat this reallocation/recomputation cycle until convergence is achieved or the number of such cycles reaches NPARTS. Note that his step is identical to the Forgy method of Section 7.2.1.

4. Discard any cluster which contains fewer than THETAN data units. The associated data units are disregarded for the remainder of the analysis.

5. Perform either a lumping or a splitting iteration (details are specified below) according to the rules:

   a. A lumping iteration is mandatory if the number of clusters is currently twice NWRDSD or more.

   b. A splitting iteration is mandatory if the number of clusters is currently half NWRDSD or less.

   c. Otherwise, alternate the processes by splitting on odd iterations and lumping on even iterations.

6. Compute new seed points as cluster centroids and perform the data unit reallocation/seed point recomputation cycles exactly as for step 3.

7. Repeat steps 4, 5, and 6 until the process converges or until these three steps have been repeated ITERMAX times.

During a *lumping* iteration all pairwise distances between cluster centroids are computed. If the distance between the two nearest centroids is less than THETAC, then the associated clusters are merged and the distance to all other centroids is computed. This process is continued up to a maximum of NCLST merges in any iteration. Of course, there may not be any merges on a lumping iteration if the centroids are separated sufficiently.

During a *splitting* iteration, a cluster provisionally is chosen for splitting if the within cluster standard deviation for any variable exceeds the product of THETAE and the standard deviation of that variable in the original data set as a whole. The data units of such a cluster are assigned to two new clusters according to whether they score above or below the mean of the splitting variable in the parent cluster. The centroids of these two clusters are computed and if the distance between them is at least 1.1 times THETAC then the split is accepted. In the original description given by Ball and Hall (1965) a cluster could not be split unless it had more than 2(THETAN + 1) data units in it; this condition was dropped in later versions.

The criterion for splitting a cluster has the effect of constraining the cluster width in *every* dimension; elongated clusters are virtually impossible to obtain. Because the seed points are taken as fixed for each reallocation at step 6, the clusters have piecewise linear boundaries. The practice of permanently discarding small clusters at step 4 eliminates outliers or "wild shots"; Ball and Hall exhibit a strong concern throughout their work for identifying and removing outliers. Because of the splitting and lumping processes the ultimate number of clusters is a surprise; the user supplies NWRDSD as a goal for the number of clusters, but the method prompts positive action toward this goal only if the current number of clusters fall outside the range of half NWRDSD to twice NWRDSD. This method is clearly the most expensive of the nearest centroid methods; unless NPARTS is set to 1 or 2, steps 1–3 alone require as much effort as is expended in the Forgy, Jancey, or basic MacQueen methods; each iteration through steps 4, 5, and 6 requires a similar amount of effort.

The question of when to stop iterating on the partition has always been somewhat perplexing for ISODATA. In the earliest documentation Ball and Hall were quite vague on this point and just seemed to keep running the program a few iterations at a time until the results were intuitively satisfactory. With the advent of the PROMENADE system, this subjective evaluation approach became an integral part of the interactive ISODATA method; at the end of each iteration the analyst chooses between accepting the current partition as satisfactory or continuing through another iteration, possibly with new parameters. In Dubes' (1970) program, convergence is achieved when successive cycles at step 6 of the method result in a change of less than 0.1 % in every variable for every centroid; in most cases this criterion will be equivalent to a full cycle with no changes in cluster membership, but it takes no account of the possibility that the convergent partition might contain clusters which should be lumped or split. On the other hand, a composite criterion based on stability of the partition plus satisfaction of lumping and splitting tests might never converge; there is no apparent barrier to the method cycling through several partitions involving different numbers of clusters. With these thoughts in mind it seems reasonable to conclude that the ISODATA method is too elaborate to be used without periodic human intervention as in the PROMENADE and OLPARS versions.

7.4   Other Approaches to Nonhierarchical Clustering

The methods presented in Sections 7.2 and 7.3 have a great deal of intuitive appeal and can be explained easily without reference to intricate mathematical models. The methods cited in this section are somewhat more complex to develop and document as well as utilize. Since adequate documentation and well-developed computer programs are available for these methods, this section is limited to brief description and commentary which will help guide the reader in a search for further information.

As was pointed out in Section 7.2.4, the nearest centroid sorting methods explicitly minimize the total within group sum of squares for the partition if data units are allocated in accordance with the Euclidean distances to existing centroids. Considerable study has been devoted to alternative criteria which may be optimized through reallocation of data units from one cluster to another. The several criteria which have been proposed are based on multivariate statistical analysis techniques, especially the methods of linear discriminant analysis and multivariate analysis of variance; Anderson (1958) and Morrison (1967) are useful references for the background material on these topics.

As discussed in Section 4.1.2 the scatter of two variables is the inner product of their centered score vectors. The total scatter matrix T is a square array in which the typical entry t_{ij} is the scatter of variables i and j computed over all data units in the data set. In a partition of the data set into k clusters, the within group scatter matrix for cluster k, W_k, has the typical entry w_{ijk} which is the scatter of variables i and j computed over all data units in cluster k; the within group scatter matrix for the partition is $W = \sum_{k=1}^{k=h} W_k$. The between group scatter matrix B has as its typical element $b_{ij} = \sum_{k=1}^{k=h} m_k \bar{x}_{ik} \bar{x}_{jk}$, where $\bar{x}_{ik}$ is the mean (centered about the grand mean in the data set) of the ith variable in the kth cluster, and m_k is the number of data units in the kth cluster. It can be shown that the three matrices satisfy the relation $T = B + W$. A particularly important element in the definition of the various clustering criteria is the determinantal equation $|B - \lambda W| = 0$; the λ_i solutions to this equation are the eigenvectors of the matrix $W^{-1}B$.

Various authors (Friedman and Rubin, 1967; Demiremen, 1969; Solomon, 1970; McRae, 1971; Scott and Symons, 1971a; Marriott, 1971) propose and discuss criteria for evaluating whether movements of individual data units result in an overall improvement of a partition. Four principal criteria have emerged from these studies.

1. Minimize trace W. The trace of a matrix is the sum of its diagonal elements. It may be shown that this criterion is the same as minimizing the total within group sum of squares of the partition, the criterion implicitly used in the methods of Sections 7.2 and 7.3.

2. Minimize the ratio of determinants $|W|/|T|$. This criterion is widely known as Wilks' lambda statistic. Since the matrix T is the same for all partitions, this criterion is equivalent to minimizing $|W|$. Another equivalent criterion is to maximize $|T|/|W|$ which may be shown to be equivalent to maximizing $|I + W^{-1}B|$ or maximizing $\prod_{i=1}^{i=n}(1 + \lambda_i)$.

3. Maximize the largest eigenvalue of $W^{-1}B$. This criterion is known as the largest root criterion and is due to the famed statistician S. N. Roy.

4. Maximize the trace of $W^{-1}B$. This criterion is known as Hotelling's trace criterion and is equivalent to maximizing $\sum_{i=1}^{i=n} \lambda_i$.

Demiremen (1969) provides a computer program which improves an initial partition by the Forgy method as described in Section 7.2.1. Criteria 1, 2, and 4 above are computed for each partition; further, a linear discriminant analysis and a multivariate analysis of variance are performed to aid the user's evaluation of each partition. These latter computations are purely descriptive and do not affect the actual clustering process. The program includes various options for scaling and orthogonalizing the data as well as an option for several distance functions.

McRae (1971) offers for general distribution another program named MIKCA which actually uses these various multivariate analysis criteria as part of the clustering process. The program performs the cluster analysis in two stages. The first stage consists of the convergent k-means method described in Section 7.2.2; this stage is terminated when convergence occurs or when a full cycle through the data set gives a worsened value of the chosen criterion (all four criteria listed above are available in the program) relative to that achieved with the partition generated in the previous cycle. The second stage consists of attempts to improve the best partition found in the first stage through reassigning individual data units from one cluster to another if such reallocation improves the value of the chosen criterion. McRae provides what he calls a "timing" parameter that permits the second stage to be applied to just the outer fringes of each cluster or to any portion of the data units up to the entire data set. The program also includes an option for standardizing the data and three choices for the distance function.

There are two serious problems associated with criteria 2, 3, and 4. The first problem is that they involve the computation of eigenvalues at each stage which severely restricts the number of variables that may be used in the problem; an eigenproblem for 20 variables is quite manageable but for 100 variables it overshadows the rest of the method in terms of computational effort. Another aspect of this problem is the need to compute the inverse of the W matrix, at least implicity; if this matrix is singular the Penrose–Moore pseudoinverse (Albert, 1971) may be used instead of the ordinary inverse but at a further cost in computational complexity. The second problem is that guidelines for making choices among the criteria are not available [as observed by Friedman and Rubin (1967, p. 1163)]; considering the extra computational cost of solving eigenproblems repeatedly, it is difficult to identify any clear-cut advantages stemming from the use of criteria 2, 3, and 4.

Scott and Symons (1971a) have shown how these various criteria can be obtained as a result of assuming each cluster to have a multivariate normal distribution and then seeking optimal partitions through maximum likelihood methods of classification; it is not clear how frequently the normality assumption is justified nor are the effects of nonnormality well understood. Wolfe (1970) has made explicit use of such likelihood methods in a program entitled NORMIX. The method assumes the data set is a mixture of multivariate normal distributions, and it seeks that partition which maximizes the likelihood function. The user can specify several different guesses for the number of clusters and the program will find a partition for each such guess. Wolfe (1971) provides complete documentation and program listings for an IBM 360/65 version of the program. With moderate effort the program can be converted to run on the CDC 6600. Both McRae's MIKCA and Wolfe's NORMIX programs are sophisticated and written competently; they are worthwhile additions to the program library in the appendices of this book.

8

PROMOTING INTERPRETATION OF CLUSTERING RESULTS

In Section 2.3 considerable discussion was devoted to philosophical observations on the utilization of cluster analysis. Particular stress was placed on the use of cluster analysis as an exploratory tool. In this latter role cluster analysis may help the analyst to generate hypotheses about the data or discern fundamental facts previously not apparent. But the mechanical results derived from submitting a set of data to some cluster analysis algorithm are themselves devoid of any inherent validity or claim to truth; such results are always in need of interpretation and are subject to being discarded as spurious or irrelevant. The product of a cluster analysis is not merely a set of clusters; the most useful outcome is increased understanding and improved organization of known facts permitting a more parsimonious description of the topic under study.

The use of cluster analysis requires the active participation of the analyst to interpret the results and judge their significance. This stage of the process is subjective, intuitive, and heuristic. When entities bearing a previously unsuspected relationship are placed side by side as a result of clustering, their juxtaposition may be sufficient in itself to spark the recognition or insight which leads to discovery; clustering can relocate an entity from its customary context so it may be seen from a new perspective. A large part of this interpretative stage is a matter of the analyst using his powers of judgement and subjective evaluation to find regularities and relations "by inspection." This chapter is devoted to describing some techniques which may aid the analyst

176

and enhance his powers of discovery. The techniques described here are simple and few in number; it should be apparent that much fruitful research remains to be done on this aspect of using cluster analysis.

8.1 Aids to Interpreting Hierarchical Classifications

The hierarchical methods of Chapter 6 give a collection of partitions ranging from each entity as a cluster unto itself to one grand cluster encompassing the entire data set. With so many alternatives it is of interest to compare the different partitions and investigate closely the nesting of clusters within each other. This section discusses three different, but complementary, aids to interpretation of hierarchical classifications; all three aids are implemented through computer programs given in Appendix G.

8.1.1 Study of Hierarchical Trees

The mechanical results of a hierarchical clustering algorithm can be described completely by reporting for each of the $n - 1$ stages the identity of the clusters merged at that stage and the value of the criterion characterizing this merge. It is difficult to make any interpretative statements from such a merge list; however the merge list contains all the information needed to draw a hierarchical tree such as those in Figs. 3.8 and 3.9. Such trees give an effective visual condensation of the clustering results which permits the analyst or any interested person to grasp rapidly the hierarchical relationships and visualize the membership of each cluster at any level of aggregation. Because of its unquestionable value as an interpretative aid, a tree is drawn on the line printer as a routine part of the output for all of the hierarchical clustering programs in Appendix E. However, the standard tree is drawn subject to a convention on the scaling of the criterion characterizing the merges: the range of the criterion is divided into 25 equal segments and all merges whose criterion values fall within a given segment of the range are treated as having occurred together. After seeing this standard tree the analyst may wish to redraw the tree or portions thereof in order to highlight certain features or study particular aspects in greater detail. The subroutine DETAIL used in conjunction with subroutine TREE provides great flexibility for drawing a tree to desired specifications. Some ways this facility may be used are discussed below.

If the two or three major groups in a data set are distinguished very well from each other, all merges but the last few may tend to appear lumped together because of the extreme dissimilarity among the major groups. Figure 8.1 depicts an example of such a situation. The major clusters are well

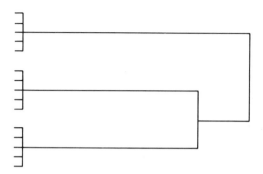

Fig. 8.1. Well-distinguished clusters.

distinguished but the figure tells nothing of the internal structure of each cluster. It would be of interest to examine each major cluster independently and draw its tree. One way to accomplish this task is to sort the entities into groups and cluster them again. However, it is not necessary to repeat all this work if the merge list generated by the original hierarchical clustering method has been saved. Subroutine DETAIL can process any segment of the merge list and thereby reproduce any portion of the tree for further study. This capacity to draw selected subtrees can also be of great value in large problems. For example, in a problem with 300 entities five pages of computer paper would be needed to draw the full tree but subtrees of up to 60 entities will fit on a single page. This facility permits detailed relationships to be depicted in segments of manageable size and the overall structure to be portrayed in terms of subtrees used as building blocks. The pertinent details of using subroutine DETAIL in this way are specified in Appendix G.

Another facility available through subroutine DETAIL is the option to stretch and shrink the scale for the criterion values characterizing cluster merges. For example, in Fig. 8.1, the analyst could choose the segments in the range of the criterion such that the structure within the subtrees is depicted on most of the page, while the connections between the major clusters are added at the right margin to pull the entire tree together. The analyst can use this facility to highlight some features and suppress others so that the tree can aid in communicating insights about the data structure.

8.1.2   PERMUTING THE SIMILARITY MATRIX

When using the hierarchical clustering approach described in Section 7.2 the analyst begins by constructing a lower triangular similarity matrix. The cluster analysis effectively rearranges the entities so that similar entities are close together, while dissimilar entities tend to be placed relatively far apart.

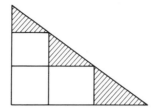

Fig. 8.2. Block diagonal structure.

One sequence of the clustered entities is provided through the placement of entities at the ends of branches in the hierarchical tree. Permuting the rows and columns of the similarity matrix to have this same sequence can often have dramatic effects: if there exist natural clusters such that entities within a group are very similar but entities from different groups are quite unlike, then permuting the similarity matrix will tend to give a block diagonal structure like that in Fig. 8.2. In the case of a correlation-like measure, the ideal outcome would involve 1s for all entries in the shaded portions of the matrix and 0s in the unshaded portion; for distance-like measures the ideal is 0s in the shaded areas and large numbers in the unshaded areas. Of course ideals rarely are realized in practice; but just rearranging the similarities in accordance with some interesting cluster analysis results often reveals substantial blocks of entities with extraordinary internal cohesion. The extent to which the off-diagonal blocks deviate from the ideal may help to identify entities belonging to the ragged fringe of a cluster or perhaps lying between two clusters. Also the opportunity to examine a carefully arranged array of similarities tends to prompt discoveries which are masked in summary statistics such as the minimum, maximum, and average similarity within a cluster. Program PERMUTE in Appendix G can be used to permute the rows and columns of a similarity matrix to any desired sequence.

8.1.3   ERROR SUM OF SQUARES ANALYSIS

When clustering data units, the analyst gives operational meaning to the term "similarity" through his choice of variables, weights, and a distance function. These choices are made with the intent of swaying the clustering in one way or another but they ultimately exert an influence which is difficult to anticipate accurately. One way to evaluate the relationship between a given hierarchical classification and each of the variables is through examination of the growth in the error sum of squares as the clustering progresses through the increasing levels of aggregation.

Suppose the data set consists of m data units. At the beginning of clustering each data unit is represented perfectly by the mean vector of the cluster to which it belongs and there is no within group error. At the highest level of aggregation there is only one cluster and it contains every data unit; the error sum of squares for each variable includes all the variance of that variable. At any stage between these two extremes the within group error sum of squares is that portion of the total variance not explained by the current set of clusters. It can be most enlightening to compare the step by step growth in the fractions of unexplained variance for all the variables in the problem; for a few variables the fractions may remain small up to the last few stages whereas for other variables the fractions may get large at a fairly early stage; the former variables may be thought of as being dominant in the results while the latter are dormant. Repeating the clustering with dormant variables eliminated should have little effect on the results; however deleting a dominant variable or making a substantial change in its weight probably will have a marked influence on the clustering results. This kind of analysis can be an especially useful device for generating pertinent changes in the set of variables and weights to be used in subsequent attempts to cluster the data.

Appendix G includes program ERROR which uses the original data and the merge list generated by the hierarchical program in Appendix E to compute the fraction of unexplained variance for each variable at each stage. This technique is suggested by Hall *et al.* (1969, Appendix 9G, pp. 11–13) and Ball (1970, pp. 73–75).

8.2   An Aid to Interpreting a Partition of Data Units into Clusters

When the analyst is interested in finding only one partition rather than a hierarchical set of partitions, the techniques discussed in Section 8.1 are diminished in value. For the present case, suppose the analyst has obtained an interesting partition of data units into clusters and he wishes to study this partition in more detail. Since the membership of each cluster is known, the analyst could examine each cluster further by extracting the score vector for each of its members from the original data set. But if the cluster consists of data units widely spread through the original sequence of the data set, it will be very difficult to gain any useful appreciation of the properties characterizing any cluster; the opportunity for forming any reasonable opinion about the character of a cluster by rummaging through the original data diminishes rapidly with increases in the number of data units and number of variables.

An obvious and almost trivial step to improve the chances for intuitive processes to contribute insights and understanding is to simply rearrange the original data to conform with the candidate partition; summary statistics such as the mean and variance for each variable in each cluster will help to identify the differences between clusters. Program POSTDU in Appendix G reads the original data set in the same way as the clustering programs of Appendixes E and F, reorders the data set according to any desired partition, and computes summary statistics for each cluster as well as the total partition. This simple device is analogous to permuting the similarity matrix as in Section 8.1.2 and can be expected to reveal obvious (after the rearrangement) features of the data set.

9

STRATEGIES FOR USING CLUSTER ANALYSIS

The previous eight chapters establish a philosophical and technical base for applying cluster analysis to real data. Once these matters are in hand it is important to develop strategies for utilizing cluster analysis effectively and economically. Unfortunately the question of alternative strategies of analysis has received little study in the literature. This chapter is devoted to discussion of some basic strategies which may be useful both as starting points for further research and as practical guides for the use of cluster analysis.

9.1 Sequential Clustering of Data Units

In most realms of applied statistics the analysis of real data is much more than a mechanical process by which the observed numbers are converted into a statement of objective reality; such effortless wisdom is even more elusive in cluster analysis. Except in the most fortunate of circumstances, the analysis proceeds through a sequence of stages where each succeeding stage contributes to improving the fit between the observed data and the analyst's conception of the process which generated the data. Taking another view, the size and complexity of a data set may prompt adoption of a sequential strategy; it may be necessary to break down the data set into subsets of manageable size, analyze the subsets sequentially, and then integrate the partial results into a composite result for the entire problem. Sequential strategies of both types are described in this section.

9.1.1 REFINEMENT AND SENSITIVITY ANALYSIS

When cluster analysis is used as a tool of discovery one cannot anticipate what combinations of variables, weights, data units, association measures, and clustering methods will lead to interesting classifications. The first cycle through a clustering problem may give a tentative partition which conflicts in important respects with the analyst's appreciation of the problem and its context. Such results immediately suggest the need to alter the formulation of the problem for another cycle through the clustering method in search of a better fit for the data set. In effect the analyst functions as the controller in the feedback system depicted schematically in Fig. 9.1. Repetitive cycles through the clustering method continue to refine the partition until the analyst concludes that he has found a balanced classification suitable for his purposes.

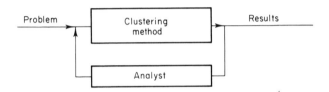

Fig. 9.1. Clustering as a feedback system.

A definite recipe cannot be provided for this sequential clustering process because it involves learning about the data set and adapting to new information at each stage. However, there are some useful elements of strategy which may be employed to make the process more productive and efficient.

1. Outliers are likely to show up as clusters consisting of only one or two data units. Such atypical features of the data set tend to distort the functioning of many clustering algorithms, especially those of Section 7.2 and Appendix F. While deviant data units ordinarily detract from the central structure of the data more than they contribute, such data units should not be discarded callously as mere errors of observation. They should be examined carefully with a view to finding a rational explanation for the deviant score profile. Outliers may provide a hint of a relevant category in the population which is poorly represented in the available data set.

2. A distinctive cluster which is well separated from the remainder of the data set is likely to be apparent in the first clustering cycle and continue to appear regularly in subsequent portions of the analysis. Rather than continue to recover the same information time after time, it is appropriate to extract such clear-cut features of the data set and focus attention on the more confused

residue. Removing a cluster part way through an analysis is somewhat analogous to a surgical operation: the cluster should be defined sufficiently well such that none of its constituent members are left behind (perhaps to be identified as outliers in later cycles) and very few (if any) data units belonging to other clusters are drawn into the departing group. The absence of the removed cluster should have no substantial effect on the identifiability of the remaining clusters.

3. In addition to clear cut separations involving individual data units and distinctive clusters, data sets sometimes include subordinate groupings which are distinct from each other but themselves consist of several clusters; Fig. 9.2 illustrates the situation with two groups of three clusters each. The

Fig. 9.2. Separation between clusters of clusters.

separation into two groups does not identify the elemental clusters; however, this distinction can be used to divide the problem into two subproblems which may be analyzed independently. Problem decompositions of this sort are valuable not only for reducing the computational burden but also for promoting insights through focusing attention on smaller and more manageable portions of the data set.

4. The simplifications achieved through identification of separable elements of the data set are dependent on the discriminating power of one or more variables; if a key variable responsible for such separations is removed from the data set, the partition obtained through clustering is sure to be changed in important respects. It is sometimes surprising how some variables seem to dominate a particular partition while other variables remain dormant; removing a dominant variable can promote the efficacy of dormant variables in unanticipated ways. The degree to which a particular partition depends on a particular variable can be estimated for certain only by trying to cluster both with and without that variable.

5. When clustering data units it may seem a little burdensome to consider all variables at all stages of clustering. Some prior knowledge about the problem and its context may permit development of a particular strategy in which various groups of variables are used sequentially; one group of variables may be used to create a primary partition of large clusters and another group of variables used to create a secondary partition through subdividing each of

the primary clusters. This sequential use of variables is one way of discovering interactions among variables without actually using cross-product terms in the computation of association measures; a final cluster is characterized as having certain values on the first group of variables, some other values on the second set of variables (which depend on the first level of clustering), and so on, such that each cluster is identified by some unique combination of scores on the variables.

6. The nearest centroid sorting methods of Section 7.2 are designed to subdivide a collection of data units into a specified number of groups. A very effective and economical divisive hierarchical clustering method is obtained merely by utilizing one of these methods sequentially to obtain several layers of subdivision; that is, a set of first level clusters is obtained by subdividing the entire data set. Then each first-level cluster is itself divided into second level clusters. The process can be continued until the analyst is satisfied with the results. A more sophisticated method may be devised by utilizing different sets of variables at each level of clustering as suggested above. Divisive hierarchical clustering by this method is an attractive alternative to the methods described in Section 6.5.4 in regard to simplicity, flexibility, efficiency, and economy. There do not seem to be any published instances in which nearest centroid sorting methods have been used in this way.

9.1.2 A SEQUENTIAL STRATEGY FOR LARGE PROBLEMS

The nearest centroid sorting methods (Section 7.2) and the sorted matrix approach to hierarchical clustering (Section 6.4) are capable of handling very large problems because they do not require central memory storage for either the data set or a similarity matrix. The stored data approach to hierarchical clustering (Section 6.3) can also handle substantial problems but is limited by the requirement to store the data set in central memory. The most severe restrictions apply to the stored similarity matrix approach to hierarchical clustering (Section 6.2); problems of only a few hundred entities can be accommodated on even the largest modern computers because of the massive storage requirements for the similarity matrix. One strategy for overcoming this limitation is to break down the data set into blocks of manageable size, cluster within each block, and then aggregate the block results into an overall partition. To illustrate, suppose the problem involves a data set of 6000 data units (a complete lower triangular similarity matrix would contain 17,997,000 entries) but the central memory of the available computer can handle problems no larger than 250 entities when using the stored matrix approach. A two-stage strategy for this case would consist of the following steps:

1. Divide the 6000 data units into 24 blocks of 250 data units each.
2. Independently cluster each block into about 10 first level clusters.

3. Collect the 240 or so first level clusters and cluster them into a set of second level clusters to obtain the final grand partition.

This two-stage strategy requires 25 runs of the clustering program with about 250 entities to be clustered in each run. If reducing the 250 data units in each block to only 10 clusters is felt to be too drastic at the first level, then a three-stage strategy could be used as follows:

1. Divide the 6000 data units into 25 blocks of 240 data units each.
2. Cluster each block into 50 first level clusters.
3. Divide the 25 blocks into 5 superblocks of 5 blocks each. Cluster the 250 first level clusters in each superblock into 50 second level clusters.
4. Aggregate the 5 superblocks and cluster the 250 second level clusters to obtain third level clusters for the final partition.

This three-stage strategy requires 31 runs of the computer program with about 250 entities in each run.

Obviously the analyst must make some decisions in this approach: first is the matter of subdividing the data set into blocks of manageable size; second is the choice of the number of stages of clustering and the degree of aggregation to be permitted at each stage. Ideally these choices should be made such that the ultimate partition is as near as possible to that which would be obtained by clustering the entire data set in a single stage. The primary consideration is to avoid grouping together data units which do not belong in the same cluster; improper merges can occur if subordinate pieces of the data set are aggregated into too few clusters. Suppose the analyst anticipated that there would be about 20 clusters in the data set of 6000 data units discussed above; the two-stage strategy reduces each block to only 10 clusters and therefore is certain to make improper merges in any block containing data units from 11 or more of the 20 clusters. On the other hand, the three-stage strategy maintains 50 distinct groups within each block at each stage; by letting the clustering remain loose up to the final stage it is most unlikely that any appreciable number of improper merges will occur. Further, as a byproduct of preventing improper merges the final partition should be influenced only slightly by the initial division of the data set into blocks of manageable size; since merges in the build-up stage(s) preceding the final stage are limited to data units which almost surely belong together, there is hardly any way to make a gross error regardless of the initial division into blocks. These observations prompt the additional conclusion that it probably is not worth the expense to deal with more than 200–300 entities at a time when using the stored matrix approach for hierarchical clustering. There appears to be little penalty involved in the loss of the facility to consider every possible pairwise merger for every data unit at every stage; as long as the minimum number of clusters in each build up stage is substantially more than the expected number of clusters, there should be an abundance of attractive merges that ultimately would occur even if all possibilities were available continuously.

9.2   Complementary Use of Several Clustering Methods

The strategies just discussed involve using a single clustering method iteratively and sequentially to obtain an adequate partition of the data set. The strategies in this section are focused on ways of using several clustering methods together in order to gain a more extensive appreciation of the structure in the data set.

9.2.1   HARMONIOUS CLASSIFICATION OF BOTH VARIABLES AND DATA UNITS

In most analyses attention is focused on clustering either data units or variables alone but not both together. When clustering data units the usual practice is to choose one set of variables, a set of associated weights, and a similarity measure to be applied uniformly for the classification of all data units; but it may be that clusters are characterized by different orientations such as in Fig. 9.3. In cluster 1 variable X_2 can vary widely as long as variable X_1 remains in a narrow range; the reverse relationship is true in cluster 2. The clusters have different descriptions in terms of the variables, so that the distance between a given data unit and each cluster centroid should be assessed using a different set of weights for each cluster. Using the same weights for all clusters implies a presumption that all the clusters have approximately the same shape and orientation. On the other hand, if one knew enough about the problem to be able to specify the unique weights for each cluster there probably would not be much need for cluster analysis. Chernoff (1970) has explored the possibilities of constructing a continuing estimate of the shape and orientation of each cluster as data units are allocated and using this information adaptively to define a unique distance measure for each cluster. Chernoff's work is directed specifically at extending MacQueen's k-means methods which were

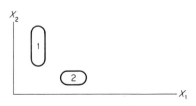

Fig. 9.3. Cluster orientations.

discussed in Sections 7.2.2 and 7.3.1. Eddy (1968) and Rohlf (1970) also consider ways of constructing a different distance measure for each cluster. All three of these discussions are somewhat exploratory in nature and describe potential developments rather than techniques presently suited to widespread use.

When clustering variables there is an implicit assumption that all the data units in the data set share some essential characteristics so that they collectively and individually represent a single population. If the data set actually includes several different clusters of data units, then measures of association between variables will reflect a mixture of effects which may not be representative of the kind of association present within any of the clusters. Figure 9.4 illustrates how mixtures of data unit clusters can conceal important within group relations. In Fig. 9.4a variables X_1 and X_2 exhibit a very strong positive correlation in cluster 1 and an equally strong negative correlation in cluster 2; however, if all the data units are taken together in one undifferentiated mass the computed correlation between the variables would be near zero. In Fig. 9.4b, the relationship within each of the three clusters is one of strong positive correlation between the variables; however, if the three clusters are taken together the observed relation is one of moderate negative correlation. In both of these cases the data set as a whole exhibits an apparent relation between the variables which is totally deceptive; far more informative would be the joint knowledge of the cluster structure for the data units and the relations between variables within each cluster. The situations depicted in Fig. 9.4 are easy to perceive in two dimensions; but in a data set of 100 variables and 1000 data units such situations may be difficult to map even with the aid of systematic clustering methods. Such examples are strong evidence that any serious attempt to cluster variables should be preceded by an exploratory clustering of data units to assess the degree of homogeneity within the data set. These remarks also apply to factor analysis, as noted in Section 9.3.1.

It appears that adequate clustering of data units requires considerable insight into the relationships among variables, especially the manner in which

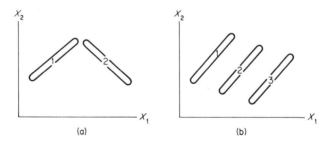

Fig. 9.4. Example of cluster structures.

the relationships vary from cluster to cluster. On the other hand, an informative cluster analysis of variables requires moderate homogeneity among data units, a requirement that can be satisfied most directly by undertaking a separate analysis for each distinct cluster of data units. Unfortunately, little prior knowledge about the classification of either variables or data units is available in most problems submitted for cluster analysis. Consequently the task of clustering often seems to be a bootstrap problem in which the data unit clusters are needed to find the clusters of variables, but variable clusters are needed to find the data unit clusters and neither set of clusters is known. A possible strategy for dealing with this situation is to undertake a sequential analysis in which data units are clustered at odd stages and variables at even stages until the two sets of clusters converge to a mutually harmonious classification of both variables and data units. The details of using such a strategy on real data remain to be developed. It may prove to be a formidable task to specify adequately these details for a batch process computer program; however, it appears that an experienced and informed analyst could achieve a simultaneous analysis of both variables and data units through use of either the PROMENADE (Hall *et al.*, 1969) or the OLPARS (Sammon, 1968) system; by combining on-line interactive control, graphical displays, and versatile analysis options (ISODATA clustering, principal components analysis, and discriminant analysis) these systems provide a powerful capacity for penetrating data analysis. However, these systems are sufficiently sophisticated and expensive to make the casual user more than a little apprehensive about whether he can use this analysis power effectively and economically.

Several other authors have studied techniques in which each cluster of data units is constructed to have a unique interpretation in terms of the variables. Litofsky (1969) and Dubin and Champoux (1970) present techniques based on the special properties of binary variables; Fisher (1968), Hartigan (1972) and Dubin (1971) propose new methods suitable for nominal and interval variables. The whole question of simultaneous clustering of variables and data units has only recently received serious study but offers considerable potential for increased effectiveness of cluster analysis.

9.2.2 SPECIALIZED USES OF VARIOUS CLUSTERING METHODS

The various cluster analysis methods discussed in Chapters 6 and 7 differ in important respects. It seems possible that the unique characteristics of these methods could complement each other so that a comprehensive analysis might involve the use of several methods, each sensitive to special features of the data set. While this topic is in need of illuminating research, there do appear to be at least three rewarding strategies available.

1. Use cheap methods first to break down the data set into manageable subsets before utilizing the more detailed and expensive methods. As discussed in Section 9.1.1, a typical cluster analysis involves a sequence of steps, the earliest of which probably should be more concerned with the differences between groups than the similarities within groups. If the analyst has a very good idea of how many groups should appear in the highest level of meaningful aggregation, then a nearest centroid sorting method from Section 7.2 may prove to be the most economical choice. On the other hand, substantial ignorance about the number of groups would make a hierarchical method from Chapter 6 more attractive because the clustering results include the widest range of possibilities for the number of clusters. Indeed, the single linkage method described in Sections 6.2.1 and 6.4.1 is suited ideally to finding clear-cut separations between groups; a characteristic of single-linkage clusters is that any entity within a cluster is more similar to some other entity within the same cluster than to any other entity in another cluster.

2. Use different methods to find different kinds of clusters. As was noted in Section 6.2.1 the single-linkage method of cluster analysis has a unique capacity to find long serpentine clusters. Most other methods delineate compact clusters which are generally convex unless the analyst introduces powers and cross products as new variables (see Section 5.1.3.2). A possibly useful strategy is to use one method to extract as many clusters of one kind as seem natural for that method and then utilize a second method to extract clusters of a different type from the residue of the data set. Perhaps it would be profitable to perform two parallel analyses so that the roles of the two methods could be reversed; the final results could differ substantially depending on whether the compact clusters were extracted first or last.

3. A partition of data units obtained at a chosen stage in a hierarchical classification can be refined using the nearest centroid sorting methods. The partitions generated by hierarchical methods are not necessarily optimal with regard to the chosen clustering criterion because early merges cannot be undone to improve the partition at later stages. However, hierarchical methods usually give very good partitions which require only modest modifications to achieve a local optimum. Thus, the results from a hierarchical method can provide an excellent initial partition from which the nearest centroid sorting methods can converge rapidly.

9.3   Cluster Analysis as an Adjunct to Other Statistical Methods

Cluster analysis is a powerful tool for discovering homogeneous groups in data sets. This characteristic can be used to advantage as an informal test of homogeneity for a given group: if distinct subgroups are not revealed by cluster analysis the data set may be taken to be substantially homogeneous.

Many of the classical methods of statistics are predicated on an assumed homogeneity within certain groups of the data set, but the means for assuring homogeneity are based primarily on careful selection of data units in accordance with prior knowledge and the principles of experimental design. Cluster analysis provides an opportunity to investigate empirically the degree of success achieved in attempts to fulfill assumptions of homogeneity.

9.3.1 CLUSTER ANALYSIS AS AN AID TO FACTOR ANALYSIS

Factor analysis is a multivariate statistical method which has enjoyed widespread use in the behavioral and social sciences. Harman (1960) and Rummel (1970) are general reference works in the field. The objective in most methods of factor analysis is to find a set of "factors" which have the following properties:

1. The factors are linear combinations of the original variables.
2. The factors are orthogonal to each other.
3. The factors are fewer in number than the original variables but they account for or "explain" most of the variance in the original data.
4. The factors are meaningful in their own right or are at least subject to some useful interpretation.

Some of the more sophisticated methods of factor analysis relax either property 1 or 2 above to permit nonlinear factors or oblique factors. The mechanics of extracting factors from a data set involve operations on either the correlation matrix or the covariance matrix. However, as discussed in Section 9.2.1 correlations obtained in heterogeneous data sets can be highly deceptive; indeed, all the problems that might be encountered when clustering variables in a data set containing distinct subgroups of data units carry over as identical problems for factor analysis. As Cattell (1965) observes, factor analysis should be applied to relatively homogeneous groups of data units, and cluster analysis is a useful device for finding such groups.

In some studies, factor analysis has been used as a prelude to cluster analysis, but considerable caution should accompany any such usage. The analyst should confirm that the factors reflect the relationships among variables which are actually observed within the clusters of data units. The most satisfactory strategy probably would be to alternate clustering and factor analysis until achieving a harmonious set of clusters and factors, much along the lines proposed in Section 9.2.1.

9.3.2 CLUSTER ANALYSIS AS AN AID TO DISCRIMINANT ANALYSIS

Discriminant analysis is a multivariate statistical method used for constructing decision procedures by which data units can be classified as members of one group or another. In the simplest problem of discriminant analysis, samples of data units are available from each of two known categories; the objective is to utilize the sample information to construct a linear discriminant function separating the groups in such a way as to minimize the number of expected misclassifications or optimize some other relevant criterion. The discriminant function may be used subsequently to classify new data units into one of the two classes. Anderson (1958) provides a thorough development of the theory behind discriminant analysis.

In a typical application of discriminant analysis, the number of groups in the data set is thought to be well known and the primary objective is to find an efficient means of distinguishing between the groups as given. The analysis sometimes fails because the resulting discriminant function separates the groups very poorly; a lack of success is usually blamed on two factors:

1. The variables describing the data units do not distinguish between the groups to a sufficient degree. Equivalently, the groups overlap in the chosen measurement space.

2. The groups cannot be separated by a function of the form adopted for the analysis.

These two factors can be reinterpreted to mean that within the measurement space defined by the available variables the defined groups are not separable and therefore the data set contains fewer identifiable groups than prior knowledge would suggest. Another alternative that could explain poor results from a discriminant analysis is that there are actually more identifiable groups in the data set than the number being sought in the analysis. The defined groups may stem from theoretical concepts or matters of convention. One particular case that can cause difficulty is when two groups are defined in accordance with a dichotomy between a central and extreme aspect of some concept; this particular kind of dichotomy was discussed earlier in Sections 3.2.2 and 3.2.3. To illustrate, consider the two-dimensional example in Fig. 9.5. Group B occupies a central place in the problem and group A consists of some boundary portions of the data set. The same kind of problem can occur readily when group B is defined by the presence of a particular attribute, while group A is merely the collection of all data units not having that particular attribute. Achieving effective discrimination between the groups in Fig. 9.5

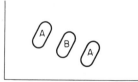

Fig. 9.5. A two-group problem with three natural groups.

while using only a single discriminant function would require a quadratic form such as an ellipse or parabola. However, if it is realized that group A consists of two distinct subgroups, then perfect discrimination could be achieved by treating the problem as having three groups and using two linear discriminant functions. Any attempt to use only one linear discriminant function to separate the two defined groups in Fig. 9.5 is doomed to dismal failure.

Of course cluster analysis is an ideal tool for discovering distinct subsets within a defined group. However it costs time and money to supplement a discriminant analysis with a cluster analysis; it would be superfluous as well to use cluster analysis if the discriminant analysis gave satisfactory discriminant functions. But if the discriminant analysis gives disappointing results, then cluster analysis should be in the first rank of diagnostic tools.

9.3.3   CLUSTER ANALYSIS AS AN AID IN THE ANALYSIS OF VARIANCE

The analysis of variance (AOV) is a statistical technique designed to test whether variations in selected independent variables cause significant variation in a dependent variable, the latter known as the response. In the simplest case of a one-way analysis of variance the effect of just one independent variable is tested. To test this effect, several groups of observations are collected on the response. The groups are taken to be homogeneous except that each group is characterized by a different level or value on the independent variable. All variations in the response not attributable to the independent variable are assumed to be due to independent random errors. Thus, the variation of the response within groups is assumed to consist solely of random error, while the variation between groups is a mixture of random error and the systematic effects due to variation in the independent variable. The AOV procedure then involves testing the null hypothesis that the response is the same in all groups.

An essential element of using AOV is insuring that the only systematic variation is due to the independent variables which characterize each group of observations. If there are additional systematic effects operating within groups (e.g., a group contains distinct subgroups) then the apparent error variance is magnified and it becomes more difficult to detect a genuine variation between groups. If there are systematic effects also operating between groups (i.e., other important variables exhibit group to group variation) then the true effect of the independent variable under investigation may be either amplified or masked. When the observations on the response are gathered in designed experiments subject to detailed control by the analyst, such problems of heterogeneous observations are relatively minor. But when the analysis is concerned with responses of human beings or systems subject to only indirect control, there may be a host of factors influencing the observed responses.

In such cases cluster analysis might be used to advantage in two ways:

1. A cluster analysis within each group either will indicate that the assumption of substantial homogeneity within groups is justified or reveal specific ways in which the assumption is violated.

2. A cluster analysis across the entire data set may reveal variations from group to group which are not due to an independent variable.

Such cluster analyses should be performed using variables other than the response and the independent variables whose effects are the topic of the analysis of variance; the object of the cluster analysis is to test (informally) the homogeneity of the groups with respect to other important aspects of the problem under study. Such additional data on other variables may be readily available or may impose an additional expense on the analysis.

The intuitive motivation for using cluster analysis with analysis of variance originally was prompted by the observation that negative results are obtained in many analyses which set out to test eminently plausible hypotheses; but such investigations often gather their experimental evidence in circumstances where uncontrollable factors may exert a strong influence on the response variable. For example, studies involving the efficacy of particular educational techniques may be based on the comparison of achievement tests for children who have been taught with the new techniques versus other children who have not had this opportunity. But the uncontrolled factors of teacher attitudes, family education, economic status, neighborhood, outlook of friends, educational backgrounds, and so forth, tend to obscure the effects on the response. Possibly a preliminary cluster analysis could help select students for the study so as to minimize the importance of uncontrolled factors; if cluster analysis cannot be used as a preliminary step, then clustering the assembled data may help to identify extraneous effects which can be taken into account and possibly removed from the data.

It should be mentioned that many users of AOV employ intuitive processes to sort out the experimental units and thereby achieve substantial homogeneity within groups. But when the numbers of effects and experimental units become large, such intuitive devices are difficult to apply consistently or efficiently; in such circumstances, cluster analysis may prove to be a useful adjunct to AOV.

9.4   Clustering with Respect to an External Criterion

The clustering methods discussed up to this point seek a set of clusters which provide a good fit to the internal structure of the data set. However, in some cases it may be of interest to construct groups which are related strongly to an additional criterion variable.

9.4.1   CLUSTERING FOR MAXIMUM PREDICTION OF A CRITERION

Suppose a data set has been partitioned into h clusters such that the kth cluster contains m_k data units; further suppose that among the variables in the data set there is a criterion variable Y, and y_{jk} is the score achieved on Y by the jth data unit in the kth cluster. Then the error sum of squares on the criterion for the given partition is

$$E = \sum_{k=1}^{k=h} \sum_{j=1}^{j=m_k} (y_{jk} - \bar{y}_k)^2, \tag{9.1}$$

where $\bar{y}_k$ is the mean score on Y in the kth cluster. The least-squares estimator of the criterion score for a data unit in the kth cluster is $\bar{y}_k$. Therefore, clustering so as to minimize E will give the optimal set of groups from which to estimate the criterion variable. If the only variable in the data set is the criterion variable, then the problem is a one-dimensional clustering problem (see Sections 3.2.1.7–9). On the other hand, when there are other relevant variables in addition to the criterion, it may be desirable to give these variables a role in shaping the partition as in the following two methods.

Sonquist and Morgan (1964) have developed a method they call automatic interaction detection (AID). The variables in the data sets are assumed to be one interval criterion variable and several categorical (nominal or ordinal) variables which are used as predictors for the criterion. The method works by hierarchical division of the data; at each stage an existing group of data units is split by dichotomizing one of the categorical predictors so that the two resulting groups are distinguished uniformly in regard to this one variable; the split actually chosen at each stage is that one which reduces the value of E in Eq. (9.1) by the largest amount of all available splits. The groups obtained at any given stage are characterized by a combination of characteristics. Figure 9.6 illustrates a hypothetical example in which the criterion variable is annual income (group means shown in parentheses), and several categorical variables are used as predictors. The lowest block on the left consists of male high school graduates while the lowest block on the right consists of single females. Notice that education is the most important second-level variable for

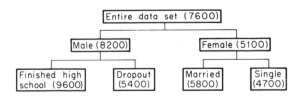

Fig. 9.6. AID example.

males, whereas marital status dominates for females. This technique can provide substantial insights to the joint relationships among the predictors as they affect the criterion.

Another method based on optimal prediction is offered by Forgy (1966). The method is a variant of Forgy's nearest centroid sorting method described in Section 7.2.1. The variables in the data set are assumed to consist of one interval criterion variable and a set of predictor variables which may involve any combination of scales; the analyst needs to supply a distance function which combines the predictor variables in a suitable way to compute the distance between a data unit and a seed point. The steps in the method are as follows:

1. Choose a set of initial seed points and assign each data unit to the nearest seed point. After all data units are assigned, compute the value of E using Eq. (9.1).

2. Taking one seed point at a time, institute a search for relocating the seed point such that when data units are reallocated to the nearest seed point the value of E will be reduced.

3. Continue step 2 until the process converges to a local minimum for E. Note that data units are always assigned to the cluster with the nearest seed point.

The heart of this method is in step 2, where it is necessary to find better seed points. Forgy (1966) reports he had programmed an experimental approach in which he chose new search directions by fitting a linear regression plane through the data units currently in the cluster; some experience had been accumulated on artificial data sets of up to four variables but it is clear from the paper that a good deal of development work remained to be accomplished. This clustering method may hold some promise but only if an efficient approach is developed for step 2; each new possibility for the location of a seed point requires a check of all the data units to see if they would change cluster membership, a much too expensive process to be repeated more than a few times with each seed point.

9.4.2   CLUSTERING WITH SCALED VARIABLES

The AID and Forgy methods just discussed are special algorithms designed to consider the criterion variable explicitly during the actual clustering of the data units. Another strategy is to use conventional clustering methods but scale the variables with the methods of Chapter 3; the predictor variables are transformed so as to be related maximally to the reference variable while retaining the unique variation associated with each predictor. For example, suppose the criterion variable is on an interval scale while the predictors are nominal variables; then one could compute class scores maximally correlated with the criterion (as discussed in Section 3.2.6.2) and use these scores to compute distances between data units and seed points in a suitable clustering method. It should be expected that such results would have a strong relation to a one-dimensional clustering of the data units based on the criterion variable alone; however the predictor variables will exert an influence on the final partition through discouraging groupings which are grossly heterogeneous in regard to any predictor.

The particular example of categorical predictors combined with an interval criterion permits a unique twist in the analysis. The optimal class scores for the categorical predictors are the class means on the criterion; these means are often available through alternative methods of estimation and therefore the criterion need not actually be measured for individual data units. For example, the criterion could be lifetime income which is estimated for various categories of workers but is not determinable for living individuals. Through this device a cluster analysis can be influenced strongly by a variable which itself could never be used directly in the analysis.

Table 9.1 summarizes the variable transformations which are appropriate for various combinations of the criterion variable and the predictors. Mixed variable problems are handled directly because the method amounts to wholesale conversion of all variables to a single scale type while using only one reference variable for all such conversions.

TABLE 9.1

SCALE CONVERSIONS FOR CLUSTERING WITH RESPECT TO AN EXTERNAL CRITERION

Predictor variables \ Criterion variable	Nominal	Ordinal	Interval
Nominal	No conversion needed	Section 3.4.1: score each category with mean rank of its members	Section 3.6.2: score each category with mean of its members on the criterion variable
Ordinal	Ignore order if desired	No conversion needed	Section 3.5.1: use category ranks as category scores
Interval	Section 3.2.1.10: use linear discriminant functions to categorize; Section 3.2.1.11: use Cochran and Hopkins method to categorize		No conversion needed

9.5   The Need for Research on Strategies

An analysis strategy is a plan for a sequence of substantive actions intended to extract some desired information from a data set. A fruitful analysis is usually guided by such a strategy, though many users of cluster analysis would be hard pressed to be explicit about the strategies they employ. Even the few strategies discussed in this chapter apparently are not well known, despite their rudimentary nature.

Unlike some other topics in statistics, the methods employed in cluster analysis have not become sanctified; there is a great degree of discretion and flexibility available to the analyst, probably a good deal more than can be utilized effectively and efficiently. Strategies of analysis help to focus attention on particular facets of the problem and highlight the methodological alternatives likely to produce the most rewarding results. Effective strategies prompt new questions about the data and suggest different ways of summarizing the data set. As illustrated in Section 9.3 such strategies may also permit cluster analysis to make surprising contributions to the effectiveness of other statistical methods. Quite clearly, innovative strategies can turn a mundane procedure into a tool of exceptional value. Strategies of cluster analysis are at a rather embryonic stage of development and refinement. It is hard to identify another research topic in cluster analysis which offers a greater potential for rewarding modest efforts with substantial achievements.

10

COMPARATIVE EVALUATION OF CLUSTER ANALYSIS METHODS

The analyst faces a perplexing problem when he is forced to choose an association measure and a clustering method for an analysis. The multitude of alternatives makes it difficult to say that a particular measure and a specific method are clearly superior selections for treating the problem at hand. Theoretical and conceptual considerations help to guide choices but are not quite sufficient because they give little insight into the performance characteristics of the alternatives when applied to real data. Nominally, every association measure or clustering method is different from every other one; but a small amount of experience with a variety of methods reveals a massive redundancy in the clustering results obtained. On any given problem, a large group of different methods will give substantially the same results while perhaps a few other methods give distinctively different results; these apparent overlaps and distinctions among methods may exhibit a radically different pattern on other problems. It often appears that most methods react to some central characteristics of the problem while a few methods detect special characteristics and produce distinctive results. Theoretical considerations, to the extent that they are known and understood, are insufficient to explain the manner in which the results of various clustering methods coincide or differ from one data set to another, yet some insight into this aspect of the various methods seems essential for making wise choices in the application of cluster analysis.

10.1   An Approach to the Evaluation of Clustering Methods

As part of the introduction to cluster analysis in Chapter 1, it was pointed out that there is an astounding number of ways to sort a collection of entities into a fixed number of groups. This fact precludes total enumeration as a cluster discovery technique and therefore assures that many partitions will go unexamined, possibly some of which might be very interesting. If one insists that the proper object of clustering algorithms is to produce the "right" partition, then this enormous variety of potential solutions gives little hope to the quest, unless the algorithm has extraordinary properties which guarantee that the "right" partition will be found. None of the algorithms in this book have such properties; nor does it appear likely that any such algorithms will ever be found. How then can one approach the question of evaluating alternative cluster analysis methods? Each method offers a candidate solution, but *a priori* every solution is just as good as any other unless the known properties of the clustering method bestow some additional credence on the solution. It seems reasonable to begin an evaluation of clustering methods by seeking to establish their characteristics as exhibited in the cluster analysis of real data.

In many clustering problems of substance there will be several different but meaningful ways of organizing the data into groups; each of these alternative clusterings may be taken as the realization of some classification principle embodied in the data and representing one of perhaps several facets of the problem. For each relevant classification principle that can be stated explicitly it seems clear that there is an "exemplary partition" which is the product of applying the principle error-free to the data set. Presumably if a cluster analysis method presented the analyst with such an exemplary partition, then he would encounter a maximum (though perhaps small) probability of discovering the associated classification principle. In addition to the exemplary partition, there also exists a multitude of other closely related partitions differing only in the placement of a few entities; such partitions are imperfect realizations of the classification principle but embody enough of the exemplary partition to offer a real opportunity for the analyst to make a useful discovery. On the other hand, many if not most of the possible partitions are so poorly related to any one of the alternative classification principles as to cause only confusion if taken seriously. With this view in mind, a clustering method is useful if it produces a partition sufficiently representative of a classification principle such that the analyst discovers that principle. Ideally, the analyst would have at his disposal a suite of methods capable of producing partitions exemplifying the full array of relevant classification principles embodied in the data set.

In many varieties of mathematical analysis, algorithms are evaluated by solving test problems having known solutions. In cluster analysis, the task of evaluation is not so direct because of the possible multiplicity of relevant, but distinct classification principles and the numerous partitions that may relate to a given principle. Within this setting the worth of a particular clustering method rests on a capacity to exemplify some principle more efficiently than other available methods; of course, if the method is the only one which can exemplify the principle it is the most efficient regardless of how poor the method may seem otherwise. In the analysis of real data, the question of evaluation involves a fundamental duality between problems and methods; on some problems most clustering methods produce similar partitions, whereas on other problems the same set of methods give a wide variety of clearly distinct partitions.

What seems to be needed is an approach to evaluation which systematically can relate the key characteristics of cluster analysis problems to the capacities of various cluster analysis methods; in other words, find the elements which make problems difficult and match them with the strengths of powerful methods. If there could be found a set of significant concept dimensions which describes problems and another such set which describes methods, then a variety of important capabilities might be within reach.

1. New strategies of analysis could be constructed, perhaps complete with decision trees to guide the sequential use of clustering methods.

2. Aspects of commonality and distinction among methods could be identified to help minimize the number of redundant partitions obtained in clustering a single data set.

3. Problem characteristics which are beyond the reach of available clustering methods could be identified and thereby help to specify needed directions of research for new clustering methods.

4. Progress could be made in the effort to distinguish between the power of a clustering method to reveal structure in a data set versus its tendencies to impose a structure. Put another way, to what extent does a given method objectively report structure in the data as it apparently exists and to what extent does it tend to force the data set into a fixed mold?

A possible approach for discovering these concept dimensions is to turn cluster analysis on itself and cluster the results obtained by applying available methods to specially constructed data sets. The similarities and differences among various clustering methods may be identified through comparison of the results obtained by clustering data sets of known characteristics, and the characteristics of various data sets may be discovered through clustering them with methods having known properties. Two important prerequisites for actually trying to pursue this approach are treated in the remainder of this chapter. Section 10.2 is devoted to a discussion of methods for measuring the similarity among clustering results (hierarchical classifications and partitions) and the subsequent clustering of methods and problems based on such measures. Section 10.3 includes preliminary lists of characteristics for both problems and methods to help guide initial attempts to evaluate cluster analysis methods through this approach.

10.2   Quantitative Assessment of Performance for Clustering Methods

When a clustering method from Chapter 6 or 7 is applied to an example data set the result is a hierarchical classification or a partition. Additional classifications or partitions may be generated by employing variations on the first method or by using a second comparable method. The clustering results obtained from alternative clustering methods can be compared and appraised by intuitive and subjective methods. However, substantial assistance may be obtained through the use of similarity measures among different clustering results. If available measures are deemed suitable, then a similarity matrix can be constructed and different clustering methods can be clustered on the basis of their performance.

10.2.1   Measuring the Similarity between Hierarchical Classifications

A hierarchical classification produced by the methods of Chapter 6 consists of a series of partitions whose clusters nest into each other; as a whole the hierarchical classification of m entities includes m related partitions ranging from m clusters of one entity each to one cluster encompassing all m entities. The relations among the entities in a hierarchical classification are numerous and complex; the task of summarizing these relations or comparing two classifications is conceptually taxing.

One approach to characterizing a hierarchical classification is to take its representation in the form of a tree and construct a corresponding similarity matrix. An entry in the matrix describing the similarity between two entities is defined as the value of the clustering criterion associated with the merger bringing the two entities into the same cluster. An example will help clarify the procedure. Figure 10.1a shows a tree depicting a hierarchical classification of six entities. Figure 10.1b presents a derived similarity matrix based on the tree. Entities 1 and 5 merge into a single cluster at a criterion value of 1. Then when entity 6 joins 1 and 5 at a criterion value of 3, one cannot distinguish (from the tree alone) whether the pair (1, 5) is more strongly associated than the pair (5, 6), and therefore the two are treated equally in the derived similarity matrix. Notice that both single-linkage and complete-linkage clustering on the derived similarity matrix will reproduce exactly the parent hierarchical tree.

The derived similarity matrix has been used in several different ways. Sokal and Rohlf (1962) compute the ordinary product moment correlation coefficient between the corresponding elements of the derived similarity matrix and the original similarity matrix from which the hierarchical tree was constructed. This correlation between matrix entries is called the "cophenetic correlation coefficient." It is fairly popular among biologists and zoologists as a measure of goodness of fit between the hierarchical classification and the original similarity matrix. Farris (1969) investigates clustering methods which seek to maximize the cophenetic correlation coefficient as the explicit clustering

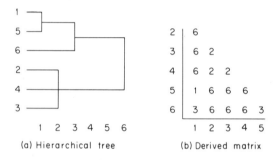

Fig. 10.1. Illustrative example.

criterion. He concludes that such methods are excessively sensitive to cluster size and subject to reversals (see Section 6.2.5, Fig. 6.4a). However, the cophenetic correlation coefficient remains useful as an evaluative tool even though it fails as an algorithmic criterion.

Borko *et al.* (1968, pp. 15–28) utilize virtually this same scheme of evaluation to compare different classifications derived from a single set of data units. The hierarchical classifications are used to construct corresponding derived similarity matrices which are compared by computing the product moment correlation between their paired entries. If there are several classifications of interest, then correlations can be computed for every pairwise comparison to build up a similarity matrix among the classifications. Recall that the product moment correlation coefficient is invariant under linear transformations; therefore the correlation does not change even if a constant is added to all entries in a matrix or if all the entries are multiplied by a constant. Because of this property, a matrix of relatively small numbers can be correlated with another matrix of relatively large numbers without concern for the apparent disparity in scale.

Hartigan (1967) chooses to characterize the comparison of similarity matrices in terms of distance rather than correlations. In particular, if $S_k(i, j)$ is taken as the value of similarity (or association measure) between entities i and j as expressed in the kth similarity matrix, then the distance between similarity matrices p and q is

$$R(S_p, S_q) = \sum_{i=2}^{i=n} \sum_{j=1}^{j=i-1} W(i, j) \, [S_p(i, j) - S_q(i, j)]^2, \qquad (10.1)$$

where n is the number of entities and $W(i, j)$ is a chosen weighting function. The indices in the summations are chosen such that i and j correspond to rows and columns, respectively, in the lower triangular similarity matrix. This distance between matrices is readily recognized to be weighted Euclidean distance computed as though the matrices are data units and their paired entries are scores on variables. Hartigan presents a complex hierarchical clustering method which attempts to minimize the distance between matrices where one matrix is the original similarity matrix and the other is a derived matrix corresponding to a hierarchical tree; the $W(i, j)$ function is also determined as part of the method.

It is clear that the distance between matrices can be computed for any pair of matrices, including those derived from two different hierarchical trees. However this distance depends on the relative magnitudes of the entries in the two matrices; if one matrix contains numbers on the order of 1000 while the numbers in the second matrix are on the order of 1, then the distance between the two matrices is virtually independent of the second matrix. Since the focus of attention is on comparing the matrices and all the entries in a matrix are of equal importance in such a comparison, it seems appropriate to let $W(i, j) = 1$ for all cases and then to rescale the entries in each matrix for comparability. A familiar method of rescaling is to standardize a collection of observations to zero mean and unit variance; but, as shown in Section 5.2.3, computing the squared Euclidean distance with scores standardized within data units merely gives a linear transformation of the correlation computed between data units. Therefore, unless some alternative scaling of the matrix entries is clearly appropriate, distances and correlations between matrices give substantially equivalent results.

10.2.2 Measuring the Similarity between Partitions

A partition is a set of mutually exclusive clusters which together encompass the entire data set. A partition is found directly from the methods of Chapter 7 or through selection of some level in a hierarchical classification obtained from the methods of Chapter 6.

Borko *et al.* (1968, pp. 62–80) suggest constructing a simple contingency table to depict the cross classification of data units in the two partitions. To illustrate, consider two different partitions of 15 data units as shown in Table 10.1. The corresponding contingency table is shown in Table 10.2. Borko *et al.* assessed the association between partitions by using the mean contingency (Φ) and chi-square statistic. Any of the measures in Section 4.2 also could be applied to this problem; the optimal prediction measures of Section 4.2.2 are probably the most appropriate since values of the measures from tables of different sizes may be compared meaningfully.

Rand (1971) offers a different approach in which the focus of attention is on the joint membership of pairs of data units in the two partitions. In particular, if two data units have the same relationship (assigned to the same cluster or assigned to different clusters) in both partitions, then they are said to be treated similarly in the two partitions; if the two data units are in the

TABLE 10.1
COMPARABLE PARTITIONS

Partition A		Partition B		
Cluster 1	Cluster 2	Cluster 1	Cluster 2	Cluster 3
4	1	4	1	2
6	2	6	3	7
9	3	8	5	11
10	5	10	9	12
11	7	14	13	
14	8		15	
	12			
	13			
	15			

TABLE 10.2
CROSS CLASSIFICATION OF
PARTITIONS

Partition A \ Partition B	1	2	3
1	4	1	1
2	1	5	3

same cluster in one partition and in different clusters in the other partition, then they are said to be treated dissimilarly. The similarity between two partitions is then computed as the ratio of the number of similar pairs to the number of all pairwise combinations of the m data units. Rand's measure of similarity may also be found in terms of a 2×2 table in which all $\frac{1}{2}m(m-1)$ pairwise combinations of data units are classified for each partition into two classes:

   0 class: the data units are in different clusters in the partition;
   1 class: the data units are in the same cluster in the partition.

Table 10.3 depicts the general form of the 2×2 table for this case. In terms of the entries in Table 10.3, Rand's measure of similarity between partitions is $(a+d)/(a+b+c+d)$, which is recognized as the simple matching coefficient, binary measure number 2 in Table 4.5. Any of the binary measures discussed in Section 4.3 can be used with this scheme of comparison between partitions. Like the contingency table approach of Borko *et al.* (1968), the method can deal with the comparison of partitions of different sizes; indeed, when applied to the example in Table 10.1 the following contingency table is obtained:

$$
\begin{array}{c}
\qquad\qquad B \\
A \; \begin{array}{|cc} 19 & 32 \\ 12 & 42 \end{array}
\end{array}
$$

The simple matching coefficient has a value of 0.581 for this comparison.

   Green and Rao (1969, p. 363) utilize a binary distance between partitions which when put into the format of Table 10.3 turns out to be $(b+c)/(a+b+c+d)$, the binary metric discussed in Section 5.3.2. This metric is the complement of the simple matching measure and therefore Green and Rao's distance is substantially equivalent to Rand's measure.

TABLE 10.3
2 × 2 TABLE FOR PAIRED
COMPARISONS BETWEEN
PARTITIONS

Partition B Partition A	1	0
1	a	b
0	c	d

If similarity between partitions is considered in terms of prediction from one partition to another, then the contingency table approach suggested by Borko *et al.* (1968) assesses the predictability of a data unit's cluster membership in the other partition. On the other hand, the paired comparisons approach exemplified in Table 10.3 assesses the ability to predict whether a pair of data units will be together or separated in one partition given they are together (or separated) in the other partition. These two notions are certainly different and no empirical comparisons are available at present. However, some substantive observations can be made on conceptual and practical grounds.

1. The contingency table approach seems to be the more interpretable of the two methods; the intuitive appreciation of the classification of a data unit in two different partitions seems more natural than an evaluation of the tendency for a pair of data units to occur separately or together.

2. The contingency table approach may be applied to larger data sets because it considers only the joint classification of the m data units whereas the paired comparisons approach must consider the joint classification of $\frac{1}{2}m(m-1)$ pairs of data units.

3. The paired comparisons approach may permit assessment of some unique varieties of association between partitions because of the much wider variety of association measures defined for 2×2 tables than defined for larger tables.

10.2.3 COMPARISONS OF METHODOLOGIES THROUGH COMPARISON OF SIMILARITY MATRICES AND PARTITIONS

The techniques just discussed in Sections 10.2.1 and 10.2.2 provide the facility to compute a measure of similarity between the clustering results achieved by two different clustering methods operating on a common set of data units. The relations among the results achieved by any number of such methods can be depicted in a similarity matrix containing the similarity measures computed for all pairwise combinations of these observed results. Manifestly, such a similarity matrix is itself suited for analysis by the hierarchical clustering methods of Section 6.2. Therefore, the clustering methods themselves can be clustered in regard to the results they produce on specified data sets. Methods falling in the same cluster are substantially redundant with regard to the common data set; methods in different clusters reveal different facets of the data set and might therefore have the potential to complement each other when used together.

The analyst may choose to compare results achieved by clustering data units while using various combinations of variables, association measures and clustering methods; the extent to which these three facets of a clustering problem interact to produce the final result could also be studied. Borko *et al.* (1968, p. 28) came close to using such an approach when they hierarchically clustered a data set 36 different ways according to a $2^2 \cdot 3^2$ factorial design in an effort to assess the effects of four factors. They analyzed their results by subjective judgement and did not see the opportunity for clustering, even though they constructed a similarity matrix among the 36 hierarchical classifications.

In addition to comparing clustering methods, these techniques can be used for comparative analysis of association measures. Each association measure can be used to compute a similarity matrix for a given data set; then a correlation can be computed between the paired entries in each pair of similarity matrices to build up a grand similarity matrix among all the association measures. Such a matrix is then ready for clustering by the hierarchical methods of Section 6.2. Green and Rao (1969, p. 363) get as far as computing the grand similarity matrix for ten association measures applied to a given data set; however, they too fail to see the opportunity for clustering and settle for a few observations on relations within the matrix.

Besides such comparisons based solely on similarity matrices computed from the data set, association measures can be compared in regard to the results they produce in conjunction with some clustering method; that is, the data set may be clustered repeatedly with a single method but using a different association measure each time. Similarities among the resulting classifications or partitions are essentially similarities among association measures as filtered through the clustering method. This latter kind of analysis can provide some useful insights into the manner in which differences between association measures are enhanced or suppressed by the action of a clustering method. Green and Rao (1969) compare eight association measures with regard to the partitions achieved at the level of eight clusters using the single linkage method of cluster analysis (see Section 6.2.1). They assess the similarity between partitions using the binary metric distance computed on the 2×2 table of paired comparisons shown in Table 10.3; however, rather than cluster these partitions, they use multidimensional scaling (Green and Carmone, 1970) to construct a two-dimensional representation of how the partitions are located relative to each other and assess the degree of clustering visually.

It seems clear that much fruitful research remains to be accomplished in the realm of quantitative assessment of the similarities between clustering results achieved with alternative methods. This observation is especially true in regard to hierarchical classifications; the derived similarity matrix is the only known way to construct a numerical representation of a given hierarchical classification and two such derived matrices can be quantitatively compared only by computing the correlation between their corresponding entries. The result is that only one aspect of similarity can be measured between two hierarchical classifications given the available tools. In contrast, there are two different ways to compare partitions and these two ways permit usage of the full array of association measures defined for nominal and binary variables.

Whether the clustering results are in the form of a hierarchical classification or a partition, it is not clear whether any particular interpretation can be attributed to a high value of similarity computed between two sets of results; there is hardly any appreciation of the different aspects of similarity that may be tapped by alternative methods of measuring similarity. Until some insight is gained in this part of the problem, there must remain doubts as to whether any given measurement of similarity bears any relevance to the problem. Put another way, at least a partial answer is needed for the question: Do informed judgements of similarity between different sets of comparable clustering results correspond in a systematic (predictable) manner with similarities computed by available methods?

10.3   Lists of Candidate Characteristics for Problems and Methods

Before setting out to evaluate alternative cluster analysis methods according to their apparent performance on example data sets, it is appropriate to have available some plausible candidates for the important characteristics of problems and methods. Admittedly it is more difficult to maintain an open mind toward unexpected discoveries while harboring a set of preferred hypotheses; on the other hand, unguided exploration of the empirical behavior exhibited by various clustering methods is far too expensive to be considered seriously as an approach to evaluation. The following sections present a variety of characteristics which apparently are of importance for describing problems and methods; the degree to which these characteristics are important, both singly and jointly, remains to be determined.

10.3.1   CANDIDATE PROBLEM CHARACTERISTICS

A problem in cluster analysis consists of a data set (possibly yet to be collected), a contextual setting, and an investigative purpose guiding the direction of the analytic effort. The emphasis in this preliminary list of problem characteristics is focused on the data set and its apparent structure properties. It remains to be seen how the elements of context and purpose can be treated with any degree of specificity outside the realm of a particular problem.

1.   *The number of variables and data units.*   The sheer size of the data set is an important element in determining the available analysis options as well as the cost and effort required for the analysis. Many methods in the literature are feasible computationally only for small problems; however, the methods chosen for description in Chapters 6 and 7 together with the strategies

of Chapter 9 are all suited for large problems. Another aspect of problem size involves the ease with which clusters may be identified. On the one hand, increased numbers of data units will reduce the errors of estimation for cluster statistics such as mean vectors and covariance matrices; on the other hand, increased numbers of data units are also accompanied by additional outliers and "between cluster" data units which tend to obscure cluster boundaries and cause a deterioration in the apparent separation between clusters. An additional consideration is that small clusters or clusters of rare entities may not be identifiable unless the data set is very large.

2. *Choice of what to cluster.* Cluster analysis techniques may be applied to data units, variables or both together. All of the methods in Chapters 6 and 7 are suitable for the clustering of data units; however, only the hierarchical methods based on a similarity matrix (Sections 6.2 and 6.4) are suitable for clustering variables. Techniques and strategies for clustering both data units and variables together are still in an early stage of development; problems requiring a harmonious analysis of both variables and data units are likely to be much more difficult than an analysis of either aspect alone.

3. *Variable types in the data set.* Chapters 3, 4, and 5 together provide a sweeping discussion of variables, scales and association measures. This discussion is organized according to the variable types used in the data set not only for convenience but also for emphasis on the fact that the choice of variables has an important impact on the variety of association measures available for the analysis. In cases involving a mixture of variable types there are several strategies of analysis available; but inevitably these methods distort some of the original data either through discarding information or imposing additional assumptions, perhaps in subtle ways.

4. *Separation of the clusters.* If the clusters are well separated then almost any method will succeed in finding them. A case in point is Fisher's (1936) famed iris data which has been widely used as test data for clustering methods. Fifty specimens from each of three species of the flower are described on four measurements. One of the species exhibits distinctive values on two of the measurements and virtually any clustering method succeeds; however, the other two species overlap and they provide a challenge. Separability is an important element in any attempt to use a decomposition strategy such as that discussed in Section 9.1.1. Another consideration is that the clusters may not be separated at all but may be arrayed in a sequence of overlapping clusters. The multiple-range tests (Duncan, 1955) used for the simultaneous comparison

Fig. 10.2. Clustering by multiple-range tests.

of treatment (group) means in the analysis of variance effectively defines overlapping clusters along a single dimension. Figure 10.2 shows the usual form for presenting the results of a multiple range test; the Xs represent treatments and the lines underneath identify groupings within which the null hypothesis of equal treatment means cannot be rejected. It is not clear how this kind of group structure can be discovered in several dimensions.

5.   *Selection and weighting of variables.*   A problem may be described by many variables, only some of which might be pertinent to a particular analysis objective; the problem is embedded in a context which may conceal important features. The fundamental task may be one of extracting the problem from its surroundings through a proper selection and combination of variables. Shepard and Carroll (1966) present a most impressive example of a nonlinear one-dimensional problem embedded in a four-dimensional space.

6.   *Number of meaningful solutions.*   In Section 10.1 it was asserted that a problem may be characterized by clusters of solutions. There may be several different *classes* of partitions or hierarchical classifications; and within a hierarchy there may be several meaningful levels of aggregation. Possibly some of these classifications or partitions can stand alone as meaningful results, but it should be interesting to identify those aspects of the problem which are discerned only as a consequence of examining joint results from several classes of solutions.

7.   *The underlying data generation process.*   A data set is merely a collection of numerical observations from some underlying data generation process which is itself the ultimate object of interest. It should be interesting to know how various such processes might become confused with each other when viewed only in terms of their observed effects. Especially valuable would be identification of distinctive features which might serve as "signatures" for different processes.

8.   *Distribution of entities among clusters.*   Problems containing clusters of widely varying sizes may be more difficult to analyze than problems in which the entities have a relatively even distribution among the clusters. Some methods show a substantial tendency to even out the distribution of entities thereby causing fragmentation of large clusters and concealment of small ones (Scott and Symons, 1971a).

10.3.2   Candidate Method Characteristics

The methods of cluster analysis have been the object of concern throughout the preceding nine chapters. Many method characteristics have been identified and discussed in regard to specific methods. In Section 7.2.3 a clear-cut opportunity is identified for a comparative analysis of the various nearest centroid sorting methods; a variety of pertinent questions are offered as a tentative guide to that comparison. The method characteristics identified in this section stand at one higher level of generality because they cut across all methods.

1. *Form of results.*   The presentation in Chapters 6 and 7 is divided according to whether the ultimate result is a hierarchical classification or a partition. The hierarchical methods may be used for clustering either variables or data units while the nearest centroid sorting methods are limited to clustering data units, principally because there is no suitable definition for a typical or average variable. The experimental two way clustering methods discussed briefly in Section 9.2.1 have specialized forms for presentation of results. It may be a fruitful tactic to construct new clustering methods to fit new forms for presenting results rather than look for additional clustering criteria which fit familiar forms.

2. *Cost and capacity factors.*   Whatever the clustering method, a computer is needed for all but the smallest of problems. But even the capacities of modern computers can be taxed by modest problems using some methods of cluster analysis. The methods described in Chapters 6 and 7 are all suitable for large problems but differ in their demands for core storage and central processor time. The hierarchical methods grow in their storage and time demands as the square of the number of entities to be clustered; however, the strategy of staged clustering presented in Section 9.1.2 makes it unnecessary to consider more than a couple hundred entities at a time and therefore gives a sharp break in this growth curve. The nearest centroid sorting methods nominally are more economical than hierarchical methods but the advantage may be diminished seriously if the analyst needs to make several guesses for the number of clusters.

3. *Interactions with association measures.*   Little study has been devoted to the joint effects of association measures and clustering methods. However, one significant result is given by Johnson (1967), who shows that the single-linkage and complete-linkage methods utilize only the ordering of the similarity values in the similarity matrix; therefore measures which are monotonic to each other are all equivalent when used with these two methods. Another significant result is that the convergent nearest centroid sorting methods give a local minimum for the trace of the within groups scatter matrix when the distance function is chosen as squared Euclidean distance (see Section 7.2.4). There are probably additional important ways in which methods and measures complement and interact with each other.

4. *Sensitivity to various cluster structures.*   Most clustering methods have special properties in regard to the cluster structure they tend to find.

The single-linkage method is suited uniquely to finding the strongest separations between clusters; this method is also set apart by its tendency to produce serpentine clusters, even when there are only compact clusters in the data set. In contrast, the nearest centroid sorting methods of Chapter 7 focus on the internal structure of the groups and may produce clusters which are not at all well separated. The complete linkage and Ward hierarchical methods also produce compact clusters which may have poor separation properties. Outliers are likely to be less of a problem for the hierarchical methods than the nearest centroid methods because the latter are likely to identify each outlier as a group unto itself and therefore aggregate the remainder of the data set into fewer clusters than originally planned. The Ward hierarchical method and others (Scott and Symons, 1971a) show a tendency to balance out the size of the clusters they produce.

   5.   *Dependence on arbitrary decisions.*   The analyst should welcome the opportunity to exercise his judgement and exert control over the clustering process, but he also should be wary of forced choices influencing the analysis but lacking relevant guidelines. One such element of arbitrariness arises in the k-means methods of Sections 7.2.2 and 7.3.1 wherein the results of the methods are dependent on the order in which the data units are presented. In some cases the requirement to specify the number of groups and an initial partition may seem arbitrary for the nearest centroid sorting methods; indeed, some authors suggest that the starting configuration be generated randomly. More subtle elements of arbitrariness may be present in other clustering methods.

10.4   The Evaluation Task Lying Ahead

The evaluation of clustering methods is particularly difficult, because the value of a cluster analysis not only depends on the mechanical workings of the algorithm but also the context, the analyst's investigative purpose, and his interpretative skills. A clustering method cannot be evaluated by simply applying the method to a set of data to see whether the "right answer" is obtained. The point is that the partition or hierarchical classification produced by a clustering method is valuable for the ideas and insights it suggests to the analyst rather than for the simple physical rearrangement of the entities in the problem. A clustering method rearranges the facts so as to sharpen the effect of the analyst's faculties for pattern recognition and pattern discovery. In this setting, evaluation is not a matter of measuring resource consumption in the performance of a fixed task; rather it is a matter of learning how the methods highlight problem characteristics embedded in the data. Once it is

214 Comparative Evaluation of Cluster Analysis Methods

known how the various methods perform in response to problem characteristics, it should be possible to devise grand strategies of analysis wherein each stage of a sequential analysis is guided by the accumulated knowledge discovered in preceding stages; the analyst will be able to exercise the methods for their complementary effects while minimizing the duplication observed in the results obtained from stage to stage. Perhaps when this condition is realized cluster analysis will fulfill its potential as an efficient and economical tool of discovery in research.

APPENDIX

CORRELATION AND NOMINAL VARIABLES

One method of calculating a measure of association between nominal variables is to assign scores to each class and then use these scores in an ordinary product moment correlation calculation. This appendix deals with the question of how to choose scores to maximize the correlation.

This topic has received sporadic attention in the statistical literature. Fisher (1940; 1963, Sections 49.2 and 49.3), Johnson (1950), Tatsuoka (1955), and Bryan (1961) use a variation of this technique for assigning scores to categorical variables in discriminant analysis. Maung (1941) shows that Fisher's method of assigning scores is equivalent to a canonical correlation method and an iterative method based on mean scores for each class; he uses the technique to assess association between eye color and hair color. Williams (1952) develops the canonical correlation approach to assigning scores. Kendall and Stuart (1961, Section 33.44 and 33.49) repeat Williams' analysis. Kshirsagar (1970) deals with significance tests as does Williams (1952). Wherry and Lane (1965) present a method based on factor analysis which has little apparent similarity to the canonical correlation method; however, computer programs for the two methods give identical results.

This appendix presents a detailed development of the canonical correlation approach outlined in Williams (1952). Williams' development is absolutely correct but sketchy at some critical points. His presentation also gives primary concern to significance testing rather than the mechanics of finding scores and correlations.

215

A.1   The Fundamental Analysis

Let A and B be two variables with p and q classes, respectively. The empirical joint distribution of the variables in a sample of $n_{..}$ observations is represented conventionally in a contingency table as in Table A.1. Using scores as interval values, the product moment correlation between A and B is

$$r = \frac{\sum\limits_{i=1}^{i=p} \sum\limits_{j=1}^{j=q} n_{ij}(a_i - \bar{a})(b_j - \bar{b})}{\left\{ \left[\sum\limits_{i=1}^{i=p} n_{i.}(a_i - \bar{a})^2 \right]\left[\sum\limits_{j=1}^{j=q} n_{.j}(b_j - \bar{b})^2 \right] \right\}^{1/2}}. \tag{A.1}$$

Since the correlation coefficient is invariant under linear transformations no loss of generality is incurred by requiring the two sets of scores to have zero mean and unit variance. That is

$$\bar{a} = \sum_{i=1}^{i=p} n_{i.}\, a_i/n_{..} = \bar{b} = \sum_{j=1}^{j=q} n_{.j}\, b_j/n_{..} = 0, \tag{A.2}$$

$$\text{var}\,(a) = \sum_{i=1}^{i=p} n_{i.}\, a_i^2/n_{..} = \text{var}\,(b) = \sum_{j=1}^{j=q} n_{.j}\, b_j^2/n_{..} = 1. \tag{A.3}$$

Substituting Eqs. (A.2) and (A.3) into Eq. (A.1) gives

$$r = \sum_{i=1}^{i=p} \sum_{j=1}^{j=q} n_{ij}\, a_i\, b_j/n_{..}. \tag{A.4}$$

Maximizing $|r|$ clearly is equivalent to maximizing r^2, the latter being a more convenient quantity to use. The optimization problem is now to choose

TABLE A.1
AUGMENTED CONTINGENCY TABLE

Category for Variable A \ Category for Variable B	1	2	$\cdots$	q	Marginal totals	Scores
1	n_{11}	n_{12}	$\cdots$	n_{1q}	$n_{1.}$	a_1
2	n_{21}	n_{22}	$\cdots$	n_{2q}	$n_{2.}$	a_2
$\vdots$	$\vdots$	$\vdots$	$\cdots$	$\vdots$	$\vdots$	$\vdots$
p	n_{p1}	n_{p2}	$\cdots$	n_{pq}	$n_{p.}$	a_p
Marginal totals	$n_{.1}$	$n_{.2}$	$\cdots$	$n_{.q}$	$n_{..}$	
Scores	b_1	b_2	$\cdots$	b_q		

scores a_i and b_j so as to maximize the square of Eq. (A.4) subject to the constraints (A.2) and (A.3). Adjoining the constraints with Lagrange multipliers m_k gives the Lagrangian

$$L = \left[\sum_{i=1}^{i=p} \sum_{j=1}^{j=q} n_{ij} a_i b_j / n_{..} \right]^2 - m_1 \sum_{i=1}^{i=p} n_{i.} a_i / n_{..}$$

$$- m_2 \sum_{j=1}^{j=q} n_{.j} b_j / n_{..} - m_3 \left[\sum_{i=1}^{i=p} n_{i.} a_i^2 / n_{..} - 1 \right] - m_4 \left[\sum_{j=1}^{j=q} n_{.j} b_j^2 / n_{..} - 1 \right] \cdot (A.5)$$

Partial differentiation of Eq. (A.5) with respect to b_j gives

$$2 \left[\sum_{i=1}^{i=p} n_{ij} a_i \right] \left[\sum_{i=1}^{i=p} \sum_{k=1}^{k=q} n_{ik} a_i b_k \right] / n_{..}^2 - m_2 n_{.j} / n_{..} - 2 m_4 n_{.j} b_j / n_{..} = 0. \quad (A.6)$$

Summing Eq. (A.6) over j gives

$$2 \left[\sum_{i=1}^{i=p} n_{i.} a_i \right] \left[\sum_{i=1}^{i=p} \sum_{k=1}^{k=q} n_{ik} a_i b_k \right] / n_{..}^2 - m_2 - 2 m_4 \sum_{j=1}^{j=q} n_{.j} b_j / n_{..} = 0. \quad (A.7)$$

By the constraints (A.2), the first and third terms of Eq. (A.7) are zero. Therefore $m_2 = 0$. Multiplying Eq. (A.6) by b_j and summing with respect to j gives

$$2 \left[\sum_{i=1}^{i=p} \sum_{j=1}^{j=q} n_{ij} a_i b_j \right]^2 / n_{..} - 2 m_4 \sum_{j=1}^{j=q} n_{.j} b_j^2 / n_{..} = 0.$$

Substituting from Eqs. (A.3) and (A.4) gives $2r^2 - 2m_4 = 0$, so $m_4 = r^2$. Substituting Eq. (A.4) into (A.6) and using the values of m_2 and m_4 gives

$$2 \left[\sum_{i=1}^{i=p} n_{ij} a_i \right] r / n_{..} - 2 r^2 n_{.j} b_j / n_{..} = 0.$$

Solving for b_j in terms of the a_i,

$$b_j = \sum_{i=1}^{i=p} n_{ij} a_i / n_{.j} r . \quad (A.8)$$

To obtain r in terms of the a_i alone, square both sides of Eq. (A.8), multiply by $n_{.j}$, and sum over j to obtain

$$\sum_{j=1}^{j=q} n_{.j} b_j^2 = n_{..} = \sum_{j=1}^{j=q} \left(\sum_{i=1}^{i=p} n_{ij} a_i \right)^2 / n_{.j} r^2 .$$

Solving for r^2 gives

$$r^2 = \sum_{j=1}^{j=q} \left(\sum_{i=1}^{i=p} n_{ij} a_i \right)^2 / n_{.j} n_{..} . \quad (A.9)$$

If the a_i scores are known and are in standard form the maximum correlation may be calculated from Eq. (A.9), and the corresponding b_j scores from Eq. (A.8). A situation in which the a_i scores are known occurs when the A variable is an interval variable. There is no conceptual problem in handling interval variables in this context since each distinct variate value defines a class. If the a_i scores are not standardized it is necessary to replace a_i in Eqs. (A.8) and (A.9) by $(a_i - \bar{a})/s_a$, where

$$\bar{a} = \sum_{i=1}^{i=p} n_{i.}\, a_i/n_{..}, \qquad s_a^2 = \sum_{i=1}^{i=p} n_{i.}(a_i - \bar{a})^2/n_{..}.$$

Making this replacement in Eq. (A.8)

$$b_j = \frac{\sum\limits_{i=1}^{i=p} n_{ij}(a_i - \bar{a})}{n_{.j}\, rs_a} = \frac{1}{rs_a}\left[\sum_{i=1}^{i=p} n_{ij}\, a_i/n_{.j} - \bar{a}\right] = \frac{\bar{a}_j - \bar{a}}{rs_a}, \qquad \text{(A.10)}$$

where $\bar{a}_j$ is the mean A score for observations in the jth class. Thus b_j is a linear transformation of $\bar{a}_j$, and since correlation is invariant under a linear transformation the b_j scores may be taken as $\bar{a}_j$ in the unstandardized case. This result is identical to that found using a regression argument in Section 3.2.6.2. That is, the correlation between a nominal variable and an interval variable is maximized by scoring the nominal classes with class means.

Replacing a_i by $(a_i - \bar{a})/s_a$ in Eq. (A.9) gives

$$r^2 = \sum_{j=1}^{j=q}\left[\sum_{i=1}^{i=p} n_{ij}(a_i - \bar{a})/s_a\right]^2 /n_{.j}n_{..},$$

and after rearranging

$$r^2 = \frac{\sum\limits_{j=1}^{j=q}\left[\sum\limits_{i=1}^{i=p} n_{ij}\, a_i\right]^2 /n_{.j}n_{..} - \bar{a}^2}{s_a^2}. \qquad \text{(A.11)}$$

Equation (A.11) may be used to calculate the maximum correlation in terms of unstandardized scores a_i.

When both sets of scores are unknown, the results to this point are still valid but incomplete. Optimization with respect to the a_i scores remains to be done. Recall that the first term of Eq. (A.5) is r^2. The b_j scores may be eliminated from this term by using Eq. (A.9) for r^2. With this substitution, partially differentiating Eq. (A.5) with respect to a_i and setting the result to zero gives

$$2\sum_{j=1}^{j=q} n_{ij} \sum_{k=1}^{k=p} n_{kj}\, a_k/n_{.j}n_{..} - m_1 n_{i.}/n_{..} - 2m_3 n_{i.}\, a_i/n_{..} = 0. \qquad \text{(A.12)}$$

Summing Eq. (A.12) over i and using the constraint (A.2) gives $m_1 = 0$. Multiplying Eq. (A.12) by a_i and summing over i gives

$$2 \sum_{j=1}^{j=q} \left[\sum_{i=1}^{i=p} n_{ij} a_i \right]^2 / n_{.j} n_{..} - 2m_3 \sum_{i=1}^{i=p} n_{i.} a_i^2 / n_{..} = 0.$$

Using constraint (A.3) and Eq. (A.9) this latter expression reduces to $m_3 = r^2$.

Up to this point the four Lagrange multipliers have been determined, the b_j are known in terms of the a_i, and r^2 is known in terms of a_i. It only remains to determine the a_i. The next few steps are concerned with achieving a particular form of Eq. (A.12) from which the determination of the a_i will follow. Substituting for m_1 and m_3 in Eq. (A.12) gives

$$2 \sum_{j=1}^{j=q} n_{ij} \sum_{k=1}^{k=p} n_{kj} a_k / n_{.j} n_{..} - 2r^2 n_{i.} a_i / n_{..} = 0. \qquad (\text{A.13})$$

Multiplying Eq. (A.13) by $(n_{..}/n_{i.})^{1/2}$, introducing $(n_{k.}/n_{k.})^{1/2}$ as a factor in the first term, and carefully arranging the results gives

$$\sum_{j=1}^{j=q} \sum_{k=1}^{k=p} \left[\frac{n_{kj}}{(n_{k.} n_{.j})^{1/2}} \frac{n_{ij}}{(n_{i.} n_{.j})^{1/2}} \right] \left[\frac{n_{k.}}{n_{..}} \right]^{1/2} a_k - r^2 \left[\frac{n_{i.}}{n_{..}} \right]^{1/2} a_i = 0. \quad (\text{A.14})$$

Noting that the order of summation is immaterial and letting $m_i = (n_{i.}/n_{..})^{1/2}$ permits Eq. (A.14) to be written

$$\sum_{k=1}^{k=p} \left[\sum_{j=1}^{j=q} \frac{n_{kj}}{(n_{k.} n_{.j})^{1/2}} \frac{n_{ij}}{(n_{i.} n_{.j})^{1/2}} \right] m_k a_k - r^2 m_i a_i = 0, \qquad (\text{A.15})$$

or more simply

$$\sum_{k=1}^{k=p} t_{ik} m_k a_k - r^2 m_i a_i = 0, \qquad (\text{A.16})$$

where t_{ik} is the term in brackets in Eq. (A.15). Note that i ranges from 1 to p, so Eq. (A.16) is but one row in a matrix equation written as

$$\mathbf{T}\{m_i a_i\} - r^2 \{m_i a_i\} = (\mathbf{T} - r^2 I)\{m_i a_i\} = 0, \qquad (\text{A.17})$$

where $\mathbf{T}$ is a $p \times p$ symmetric matrix whose typical element is t_{ik} and $\{m_i a_i\}$ is a p-vector. Equation (A.17) is readily recognized as an eigenproblem in which r^2 is an eigenvalue (latent root) and $\{m_i a_i\}$ is an eigenvector (latent vector) of $\mathbf{T}$. Thus the determination of the maximum correlation is a matter of determining a root of $\mathbf{T}$. Incidentally, the analysis is symmetric so that the a_i could be determined in terms of the b_j. In this latter case $\mathbf{T}$ would be a $q \times q$ symmetric matrix. With this fact in mind one can always construct $\mathbf{T}$ so that its dimension is the smaller of p and q.

One solution of Eqs. (A.15) and (A.17) is

$$r^2 = a_1 = a_2 = \cdots = a_p = 1,$$

as may be readily verified by direct substitution in Eq. (A.15). This solution gives the largest root of **T** and corresponds to the trivial case of collapsing each variable to a single class as a result of assigning the same score to every category. The **T** matrix may be deflated to remove this extraneous solution by subtracting the outer product of the eigenvector. That is

$$\hat{\mathbf{T}} = \mathbf{T} - \{m_i a_i\}\{m_i a_i\}^{\mathrm{T}} = \mathbf{T} - \{m_i\}\{m_i\}^{\mathrm{T}}.$$

The typical element of $\hat{\mathbf{T}}$ is

$$\hat{t}_{ik} = \sum_{j=1}^{j=q} \frac{n_{kj}}{(n_{k.}\,n_{.j})^{1/2}}\, \frac{n_{ij}}{(n_{i.}\,n_{.j})^{1/2}} - \frac{(n_{i.}\,n_{k.})^{1/2}}{n_{..}}. \tag{A.18}$$

The largest root of $\hat{\mathbf{T}}$ is the maximum squared correlation between the variables A and B. The corresponding eigenvector is $\{m_i a_i\}$. These quantities may be determined very efficiently by any principal axis factor analysis or eigenvalue method.

Extracting the a_i scores from $\{m_i a_i\}$ requires a little caution. Recall that the a_i scores must satisfy the constraints (A.2) and (A.3). In general the eigenvectors calculated by standard computer programs will have to be subjected to a linear transformation before dividing each term by m_i to obtain a_i. The standardizing constraints (A.2) and (A.3) may be written in terms of $m_i a_i$ as

$$\sum_{i=1}^{i=p} n_{i.}\,a_i = \sum_{i=1}^{i=p} (n_{i.})^{1/2} m_i a_i = 0, \tag{A.19}$$

$$\sum_{i=1}^{i=p} n_{i.}\,a_i^2/n_{..} = \sum_{i=1}^{i=p} (m_i a_i)^2 = 1. \tag{A.20}$$

Suppose $\{e_i\}$ is the eigenvector corresponding to the largest root of $\hat{\mathbf{T}}$. A linear transformation from e_i to $m_i a_i$ is

$$m_i a_i = (e_i - c)/d. \tag{A.21}$$

Substituting Eq. (A.21) into Eqs. (A.19) and (A.20) results in determination of c and d as

$$c = \sum_{i=1}^{i=p} (n_{i.})^{1/2} e_i \Big/ \sum_{i=1}^{i=p} (n_{i.})^{1/2},$$

$$d = \left[\sum_{i=1}^{i=p} (e_i - c)^2\right]^{1/2}.$$

The a_i scores may then be determined after making the transformation of Eq. (A.21).

A.2 The Problem of Isolated Cells

An isolated cell in a contingency table is a cell which is the only nonnull entry in both its row and column. That is, if the cell is located at the intersection of row i and column j then $n_{ij} = n_{i.} = n_{.j}$, $n_{ik} = 0$ for $k \neq j$, and $n_{kj} = 0$ for $k \neq i$. The presence of one or more such cells in a contingency table results in extraneous solutions to the maximum correlation problem. Surprisingly, this problem is not mentioned in any of the literature on the use of scores in contingency tables, yet it has been a persistent source of difficulties in studies with real data.

The essential points of the analysis are clearest when the contingency table contains only one isolated cell. The problem will be examined in this simplified form and then the solution extended to the case of k isolated cells.

Without loss of generality rows and columns may be exchanged so that the isolated cell is at the intersection of row 1 and column 1. Then for the case $i \neq 1$, Eq. (A.15) reduces to

$$\sum_{k=2}^{k=p} \left[\sum_{j=2}^{j=q} \frac{n_{kj}}{(n_{k.} n_{.j})^{1/2}} \frac{n_{ij}}{(n_{i.} n_{.j})^{1/2}} \right] m_k a_k - r^2 m_i a_i = 0,$$

because $n_{1j} = 0$ for $j \neq 1$, and $n_{i1} = 0$ for $i \neq 1$. This latter equation is easily recognized to be the maximum correlation problem with the first row and first column deleted; and as shown earlier the maximum r^2 solution is

$$r^2 = a_2 = a_3 = \cdots = a_p = 1.$$

For the case $i = 1$, Eq. (A.15) reduces to

$$m_1 a_1 - r^2 m_1 a_1 = 0,$$

because the only nonzero term in the sum of terms is that one for which $i = j = k = 1$; and since $n_{11} = n_{.1} = n_{1.}$ the fractions reduce to unity. Since $r = 1$, the score a_1 is arbitrary and may be chosen to satisfy constraint (A.2) or (A.3). The complete form of this extraneous solution is then

$$r^2 = 1,$$
$$b_1 = a_1 = \text{arbitrary},$$
$$b_j = a_i = 1, \qquad i = 2, \ldots, p, \quad j = 2, \ldots, q.$$

In practical terms the solution dichotomizes the table into a contrast of the isolated cell versus the remainder. The solution $r^2 = 1$ occurs regardless of the magnitude of n_{11}. This result is a manifestation of the fact that the row and column associated with the isolated cell are orthogonal to the other rows and columns.

This outcome is lacking distinctly in intuitive appeal. The existence of one isolated cell containing a single observation (out of perhaps thousands of observations) insures $r^2 = 1$ regardless of the structure in the remainder of the table. One possible remedy is merely to delete the isolated cell (with its row and column) and calculate the correlation on the reduced table. But this approach takes no account of the perfect predictability inherent in the isolated cells. If any significant number of observations fall in isolated cells this procedure could grossly underestimate r^2. A reasonable alternative giving due credit to isolated cells is developed below.

Let r_d^2 be the squared correlation obtained from the contingency table with the isolated cell deleted. The squared correlation may be considered a ratio of between group sum of squares to total sum of squares, that is

$$r_d^2 = B/T.$$

The a_i scores determined along with r_d^2 in the reduced table have zero mean

$$\sum_{i=2}^{i=p} n_{i.}\, a_i = 0, \tag{A.22}$$

and unit variance

$$\sum_{i=2}^{i=p} n_{i.}\, a_i^2 / (n_{..} - n_{11}) = T/(n_{..} - n_{11}) = 1. \tag{A.23}$$

If the isolated cell is reintroduced in the contingency table with score a_1 it makes no contribution to the within group sum of squares W because of the associated perfect predictability. And since $T = B + W$ any increase in T due to the isolated cell is matched by an equal increase in B. Therefore a corrected squared correlation should have the form

$$r_c^2 = \frac{B + C}{T + C}.$$

This expression also may be written as

$$r_c^2 = \frac{r_d^2\, T + C}{T + C} = \frac{T}{T + C}(r_d^2) + \frac{C}{T + C}(1), \tag{A.24}$$

which is merely a weighted average of the squared correlation for observations in the reduced table (r_d^2) and the squared correlation for observations in the isolated cell (1). It seems most reasonable to weight the two squared correlations in direct proportion to the number of observations associated with each term. Since $T = n_{..} - n_{11}$, this proportional weighting is achieved by choosing $C = n_{11}$. In this case

$$r_c^2 = \frac{n_{..} - n_{11}}{n_{..}}\, r_d^2 + \frac{n_{11}}{n_{..}}$$

This choice for C seems to be universally appropriate. However, it may be that some peculiar circumstances will indicate an alternative solution. Accordingly C is retained as a parameter in the remainder of this analysis.

Once C has been chosen, it remains to determine a set of a_i scores which can be associated with r_c^2. Because the approach taken here is modification of r_d^2 as in Eq. (A.24), the a_i scores determined in the reduced table should be preserved; the isolated cell is then to be given a score which places it among the other classes so as to reproduce the corrected correlation r_c. Introducing the score a_1 gives the new mean

$$\bar{a} = \sum_{i=1}^{i=p} n_{i.}\, a_i/n_{..} = n_{11}\, a_1/n_{..}, \tag{A.25}$$

and variance

$$\sum_{i=1}^{i=p} n_{i.}(a_i - \bar{a})^2/n_{..} = (T + C)/n_{..},$$

or

$$\sum_{i=1}^{i=p} n_{i.}\, a_i^2 - n_{..}\, \bar{a}^2 = T + C.$$

Using Eqs. (A.23) and (A.25), the last expression becomes

$$T + n_{11}\, a_1^2 - n_{11}^2\, a_1^2/n_{..} = T + C.$$

Therefore

$$a_1 = \left[\frac{n_{..}\, C}{n_{11}(n_{..} - n_{11})} \right]^{1/2}. \tag{A.26}$$

If $C = n_{11}$ then Eq. (A.26) is even simpler. For each $C > 0$, both a_1 and $-a_1$ will prove satisfactory as scores, at least in regard to reproducing r_c^2 for a given value of C. If the sign is important the choice probably will have to be made in light of information peculiar to the particular problem and its context.

Now suppose there are k isolated cells in a $p \times q$ contingency table, the rows and columns having been exchanged as necessary to put the isolated cells in the first k diagonal cells of the table. When the squared correlation is calculated for the reduced table (first k rows and columns deleted) the associated a_i scores have zero mean and unit variance, that is

$$\sum_{i=k+1}^{i=p} n_{i.}\, a_i = 0, \tag{A.27}$$

$$\sum_{i=k+1}^{i=p} n_{i.}\, a_i^2 = n_{..} - \sum_{i=1}^{i=k} n_{i.} = T. \tag{A.28}$$

For each isolated cell a correction C_i is applied to the sums of squares B and T to give the corrected correlation

$$r_c^2 = \frac{B + \sum_{i=1}^{i=k} C_i}{T + \sum_{i=1}^{i=k} C_i} .$$

This expression may be rewritten as

$$r_c^2 = \frac{T}{T + \sum_{i=1}^{i=k} C_i} (r_d^2) + \frac{\sum_{i=1}^{i=k} C_i}{T + \sum_{i=1}^{i=k} C_i} (1). \tag{A.29}$$

Substituting $C_i = n_{i.}$ gives

$$r_c^2 = \frac{n_{..} - \sum_{i=1}^{i=k} n_{i.}}{n_{..}} (r_d^2) + \frac{\sum_{i=1}^{i=k} n_{i.}}{n_{..}} (1). \tag{A.30}$$

If the analysis merely requires the correlations and not the scores, then Eq. (A.30) gives the desired result with little more effort than that required for a table without isolated cells.

Determination of scores corresponding to the r_c^2 of Eq. (A.29) involves a little more effort. Suppose the isolated cells are reintroduced one at a time beginning with k and descending to 1. After addition of cell j the mean over the observations in the table is

$$\bar{a}_j = \sum_{i=j}^{i=k} n_{i.} a_i / (n_{..} - \sum_{i=1}^{i=j-1} n_{i.}), \tag{A.31}$$

all terms for $i > k$ being neglected by virtue of Eq. (A.27). The total sum of squares after addition of cell j is then

$$T + \sum_{i=j}^{i=k} C_i = \sum_{i=j}^{i=k} n_{i.} (a_i - \bar{a}_j)^2$$

$$= T + \sum_{i=j}^{i=k} n_{i.} a_i^2 - (n_{..} - \sum_{i=1}^{i=j-1} n_{i.}) \bar{a}_j^2,$$

by using Eq. (A.28). Eliminating the common term T then gives

$$\sum_{i=j}^{i=k} C_i = \sum_{i=j}^{i=k} n_{i.} a_i^2 - (n_{..} - \sum_{i=1}^{i=j-1} n_{i.}) \bar{a}_j^2. \tag{A.32}$$

Equation (A.32) may be used to determine a_j from C_j, given $C_{j+1}, \ldots, C_k$. Writing the left side of Eq. (A.32) as $C_j + \sum_{i=j+1}^{i=k} C_i$ and using this same equation to evaluate $\sum_{i=j+1}^{i=k} C_i$ gives

$$C_j = n_{j.}\, a_j^2 + \left(n_{..} - \sum_{i=1}^{i=j} n_{i.}\right)\bar{a}_{j+1}^2 - \left(n_{..} - \sum_{i=1}^{i=j-1} n_{i.}\right)\bar{a}_j^2. \qquad (A.33)$$

It is necessary to expand the last term of Eq. (A.33) by substituting for $\bar{a}_j^2$ from Eq. (A.31) to obtain

$$\frac{\left[\sum_{i=j}^{i=k} n_{i.}\, a_i\right]^2}{n_{..} - \sum_{i=1}^{i=j-1} n_{i.}} = \frac{n_{j.}^2\, a_j^2 + 2n_{j.}\, a_j \sum_{i=j+1}^{i=k} n_{i.}\, a_i + \left[\sum_{i=j+1}^{i=k} n_{i.}\, a_i\right]^2}{n_{..} - \sum_{i=1}^{i=j-1} n_{i.}}.$$

Equation (A.33) may then be rewritten as

$$C_j = \frac{n_{j.}\left(n_{..} - \sum_{i=1}^{i=j} n_{i.}\right)}{n_{..} - \sum_{i=1}^{i=j-1} n_{i.}} \left[a_j^2 + \frac{\left(n_{..} - \sum_{i=1}^{i=j-1} n_{i.}\right)\bar{a}_{j+1}^2}{n_{j.}} \right.$$
$$\left. - \frac{2a_j \sum_{i=j+1}^{i=k} n_{i.}\, a_i}{n_{..} - \sum_{i=1}^{i=j} n_{i.}} - \frac{\left(\sum_{i=j+1}^{i=k} n_{i.}\, a_i\right)^2}{n_{j.}\left(n_{..} - \sum_{i=1}^{i=j} n_{i.}\right)} \right]. \qquad (A.34)$$

The sum of terms in brackets is then rewritten with the aid of Eq. (A.31) as

$$a_j^2 + \frac{n_{..} - \sum_{i=1}^{i=j-1} n_{i.}}{n_{j.}}\,\bar{a}_{j+1}^2 - 2a_j\,\bar{a}_{j+1} - \frac{\left(n_{..} - \sum_{i=1}^{i=j} n_{i.}\right)}{n_{j.}}\,\bar{a}_{j+1}^2,$$

or

$$a_j^2 - 2a_j\,\bar{a}_{j+1} + \bar{a}_{j+1}^2 = (a_j - \bar{a}_{j+1})^2.$$

Using this result in Eq. (A.34) and solving for a_j gives

$$a_j = \bar{a}_{j+1} \pm \left[\frac{C_j\left(n_{..} - \sum_{i=1}^{i=j-1} n_{i.}\right)}{n_{j.}\left(n_{..} - \sum_{i=1}^{i=j} n_{i.}\right)} \right]^{1/2}. \qquad (A.35)$$

For the case of one cell, $j = k = 1$, so that $\bar{a}_2 = 0$ and Eq. (A.35) reduces to Eq. (A.26). If C_j is set to n_j, then the radical is merely the ratio of the number of observations in the table after addition of cell j to the number before.

For k isolated cells there is a certain ambiguity in the scores obtainable from Eq. (A.35). Indeed, there are at least $2^k k!$ sets of scores satisfying Eq. (A.35). The determination of each score involves a choice of sign for the square root term so that there are 2^k possible sets for a given sequence of cells. But the scores depend on this sequence and there are $k!$ different sequences. Of course, one could adopt a convention for sequence and sign choice. The danger here is that the convention would probably be part of a computer program and tend to become enshrined as an essential aspect of the procedure while other equally viable alternatives languish in neglect. With this view in mind, the best policy probably is to treat reporting of scores for isolated cells as exceptional rather than routine. It undoubtedly will require substantial external information to make an informed selection of the most pleasing set of scores, whenever such a selection is possible.

A.3   Deflating the Squared Correlation

Suppose for the moment that there are no isolated cells in the contingency table. The procedure described earlier in this appendix gives the maximum possible squared correlation between the variables. There is no set of scores which can give a larger number. This fact raises the suspicion that the result may be somewhat inflated, at least as compared with the ordinary product-moment correlation calculated with interval variables. To clarify the point, take two interval variables and calculate the squared correlation the usual way. Then treat the variables as though they are on nominal scales by letting each distinct interval value define a class. Now calculate the squared correlation using the canonical correlation method. The latter result certainly will not be less than the first number (at the very least the observed values may be used as scores) and may be a great deal larger. Indeed, if each class has a single member (as might occur easily with truly interval data) the contingency table will consist exclusively of isolated cells and give a squared correlation of 1.0.

The usual product-moment correlation coefficient deals with the association between variables in the interval–interval case; the methods of this appendix provide correlations for the interval–nominal and nominal–nominal cases. The three results are genuine correlations since they all may be calculated by the use of well-defined numbers in the usual correlation formula. But, the latter two cases bear an evident upward bias as compared to the first. How can the three methods be rendered more comparable?

One approach is to inflate the usual result for the interval–interval case. The device of letting each distinct interval value define a class probably overshoots the mark, however; the multitude of small classes inevitably gives a large correlation. Such an approach also discards the order information originally present in the interval data. Preserving the order would give a result intermediate between these two extremes but presents formidable computational problems (Bradley *et al.*, 1962).

The alternative is to deflate the correlation obtained for the nominal–interval and nominal–nominal cases. Since the optimal scores reproduce the maximum correlation, any other set of scores can give no larger value and may give a much smaller one. What other set of scores should be chosen? If the variables were ordinal instead of nominal the obvious choice would be ranks. Ordinal to interval scale conversions are discussed in Section 3.2.5 and it appears that product-moment correlation is surprisingly robust to the use of ranks. In the present case the procedure giving the maximum correlation produces both an ordering and spacing of the classes. By adopting this order and rescoring the classes with the resultant ranks, the correlation is deflated but the optimal predictive ordering is retained. It seems reasonable to allow the ordering to be tailored to each pairwise comparison of variables; otherwise an ordering would have to be imposed *a priori* for consistent use in all comparisons. The nominal nature of the variable implies that no particular ordering has any status over another except in regard to predicting the behavior of another variable. To imply that the ordering may not be changed from one comparison variable to another is to impose a constraint in conflict with the fundamental nature of nominal variables.

The computational details for this procedure are relatively simple. In the interval–nominal case, find the optimal scores as class means; replace these scores by their ranks and calculate the correlation from Eq. (A.1) or an equivalent form. In the nominal–nominal case, find optimal scores for the row variables and replace them with ranks; then calculate class means for the columns, replace them with ranks and calculate the correlation using the two sets of ranks.

At the beginning of this section the case of isolated cells was set aside. The procedure of using ranks can be applied without difficulty in the reduced table (isolated cells deleted) to obtain an associated r_d^2. The question is what to do with the isolated cells? A workable procedure with considerable appeal is to simply use Eq. (A.30), computing r_d^2 with ranks. Ranks for the isolated classes are best left undetermined just as in the case of optimal scores.

B

PROGRAMS FOR SCALE CONVERSIONS

B.1 Partitions of the Truncated Normal Distribution

In Section 3.2.1.5 a method is described for categorizing interval variables by partitioning an assumed underlying truncated normal distribution into equal length categories. The computer program CUTS and its subroutine ERF are provided in this appendix to simplify investigation of alternative truncation points.

The program CUTS reads input and computes the following quantities:

1. Cut points between classes,
2. Proportion of distribution in each class,
3. Cumulative proportions.

The subroutine ERF approximates the value of the cumulative distribution function (CDF) for the standard normal distribution; this subroutine is commonly available among University of Texas engineering students and its origin is unknown. All computations in both CUTS and ERF are in terms of standardized variates (zero mean and unit variance). Input for each case consists of a single card:

Columns	Entry
1–10	PLO: lower truncation point in standard form (number of standard deviations from the mean)
11–20	PHI: upper truncation point in standard form
21–30	N: maximum number of classes of interest (right justified)

The program gives partitions for all numbers of classes from 2 to N. Computation is terminated by reading a value of PLO which is less than -5. Sample output is shown in Fig. 3.5.

The program as listed is dimensioned for 20 classes. These dimensions may be increased without further alteration to the program.

B.2 Iterative Improvement of a Partition

In Section 3.2.1.9 a method is described for categorizing interval data through a one-dimensional iterative clustering procedure. This section includes discussion and listing of the implementing program.

Program DIVIDE is the main program. Specifications for input of data are given by comment cards at the beginning of DIVIDE. Through use of the ISORT parameter the user has the option of arranging the data according to some prescribed sequence or first letting the program rearrange the data into ascending sequence. Subroutine SORT is used to obtain the ascending sequence if desired. A prescribed sequence can be used to take account of constraints on the groups; see Fisher (1958) for an example in which the data have a time sequence and the objective is to find epochs in which the data is most homogeneous. The IPART parameter gives three different options for the initial partition of the data.

1. Equal membership partition: each group has the same number of members except the last group which is assigned all extra data units.

2. Equal interval partition: the range of the data is computed and divided into NG intervals of equal length; the data units in each interval form a group; the data units must be sorted in ascending sequence for this option.

3. Specified partition: the user indicates the number of data units in each group of the initial partition; this option is useful for improving the result of a one-dimensional hierarchical clustering or testing some other proposed grouping.

The program implements the algorithm described in Section 3.2.1.9. Subroutine TEST computes the effect of moving a data unit which is next to a class boundary into the adjacent class. When no such moves can improve the partition further, the error sum of squares [W function of Eq. (3.1)] for each class is computed in function PSUMSQ. The output for each group consists of:

1. The sequence numbers for the first and last data units in the group.

2. The scores on the original interval variable for the first and last data units.

3. The error sum of squares.

The total error sum of squares for all groups is also printed.

The program as listed is dimensioned for 1000 data units and 20 groups; either dimension may be changed as desired without further alteration to the program. If the parameter NG is set larger than 20, then input card 1 must be continued on an additional card to accommodate additional group specifications for the initial partition.

Program CUTS

```
      PROGRAM CUTS(INPUT,OUTPUT)
C   CALCULATES CUMMULATIVE FRACTION OF OBSERVATIONS IN 2 THROUGH N
C   CATEGORIES, THE CATEGORIES OBTAINED BY UNIFORMLY DIVIDING THE RANGE
C   OF THE STANDARD NORMAL DEVIATE Z INTO N EQUAL LENGTH SEGMENTS.  THE
C   RANGE OF Z IS TRUNCATED TO THE BOUNDS ZLO AND ZHI.
      DIMENSION Z(20),P(20),PROP(20)
5     READ 1000,ZLO,ZHI,N
      IF(ZLO.LT.-5.) STOP
      PRINT 2000,ZLO,ZHI,N
      RANGE=ZHI-ZLO
      PLO=ERF(ZLO)
      PHI=ERF(ZHI)
      PNORM=PHI-PLO
      DO 50 I=2,N
      SPACE=RANGE/I
      P(I)=PHI
      I1=I-1
      DO 10 J=1,I1
      Z(J)=ZLO+SPACE*J
10    P(J)=ERF(Z(J))
      PROP(1)=(P(1)-PLO)/PNORM
      DO 20 J=2,I
20    PROP(J)=(P(J)-P(J-1))/PNORM
40    PRINT 2100,I
      PRINT 2200,(Z(J),J=1,I1)
      PRINT 2400,(PROP(J),J=1,I)
      DO 45 J=2,I
45    PROP(J)=PROP(J)+PROP(J-1)
      PRINT 2300,(PROP(J),J=1,I)
50    CONTINUE
      GO TO 5
1000  FORMAT(2F10.0,I10)
2000  FORMAT(1H1,////////,12X,11H RANGE FROM,F5.2,3H TO,F5.2,18H DIVIDE(
      AINTO 2 TO,I3,8H CLASSES)
2100  FORMAT(1H0,12X,I3,7H GROUPS)
2200  FORMAT(12X,11H CUT POINTS,(/,13X,10F7.3))
2300  FORMAT(12X,24H CUMMULATIVE PROPORTIONS,(/,13X,10F7.4))
2400  FORMAT(12X,25H PROPORTION IN EACH CLASS,(/,13X,10F7.4))
      END
```

Function ERF

```
      FUNCTION ERF(Z)
C   THIS ROUTINE INTEGRATES THE NORMAL DISTRIBUTION FUNCTION FROM MINUS
C   INFINITY TO Z WHERE Z THE NORMALLY DISTRIBUTED RANDOM VARIABLE IN
C   STANDARD FORM.  THE MAXIMUM ERROR OF THE APPROXIMATION IS 0.0000003
      IF(Z.GT.4.17) GO TO 2
      IF(Z.LT.-4.17) GO TO 3
      ZZ=Z
```

```
      IF(Z.LT.0.0) ZZ=-Z
      T=ZZ/1.4142142
      D=(((((((.430638E-4*T+.2765672E-3)*T+.1520143E-3)*T+.92705272E-2)*T
     A+.42282012E-1)*T+.70523078E-1)*T+1.0)**2
      D=D*D
      D=D*D
      D=D*D
      ERF=.5-.5/D
      IF(Z)4,5,6
    4 ERF=.5-ERF
      GO TO 7
    5 ERF=.5
      GO TO 7
    6 ERF=.5+ERF
      GO TO 7
    2 ERF=1.0
      RETURN
    3 ERF=0.0
    7 CONTINUE
      RETURN
      END
```

Program DIVIDE

```
      PROGRAM DIVIDE(INPUT,OUTPUT,TAPE5=INPUT,TAPE6=OUTPUT,TAPE1)
      DIMENSION DATA(1000),NUMBR(20),TOTAL(20),LAST(20),TITLE(20),
     AFMT(20),ICHNG(20)
      INTEGER START,END
C--------------------------------------------------------------------
C     INPUT SPECIFICATIONS
C     CARD 1   RUN PARAMETERS
C        COL      1   KRUN=REPEAT PARAMETER
C                     KRUN.NE.1=NEW DATA FOR THIS PASS
C                     KRUN=1=USE DATA FROM PREVIOUS PASS
C                     KRUN=2=NO MORE PROCESSING DESIRED.  END OF JOB.
C        COLS 2-5     NE=NUMBER OF ITEMS IN THE DATA SET (MAX 1000)
C        COLS 6-7     NG=NUMBER OF GROUPS DESIRED (MAX 20)
C        COL      8   NTAPE=INPUT UNIT FOR DATA (5=CARDS)
C        COL      9   ISORT=SORT PARAMETER
C                     ISORT=1=SORT DATA IN ASCENDING SEQUENCE
C                     ISORT.NE.1=USE DATA IN THE INPUT SEQUENCE
C        COL     10   IPART=INITIAL PARTITON PARAMETER
C                     IPART=1=EQUAL MEMBERSHIP PARTITION
C                     IPART=2=EQUAL INTERVAL PARTITION
C                     IPART=3=NUMBER IN EACH GROUP SPECIFIED ON REMAINDER OF
C.                        THIS CARD
C        COLS 11-70   (NUMBR(I),I=1,NG)=NUMBER OF ENTITIES IN EACH GROUP IN
C                        2013 FORMAT IF IPART=3
C
C     THE FOLLOWING CARDS ARE READ ONLY IF KRUN.NE.1
C
C     CARD 2   TITLE CARD
C     CARD 3   INPUT FORMAT FOR DATA
C     CARD 4 AND FOLLOWING.   DATA (IF ON CARDS)
C
C     FOR ADDITIONAL RUNS WITH THE SAME DATA INCLUDE ADDITIONAL RUN
C     PARAMETER CARDS AFTER THE DATA.  FOR RUNS WITH ADDITIONAL DATA SETS
C     INCLUDE FULL SET OF INPUT CARDS.
C--------------------------------------------------------------------
C
   10 READ(5,1000) KRUN,NE,NG,NTAPE,ISORT,IPART,(NUMBR(I),I=1,NG)
      IF(KRUN.EQ.1) GO TO 20
```

```
          IF(KRUN.EQ.2) STOP
          READ(5,1100) TITLE
          READ(5,1100) FMT
20        WRITE(6,2000) TITLE
          WRITE(6,2100) KRUN,NE,NG,NTAPE,ISORT,IPART
          IF(KRUN.EQ.1) GO TO 30
          WRITE(6,2300) FMT
          READ(NTAPE,FMT) (DATA(I),I=1,NE)
30        IF(ISORT.EQ.1) CALL SORT(DATA,NE)
          NG1=NG-1
C  SET UP INITIAL PARTITION
          GO TO (40,70,110),IPART
C  EQUAL MEMBERSHIP PARTITION
40        NPERG=NE/NG
          NSUM=0
          DO 50 I=1,NG1
          NSUM=NSUM+NPERG
50        NUMBR(I)=NPERG
          GO TO 100
C  EQUAL INTERVAL PARTITION
70        SPACE=(DATA(NE)-DATA(1))/NG
          TOP=SPACE
          I=1
          NUMBR(1)=0
          DO 90 J=1,NE
80        IF(DATA(J).LE.TOP) GO TO 90
C  INSURE THAT EVERY GROUP HAS AT LEAST ONE MEMBER
          IF(NUMBR(I).EQ.0) GO TO 90
          I=I+1
          IF(I.EQ.NG) GO TO 95
          TOP=TOP+SPACE
          NUMBR(I)=0
          GO TO 80
90        NUMBR(I)=NUMBR(I)+1
95        NSUM=J-1
100       NUMBR(NG)=NE-NSUM
110       WRITE(6,2200) (NUMBR(I),I=1,NG)
C  DETERMINE IDENTITY OF LAST ELEMENT IN EACH GROUP AND SET ICHNG
          LAST(1)=NUMBR(1)
          ICHNG(1)=1
          DO 120 I=2,NG
          LAST(I)=LAST(I-1)+NUMBR(I)
120       ICHNG(I)=1
C  FIND TOTAL IN EACH GROUP
          K1=1
          K2=LAST(1)
          DO 125 I=1,NG
          TOTAL(I)=0.
          DO 124 J=K1,K2
124       TOTAL(I)=TOTAL(I)+DATA(J)
          K1=K2+1
125       K2=LAST(I+1)
C  ITERATE ON THE PARTITION TO MAKE IMPROVEMENTS
130       IMPRV=0
          DO 160 LOWER=1,NG1
C  SKIP PAIRS THAT DID NOT CHANGE LAST TIME
          IF(ICHNG(LOWER).LE.0.AND.ICHNG(LOWER+1).LE.0) GO TO 130
C  TRY SHIFTING THE CLASS BOUNDARY UPWARD
          CALL TEST(DATA,NUMBR,TOTAL,LAST,NCHNG,1,LOWER)
          IF(NCHNG.EQ.0) GO TO 150
140       ICHNG(LOWER)=2
          ICHNG(LOWER+1)=2
          IMPRV=1
          GO TO 160
```

```
C  TRY SHIFTING THE CLASS BOUNDARY DOWNWARD
150    CALL TEST(DATA,NUMBR,TOTAL,LAST,NCHNG,-1,LOWER)
       IF(NCHNG.NE.0) GO TO 140
160    CONTINUE
       IF(IMPRV.EQ.0) GO TO 180
C  PREPARE FOR ANOTHER ROUND
       DO 170 I=1,NG
170    ICHNG(I)=ICHNG(I)-1
       GO TO 130
C  FINISHED.  PRINT RESULTS
180    WRITE(6,2400)
       K1=1
       K2=LAST(1)
       PARTSS=PSUMSQ(DATA,K1,K2)
       SUMSQ=PARTSS
       WRITE(6,2500) K1,K1,K2,DATA(K1),DATA(K2),PARTSS
       DO 190 I=2,NG
       K1=LAST(I-1)+1
       K2=LAST(I)
       PARTSS=PSUMSQ(DATA,K1,K2)
       SUMSQ=SUMSQ+PARTSS
190    WRITE(6,2500)  I,K1,K2,DATA(K1),DATA(K2),PARTSS
       WRITE(6,2600) SUMSQ
       GO TO 10
1000   FORMAT(I1,I4,I2,3I1,20I3)
1100   FORMAT(20A4)
2000   FORMAT(1H1,20A4)
2100   FORMAT(7H KRUN =,I6,/,5H NE =,I8,/,5H NG =,I8,/,8H NTAPE =,I5,/,
      A8H ISORT =,I5,/,8H IPART =,I5)
2200   FORMAT(20H INITIAL GROUP SIZES,/,1X,20I4)
2300   FORMAT(7H FORMAT,20A4)
2400   FORMAT(16HOFINAL PARTITION,/,6HOGROUP,4X,12HOBSERVATIONS,10X,
      A16H RANGE OF VALUES,10X,11HSUM SQUARES)
2500   FORMAT(1X,I5,3X,I5,3H TO,I5,4X,E12.4,3H TO,E12.4,3X,E12.4)
2600   FORMAT(51X,5HTOTAL,E12.4)
       END
```

Subroutine TEST

```
       SUBROUTINE TEST(DATA,NUMBR,TOTAL,LAST,NCHNG,KIND,LOWER)
C  THIS SUBROUTINE TESTS WHETHER TRANSFERRING POINT *MOVE* FROM ONE
C  GROUP TO ANOTHER WILL INCREASE THE BETWEEN GROUP SUM OF SQUARED
C  DEVIATIONS.
       DIMENSION NUMBR(1),TOTAL(1),LAST(1),DATA(1)
       INTEGER UPPER
       NCHNG=0
       UPPER=LOWER +1
C  CALCULATE CONTRIBUTION TO BETWEEN GROUP SUM OF SQUARES MADE
C  BY GROUPS *UPPER* AND *LOWER*.
       BASE=(TOTAL(UPPER)**2)/NUMBR(UPPER)+(TOTAL(LOWER)**2)/NUMBR(LOWER)
10     IF(KIND.EQ.-1) GO TO 20
C  TRY MOVING HIGHEST ELEMENT IN GROUP *LOWER* TO GROUP *UPPER*.
       MOVE=LAST(LOWER)
       DELTA=DATA(MOVE)
       GO TO 30
C  TRY MOVING LOWEST ELEMENT IN GROUP *UPPER* TO GROUP *LOWER*.
20     MOVE=LAST(LOWER)+1
       DELTA=-DATA(MOVE)
30     NUP=NUMBR(UPPER)+KIND
       NLOW=NUMBR(LOWER)-KIND
       TOTUP=TOTAL(UPPER)+DELTA
```

```
      TOTLOW=TOTAL(LOWER)-DELTA
      TRIAL=(TOTUP**2)/NUP+(TOTLOW**2)/NLOW
      IF(TRIAL.LT.BASE) RETURN
C  THE MOVE WAS A SUCCESS.  UPDATE
      BASE=TRIAL
      NUMBR(UPPER)=NUP
      NUMBR(LOWER)=NLOW
      TOTAL(UPPER)=TOTUP
      TOTAL(LOWER)=TOTLOW
      LAST(LOWER)=LAST(LOWER)-KIND
      NCHNG=NCHNG+1
      GO TO 10
      END
```

Subroutine SORT

```
      SUBROUTINE SORT(X,N)
C  SORTS ARRAY *X* CONSISTING OF *N* ITEMS INTO ASCENDING SEQUENCE
C  SOURCE
C    FUNDERLIC, R.E. (ED), THE PROGRAMMERS HANDBOOK, REPORT NUMBER
C    K-1729, UNION CARBIDE CORPORATION, NUCLEAR DIVISION, OAK RIDGE,
C    TENNESSEE, FEBRUARY 1968, PP 285-287.
C  BASED ON THE ALGORITHM OF
C    SHELL, D.L., A HIGH SPEED SORTING PROCEDURE, COMMUNICATIONS OF THE
C    ACM, VOLUME 2, JULY 1959, PP 30-32.
      DIMENSION X(1)
      M=N
10    M=M/2
      IF(M.EQ.0) RETURN
      K=N-M
      J=1
20    I=J
30    IM=I+M
      IF(X(I).LE.X(IM)) GO TO 40
      TEMP=X(I)
      X(I)=X(IM)
      X(IM)=TEMP
      I=I-M
      IF(I.GE.1) GO TO 30
40    J=J+1
      IF(J.GT.K) GO TO 10
      GO TO 20
      END
```

Function PSUMSQ

```
      FUNCTION PSUMSQ(DATA,K1,K2)
      DIMENSION DATA(1)
      SUM=0.
      SQS=0.
      DO 10 I=K1,K2
      SUM=SUM+DATA(I)
10    SQS=SQS+DATA(I)**2
      PSUMSQ=SQS-(SUM**2)/(K2-K1+1)
      RETURN
      END
```

APPENDIX

C

PROGRAMS FOR ASSOCIATION MEASURES
AMONG NOMINAL AND INTERVAL VARIABLES

The program described in this appendix is written to compute matrices of association measures among interval variables, nominal variables and mixtures thereof. The program computes the $\binom{N}{2} = \frac{1}{2}N(N-1)$ elements in the lower triangular matrix of symmetric association measures among N variables. Theoretical aspects of these measures are discussed in Chapter 4.

C.1 General Design Features

The program in this appendix and several other programs in later appendices have a special design structure which entails the following features.

1. The main program (DRIVER) is a dummy program which assigns input/output units and establishes the dimension of a single array (X) used for all problem-dependent storage.

2. The central program segment is a subroutine (GCORR in this appendix). It includes a substantial block of comment cards describing important aspects of program utilization including input procedures and deck setup. This subroutine reads the input parameters and computes the storage requirement for each array needed in subsequent processing. The necessary space for each array is allocated within the X array. The computed requirement for storage is compared to that available in the X array and execution is stopped if there is insufficient capacity. Through this device the user may change the program

capacity by altering the dimension of only one array. Further the required storage is printed for each run so that subsequent runs can be tailored to the problem requirements, an important element in multiprocessor systems where requested field length is a determinant of execution priority.

3. Alternative analysis procedures (in this appendix, the measures of association) are implemented through deck assembly. That is, each alternative involves one or more subroutines conforming to a standard interface with the central program segment. Each of the several analysis options may be utilized on a single set of data merely by changing subroutines in the job deck.

C.2   Deck Setup and Utilization

Every job run with this program requires the following program segments in the job deck:

1. User supplied segments: (a) DRIVER, (b) USER.
2. Segments which are the same for all jobs: (a) GCORR, (b) NCAT, (c) INPTR, (d) CORXX.
3. Segments for which there are alternatives: (a) CORKX, (b) CORKK.

Program DRIVER is a dummy main program; an example version is given in the comment cards at the beginning of subroutine GCORR and should suffice in most cases. Subroutine USER is intended to read a full set of data for one data unit at a time. The example given in the comment cards of subroutine GCORR is more elaborate than would be required for most problems but illustrates the variety of operations that may be employed with this subroutine. Program DRIVER and subroutine USER are the only portions of the program that may require alteration to fit the requirements of a particular problem. The remaining subroutines may be compiled and put on tape for general purpose use.

Subroutine GCORR is the central program segment discussed as feature 2 of the general design features in Section C.1. Subroutine NCAT reads the number of categories for each categorical variable and finds the two largest such numbers. The number of classes specified for a particular variable must be at least as large as the largest number used for a class label. Null classes are permitted. To illustrate, suppose a nominal variable consists of seven classes but only classes 1, 2, 4, and 6 occur in the data set. If these four numbers are used as data, the specified number of classes must be at least 6 even though classes 3 and 5 are null. If subroutine USER is programmed to renumber the classes so that the largest number is 4 (e.g., replace each 6 by a 3) then the number of classes may be set to 4.

Subroutine INPTR calls subroutine USER, controls storage of the data, and computes the mean and variance for interval variables (if any). Subroutine

CORXX calculates the squared product moment correlation between two interval variables; the squared correlation is used because the cluster analysis algorithms of this book cannot make use of the distinction between positive and negative association; it is also more convenient to compute the square of the correlation than its absolute value.

The choice of association measures for sets of mixed or nominal variables is effected through selection of the various versions of subroutines CORKX and CORKK. Subroutine CORKX computes a measure of association between a nominal and an interval variable, whereas subroutine CORKK computes a measure of association between two nominal variables. Four alternatives are available:

1. Optimal class scores are computed as discussed in Sections 4.2.3 and 4.4.3 and Appendix A. These scores are used to compute the squared-product moment correlation. This alternative may be used with either mixed variables or nominal variables. Subroutines CORKX and CORKK include comment cards to identify the appropriate version. Subroutine CORKK requires subroutine EIGEN to compute eigenvalues and eigenvectors.

2. Optimal class scores are replaced by their ranks and the latter are used as scores to compute the squared product moment correlation. Special versions of subroutines CORKX and CORKK implement this alternative. Subroutines EIGEN and VSORT (used to order the optimal scores) are required. This alternative also may be used with either mixed or nominal variables.

3. The Goodman and Kruskal lambda measure is the third version of CORKK. The measure is defined only for nominal variables and has no counterpart for interval variables; therefore it should be used only with nominal variables (i.e., set NVX = 0 in the input). The measure is discussed in Section 4.2.2.

4. The D measure of Eq. (4.11) is in the fourth version of subroutine CORKK. This measure also is limited to nominal variables. Neither of these last two alternatives require additional subroutines.

Each version of subroutine CORKK includes an identifying title in a DATA statement and this title is written automatically on the line printer output. All versions of subroutine CORKK include an option (see parameter NPRNT in subroutine GCORR) to print the contingency table for each pairwise comparison. The first two versions which involve class scores also include an option to print these scores.

Subroutine GCORR

```
      SUBROUTINE GCORR(X,LIMIT)
C
C     THIS PROGRAM CONTROLS COMPUTATION OF A SIMILARITY MATRIX REFLECTING
C     ASSOCIATION AMONG VARIABLES.  VARIOUS SIMILARITY COEFFICENTS ARE
C     AVAILABLE THROUGH THE MANY VERSIONS OF SUBROUTINF *CORKK*.
```

```
C   SUBROUTINES *CORKX* AND *CORXX* ARE DEFINED ONLY FOR CORRELATION.
C
C   THE USER PROVIDES TWO JOB DEPENDENT DECK SEGMENTS.  THE FIRST IS THE
C   MAIN PROGRAM WHICH PERFORMS THE FOLLOWING FUNCTIONS.
C      1.  ASSIGNS NECESSARY INPUT/OUTPUT UNITS
C      2.  SPECIFIES THE DIMENSION OF THE *X* ARRAY AND SETS THIS
C          DIMENSION EQUAL TO *LIMIT*.
C      3.  CALLS *SUBROUTINE GCORR*.
C      4.  IDENTIFIES THE RUN BY WRITING TITLES ON DESIRED OUTPUT UNITS
C   THE FOLLOWING EXAMPLE SHOULD SUFFICE IN MOST CASES.
C
C      PROGRAM DRIVER(INPUT,OUTPUT,PUNCH,TAPE5=INPUT,TAPE6=OUTPUT,
C     ATAPE7=PUNCH,TAPE1)
C      DIMENSION X(7000)
C      LIMIT=7000
C      WRITE(6,200)
C      WRITE(7,200)
C 200 FORMAT(25H ABC DATA SET.  SUBSET G.)
C      CALL GCORR(X,LIMIT)
C      END
C
C   THE SECOND USER SUPPLIED DECK SEGMENT IS *SUBROUTINE USER*.  THIS
C   SUBROUTINE IS USED TO READ THE RAW DATA.  EACH CALL TO *SUBROUTINE
C   USER* SHOULD RESULT IN READING OF A FULL SET OF DATA FOR ONE SUBJECT.
C   AMONG THE MANY OPERATIONS THE USER MAY PERFORM IN THIS SUBROUTINE ARE
C   THE FOLLOWING.
C      1.  MERGE RECORDS FROM SEVERAL FILES TO GENERATE A SINGLE SET OF
C          DATA FOR EACH SUBJECT.
C      2.  APPLY TRANSFORMATIONS TO ANY VARIABLES.
C      3.  GENERATE NEW VARIABLES AS FUNCTIONS OF THOSE THAT WERE READ IN.
C   CATEGORICAL VARIABLES MUST BE CODED AS POSITIVE INTEGERS
C   1,2,3,...ETC.   THE LARGEST INTEGER USED AS A LABEL IS THE NUMBER OF
C   CATEGORIES TO BE LISTED ON CARD 2.  NULL CATEGORIES ARE PERMITTED.
C   THE FOLLOWING EXAMPLE ILLUSTRATES SOME USES.
C
C      SUBROUTINE USER(X,K)
C      DIMENSION X(1),K(1)
C      READ(1,100) X(1),K(3),(X(I),I=2,7)
C      READ(2) K(1),K(2),X(8)
C      READ(5,200) K(4),K(5),X(9)
C      K(1)=K(1)+1
C      X(10)=3.*X(1)**2+ALOG(X(4))
C 100 FORMAT(F4.1,I1,7X,4F4.2,1X,2F2.0)
C 200 FORMAT(23X,I1,7X,I2,12X,F6.2)
C      RETURN
C      END
C
C--------------------------------------------------------------------
C   INPUT SPECIFICATIONS
C
C   CARD 1    PARAMETER CARD
C      COLS  1- 5  NS=NUMBER OF SUBJECTS OR DATA UNITS
C      COLS  6-10  NVX=NUMBER OF INTERVAL (X) VARIABLES
C      COLS 11-15  NVK=NUMBER OF CATEGORICAL (K) VARIABLES
C      COLS 16-20  NTOUT=OUTPUT UNIT FOR SIMILARITY MATRIX
C                  NTOUT=7  RESULTS ARE PUNCHED ON CARDS
C                  NTOUT.NE.7  USER SHOULD SET UP A FILE (PERHAPS A SCRATC
C                            FILE) FOR WRITING THE SIMILARITY MATRIX.
C                  THIS OUTPUT IS IN ADDITION TO THAT ON THE LINE PRINTER.
C      COLS 21-25  NPRNT=PRINT OPTION FOR CATEGORICAL VARIABLES
C                  NPRNT=1  PRINT ONLY THE RESULTING SIMILARITY MATRIX
C                  NPRNT=2  PRINT SCORES FOR THE CATEGORIES
C                  NPRNT=3  PRINT SCORES FOR THE CATEGORIES AND PRINT
C                            CONTINGENCY TABLES IN *SUBROUTINE CORKK*.
```

```
C     COLS 26-30 KSQR=OPTION FOR OUTPUT
C                KSQR=1  OUTPUT IS ABSOLUTE VALUE OF SIMILARITY MEASURE
C                KSQR.NE.1  OUTPUT IS SQUARED SIMILARITY MEASURE
C                IF THIS PROGRAM IS USED TO COMPUTE THE LAMBDA OR D
C                MEASURE (EQUATION (4-11)), DO NOT SET KSQR=1 AS THE
C                APPROPRIATE VERSIONS OF CORKK COMPUTE THE PROPER
C                COEFFICIENT, NOT ITS SQUARE.
C
C     CARD(S) 2  NUMBER OF CATEGORIES FOR EACH CATEGORICAL VARIABLE IN
C     FORMAT (40I2)
C*****DO NOT INCLUDE CARD 2 IF NVK=0*****
C
C     CARD(S) 3  RAW DATA--INPUT SUCH THAT EACH LOGICAL RECORD IS ONE
C     SUBJECT.  SINCE INTERVAL AND CATEGORICAL VARIABLES WILL OFTEN BE
C     RECORDED IN MIXED SEQUENCE, INPUT IS THROUGH *SUBROUTINE USER*.
C
C--------------------------------------------------------------------------
C
C     WHEN THE SIMILARITY MEASURE IS PRODUCT MOMENT CORRELATION
C     DICHOTOMOUS (BINARY) VARIABLES MAY BE TREATED AS EITHER INTERVAL (X)
C     OR CATEGORICAL (K) VARIABLES.  THIS FACT RESULTS FROM THE
C     INVARIANCE OF CORRELATIONS UNDER LINEAR TRANSFORMATIONS OF THE
C     VARIABLES.
C
C     ANY POSITIONING OF UNIT *NTOUT* MUST BE HANDLED IN *PROGRAM DRIVER*
C     OR BY CONTROL CARDS. THIS INCLUDES TAPE REWIND AND SKIPPING OF FILES.
C
C     X=GENERAL PURPOSE STORAGE ARRAY PARTITIONED AS FOLLOWS
C     X(N1)  TO X(N2-1)  NVX*NS WORDS--RAW DATA FOR INTERVAL VARIABLES
C     X(N2)  TO X(N3-1)  NVX WORDS--MEAN FOR EACH INTERVAL VARIABLE
C     X(N3)  TO X(N4-1)  NVX WORDS--VARIANCE FOR EACH INTERVAL VARIABLE
C     X(N4)  TO X(N5-1)  NVX+NVK-1 WORDS--STORAGE AREA FOR BUILDUP OF THE
C                        MATRIX OF SIMILARITIES, ONE ROW AT A TIME.
C     X(N5)  TO X(N6-1)  NVK*NS WORDS--RAW DATA FOR CATEGORICAL VARIABLES
C     X(N6)  TO X(N7-1)  NVK WORDS--NUMBER OF CATEGORIES FOR EACH
C                        CATEGORICAL VARIABLE
C     X(N7)  TO X(N8-1)  L1*L2 WORDS--CONTINGENCY TABLE FOR CORRELATION
C                        BETWEEN CATEGORICAL VARIABLES IN *CORKK*.  IN
C                        *CORKX*, THE SUM OF X VALUES IN EACH CATEGORY.
C     X(N8)  TO X(N9-1)  (L2+1)*L2/2 WORDS--LOWER TRIANGULAR *THAT* MATRIX
C                        IN *CORKK*.
C     X(N9)  TO X(N10-1) L1 WORDS--MARGINAL TOTALS FOR ONE CATEGORICAL
C                        VARIABLE IN *CORKK*.  IN *CORKX*, THE NUMBER OF
C                        SUBJECTS IN EACH CATEGORY.
C     X(N10) TO X(N11-1) L2 WORDS--MARGINAL TOTALS FOR OTHER CATEGORICAL
C                        VARIABLE IN *CORKK*.
C     X(N11) TO X(N12)   L2*L2 WORDS--MATRIX OF EIGENVECTORS IN *EIGEN*.
C                        SCORES FOR CATEGORIES IN BOTH *CORKX* AND *CORKK*.
C                        USED ONLY IF NECESSARY
C     BE SURE THAT *LIMIT* IS THE DIMENSION OF THE *X* ARRAY.  IF THE
C     REQUIRED STORAGE EXCEEDS *LIMIT* THE PROGRAM WILL ABORT THE JOB.
C--------------------------------------------------------------------------
      COMMON/AB/ KSCOR,TITLE(20)
      DIMENSION X(1)
      READ(5,1000) NS,NVX,NVK,NTOUT,NPRNT,KSQR
      WRITE(6,2000) NS,NVX,NVK,NTOUT,NPRNT,KSQR
C     RESERVE STORAGE FOR PROGRAM VARIABLES
      N1=1
      N2=N1+NVX*NS
      N3=N2+NVX
      N4=N3+NVX
      N5=N4+NVX+NVK-1
      N6=N5+NVK*NS
      N7=N6+NVK
```

```
      CALL NCAT(X(N6),NVK,L1,L2)
      N8=N7+L1*L2
      GO TO (1,1,3),KSCOR
1     N9=((L2+1)*L2)/2
      IF(KSCOR.EQ.2) GO TO 2
      N10=2*L1
      N9=N8+MAXO(N9,N10)
      GO TO 4
2     N9=N8+N9
      GO TO 4
3     N9=N8+L2
4     N10=N9+L1
      N11=N10+L2
      N12=N11
      IF(KSCOR.EQ.3) GO TO 5
      IF(KSCOR.EQ.2.AND.NRPNT.EQ.1) GO TO 5
      L1=L1+L2-1
      L2=L2*L2-1
      N12=N11+MAXO(L1,L2)
C     CHECK FOR SUFFICIENT STORAGE CAPACITY
5     WRITE(6,2100) N12,LIMIT
      IF(N12.GT.LIMIT) STOP
C     READ RAW DATA AND CALCULATE INTERVAL STATISTICS
      CALL INPTR(X(N1),X(N2),X(N3),NS,NVX,NVK)
C     THE CORRELATION MATRIX IS BUILT BY ROWS IN LOWER TRIANGULAR FORM
C     WITHOUT THE MAIN DIAGONAL
      NROW=1
      L2=N4-1
      WRITE(6,2400) TITLE
C     CALCULATE ROWS INVOLVING ONLY X-X CORRELATIONS
      IF(NVX-1) 100,40,10
C     THERE ARE TWO OR MORE INTERVAL VARIABLES
10    DO 30 I=2,NVX
      L1=I-1
      DO 20 J=1,L1
      L2J=L2+J
      X(L2J)=CORXX(X(N1),X(N2),X(N3),NS,NVX,I,J)
      IF(KSQR.EQ.1) X(L2J)=SQRT(X(L2J))
20    CONTINUE
C     WRITE A ROW OF THE LOWER TRIANGULAR MATRIX
      NROW=NROW+1
      WRITE(6,2600) NROW
      WRITE(6,2500) (J,X(L2+J),J=1,L1)
30    WRITE(NTOUT,2200) (X(L2+J),J=1,L1)
C     CALCULATE ROWS INVOLVING BOTH K-X AND K-K CORRELATION
40    IF(NVK.EQ.0) GO TO 150
      DO 90 I=1,NVK
      DO 50 J=1,NVX
      L2J=L2+J
      X(L2J)=CORKX(X(N1),X(N2),X(N3),X(N5),X(N6),X(N7),X(N9),X(N11),
     ANPRNT,NS,NVX,NVK,I,J)
      IF(KSQR.EQ.1) X(L2J)=SQRT(X(L2J))
50    CONTINUE
      IF(I.EQ.1) GO TO 80
      L1=I-1
      DO 70 J=1,L1
      L2J=L2+J+NVX
      X(L2J)=CORKK(X(N5),X(N6),X(N7),X(N8),X(N9),X(N10),X(N11),NPRNT,
     ANS,NVK,I,J)
      IF(KSQR.EQ.1) X(L2J)=SQRT(X(L2J))
70    CONTINUE
80    L1=NROW
      NROW=NROW+1
      WRITE(6,2600) NROW
```

```
          WRITE(6,2500) (J,X(L2+J),J=1,L1)
90        WRITE(NTOUT,2200) (X(L2+J),J=1,L1)
          GO TO 150
C   THERE ARE NO INTERVAL VARIABLES
100       DO 140 I=2,NVK
          L1=I-1
          DO 120 J=1,L1
          L2J=L2+J
          X(L2J)=CORKK(X(N5),X(N6),X(N7),X(N8),X(N9),X(N10),X(N11),NPRNT,
          ANS,NVK,I,J)
          IF(KSQR.EQ.1) X(L2J)=SQRT(X(L2J))
120       CONTINUE
          NROW=NROW+1
          WRITE(6,2600) NROW
          WRITE(6,2500) (J,X(L2+J),J=1,L1)
140       WRITE(NTOUT,2200) (X(L2+J),J=1,L1)
150       CONTINUE
          ENDFILE NTOUT
          WRITE(6,2300)
          RETURN
1000      FORMAT(6I5)
2000      FORMAT(5H1NS =,I8,/,6H NVX =,I7,/,6H NVK =,I7,/,8H NTOUT =,I5,/,
          A8H NPRNT =,I5,/,7H KSQR =,I6,/,4H0VAR,10X,7HNO CATS)
2100      FORMAT(19H1REQUIRED STORAGE =,I10,6H WORDS,/,
          A         19H0ALLOTTED STORAGE =,I10,6H WORDS)
2200      FORMAT(8F10.7)
2300      FORMAT(10H0FINISHED.)
2400      FORMAT(1H1,20A4)
2500      FORMAT(1X,8(I5,F10.7))
2600      FORMAT( 9H0VARIABLE, I5)
          END
```

Subroutine INPTR

```
          SUBROUTINE INPTR(X,XMEAN,VAR,K,NS,NVX,NVK)
C   THIS SUBROUTINE CONTROLS INPUT OF RAW DATA AND CALCULATES BOTH THE
C   MEAN AND VARIANCE FOR EACH INTERVAL VARIABLE
C
          DIMENSION X(1),XMEAN(1),VAR(1),K(1)
C         READ IN FULL DATA ARRAY
          NOWX=1
          NOWK=1
          DO 10 J=1,NS
          CALL USER(X(NOWX),K(NOWK))
          NOWX=NOWX+NVX
          NOWK=NOWK+NVK
10        CONTINUE
          IF(NVX.EQ.0) RETURN
C   CALCULATE MEAN AND VARIANCE FOR INTERVAL VARIABLES
          DO 20 I=1,NVX
          XMEAN(I)=0.
20        VAR(I)=0.
          NOWX=1
          DO 30 J=1,NS
          DO 30 I=1,NVX
          XMEAN(I)=XMEAN(I)+X(NOWX)
          VAR(I)=VAR(I)+X(NOWX)**2
30        NOWX=NOWX+1
          DO 40 I=1,NVX
          XMEAN(I)=XMEAN(I)/NS
40        VAR(I)=VAR(I)/NS-XMEAN(I)**2
          RETURN
          END
```

Subroutine NCAT

```
      SUBROUTINE NCAT(KAT,NVK,L1,L2)
C THIS SUBROUTINE READS IN THE NUMBER OF CATEGORIES FOR EACH
C CATEGORICAL VARIABLE AND DETERMINES THE DIMENSIONS OF WORKING
C AREAS NEEDED FOR CALCULATIONS INVOLVING CATEGORICAL VARIABLES
C L1=LARGEST NUMBER OF CATEGORIES DECLARED FOR ANY VARIABLE
C L2=SECOND LARGEST NUMBER OF CATEGORIES DECLARED FOR ANY VARIABLE
      DIMENSION KAT(1)
      IF(NVK.EQ.0) GO TO 70
      READ(5,1000) (KAT(I),I=1,NVK)
      WRITE(6,2000) (I,KAT(I),I=1,NVK)
2000  FORMAT(1X,I3,15X,I2)
      IF(NVK.EQ.1) GO TO 60
C START SEARCH FOR LARGEST AND NEXT LARGEST NUMBER OF CATEGORIES
      IF(KAT(2).GT.KAT(1)) GO TO 10
      L1=KAT(1)
      L2=KAT(2)
      GO TO 20
10    L1=KAT(2)
      L2=KAT(1)
20    IF(NVK.EQ.2) RETURN
      DO 40 I=3,NVK
      IF(KAT(I).LE.L2) GO TO 40
      IF(KAT(I).LE.L1) GO TO 30
      L2=L1
      L1=KAT(I)
      GO TO 40
30    L2=KAT(I)
40    CONTINUE
      RETURN
60    L1=KAT(1)
      L2=0
      RETURN
70    L1=0
      L2=0
      RETURN
1000  FORMAT(40I2)
      END
```

Subroutine EIGEN

```
      SUBROUTINE EIGEN(A,R,N,MV)
C
C
C        SUBROUTINE EIGEN
C
C        PURPOSE
C           COMPUTE EIGENVALUES AND EIGENVECTORS OF A REAL SYMMETRIC
C           MATRIX
C
C        USAGE
C           CALL EIGEN(A,R,N,MV)
C
C        DESCRIPTION OF PARAMETERS
C           A - ORIGINAL MATRIX (SYMMETRIC), DESTROYED IN COMPUTATION.
C               RESULTANT EIGENVALUES ARE DEVELOPED IN DIAGONAL OF
```

```
C                    MATRIX A IN DESCENDING ORDER.
C             R - RESULTANT MATRIX OF EIGENVECTORS (STORED COLUMNWISE,
C                    IN SAME SEQUENCE AS EIGENVALUES)
C             N - ORDER OF MATRICES A AND R
C             MV- INPUT CODE
C                 .NE.1   COMPUTE EIGENVALUES AND EIGENVECTORS
C                 .EQ.1   COMPUTE EIGENVALUES ONLY (R NEED NOT BE
C                         DIMENSIONED BUT MUST STILL APPEAR IN CALLING
C                         SEQUENCE)
C
C        REMARKS
C             ORIGINAL MATRIX A MUST BE REAL SYMMETRIC (STORAGE MODE=1)
C             STORED EITHER AS THE UPPER TRIANGULAR PART BY COLUMNS OR
C             THE LOWER TRIANGULAR PART BY ROWS (INCLUDING MAIN DIAGONAL)
C             MATRIX A CANNOT BE IN THE SAME LOCATION AS MATRIX R
C
C        SUBROUTINES AND FUNCTION SUBPROGRAMS REQUIRED
C             NONE
C
C        METHOD
C             DIAGONALIZATION METHOD ORIGINATED BY JACOBI AND ADAPTED
C             BY VON NEUMANN FOR LARGE COMPUTERS AS FOUND IN ≠MATHEMATICAL
C             METHODS FOR DIGITAL COMPUTERS≠, EDITED BY A. RALSTON AND
C             H.S. WILF, JOHN WILEY AND SONS, NEW YORK, 1962, CHAPTER 7
C
C        ........................................................
C
      DIMENSION A(1),R(1)
C
C        ........................................................
C
C        IF A DOUBLE PRECISION VERSION OF THIS ROUTINE IS DESIRED, THE
C        C IN COLUMN 1 SHOULD BE REMOVED FROM THE DOUBLE PRECISION
C        STATEMENT WHICH FOLLOWS.
C
C     DOUBLE PRECISION A,R,ANORM,ANRMX,THR,X,Y,SINX,SINX2,COSX,
C    1                 COSX2,SINCS,RANGE
C
C        THE C MUST ALSO BE REMOVED FROM DOUBLE PRECISION STATEMENTS
C        APPEARING IN OTHER ROUTINES USED IN CONJUNCTION WITH THIS
C        ROUTINE.
C
C        THE DOUBLE PRECISION VERSION OF THIS SUBROUTINE MUST ALSO
C        CONTAIN DOUBLE PRECISION FORTRAN FUNCTIONS.  SQRT IN STATEMENTS
C        40, 68, 75, AND 78 MUST BE CHANGED TO DSQRT.  ABS IN STATEMENT
C        62 MUST BE CHANGED TO DABS. THE CONSTANT IN STATEMENT 5 SHOULD
C        BE CHANGED TO 1.0D-12.
C
C        ........................................................
C
C        GENERATE IDENTITY MATRIX
C
    5 RANGE=1.0E-12
      IF(MV-1) 10,25,10
   10 IQ=-N
      DO 20 J=1,N
      IQ=IQ+N
      DO 20 I=1,N
      IJ=IQ+I
      R(IJ)=0.0
      IF(I-J) 20,15,20
   15 R(IJ)=1.0
   20 CONTINUE
```

```
C
C          COMPUTE INITIAL AND FINAL NORMS (ANORM AND ANORMX)
C
   25 ANORM=0.0
      DO 35 I=1,N
      DO 35 J=I,N
      IF(I-J) 30,35,30
   30 IA=I+(J*J-J)/2
      ANORM=ANORM+A(IA)*A(IA)
   35 CONTINUE
      IF(ANORM) 165,165,40
   40 ANORM=1.414E0* SQRT(ANORM)
      ANRMX=ANORM*RANGE/ FLOAT(N)
C
C          INITIALIZE INDICATORS AND COMPUTE THRESHOLD, THR
C
      IND=0
      THR=ANORM
   45 THR=THR/ FLOAT(N)
   50 L=1
   55 M=L+1
C
C          COMPUTE SIN AND COS
C
   60 MQ=(M*M-M)/2
      LQ=(L*L-L)/2
      LM=L+MQ
   62 IF( ABS(A(LM))-THR) 130,65,65
   65 IND=1
      LL=L+LQ
      MM=M+MQ
      X=0.5*(A(LL)-A(MM))
   68 Y=-A(LM)/ SQRT(A(LM)*A(LM)+X*X)
      IF(X) 70,75,75
   70 Y=-Y
   75 SINX=Y/ SQRT(2.0*(1.0+( SQRT(1.0-Y*Y))))
      SINX2=SINX*SINX
   78 COSX= SQRT(1.0-SINX2)
      COSX2=COSX*COSX
      SINCS =SINX*COSX
C
C          ROTATE L AND M COLUMNS
C
      ILQ=N*(L-1)
      IMQ=N*(M-1)
      DO 125 I=1,N
      IQ=(I*I-I)/2
      IF(I-L) 80,115,80
   80 IF(I-M) 85,115,90
   85 IM=I+MQ
      GO TO 95
   90 IM=M+IQ
   95 IF(I-L) 100,105,105
  100 IL=I+LQ
      GO TO 110
  105 IL=L+IQ
  110 X=A(IL)*COSX-A(IM)*SINX
      A(IM)=A(IL)*SINX+A(IM)*COSX
      A(IL)=X
  115 IF(MV-1) 120,125,120
  120    ILR=ILQ+I
      IMR=IMQ+I
      X=R(ILR)*COSX-R(IMR)*SINX
      R(IMR)=R(ILR)*SINX+R(IMR)*COSX
```

```
      R(ILR)=X
  125 CONTINUE
      X=2.0*A(LM)*SINCS
      Y=A(LL)*COSX2+A(MM)*SINX2-X
      X=A(LL)*SINX2+A(MM)*COSX2+X
      A(LM)=(A(LL)-A(MM))*SINCS+A(LM)*(COSX2-SINX2)
      A(LL)=Y
      A(MM)=X
C
C         TESTS FOR COMPLETION
C
C         TEST FOR M = LAST COLUMN
C
  130 IF(M-N) 135,140,135
  135 M=M+1
      GO TO 60
C
C         TEST FOR L = SECOND FROM LAST COLUMN
C
  140 IF(L-(N-1)) 145,150,145
  145 L=L+1
      GO TO 55
  150 IF(IND-1) 160,155,160
  155 IND=0
      GO TO 50
C
C         COMPARE THRESHOLD WITH FINAL NORM
C
  160 IF(THR-ANRMX) 165,165,45
C
C         SORT EIGENVALUES AND EIGENVECTORS
C
  165 IQ=-N
      DO 185 I=1,N
      IQ=IQ+N
      LL=I+(I*I-I)/2
      JQ=N*(I-2)
      DO 185 J=I,N
      JQ=JQ+N
      MM=J+(J*J-J)/2
      IF(A(LL)-A(MM)) 170,185,185
  170 X=A(LL)
      A(LL)=A(MM)
      A(MM)=X
      IF(MV-1) 175,185,175
  175 DO 180 K=1,N
      ILR=IQ+K
      IMR=JQ+K
      X=R(ILR)
      R(ILR)=R(IMR)
  180 R(IMR)=X
  185 CONTINUE
      RETURN
      END
```

Subroutine VSORT

```
      SUBROUTINE VSORT(S,ITAG,ITEMS)
      DIMENSION S(1),ITAG(1)
C ADAPTED FROM
C   FUNDERLIC, R.E. (ED), THE PROGRAMMERS HANDBOOK, AEC R+D REPORT
C   K-1729, UNION CARBIDE CORPORATION, OAK RIDGE, TENNESSEE, FEB 1968,
```

```
C     PP 286-288.
C
C     BASED ON THE PROCEDURE DESCRIBED IN
C        SHELL, D.L., A HIGH SPEED SORTING PROCEDURE, COMMUNICATIONS ACM,
C        VOLUME 2, JULY 1959, PP 30-32.
C     THIS SUBROUTINE SORTS INTO ASCENDING SEQUENCE THE ELEMENTS STORED IN
C     *S*.   INSTEAD OF ACTUALLY MOVING THESE ELEMENTS,   THEY ARE LEFT IN
C     PLACE AND THEIR NEW SEQUENCE IS STORED IN *ITAG*.   THAT IS, IF
C     L=ITAG(I) THEN S(L) IS THE I-TH ELEMENT IN THE SORTED SEQUENCE.
C     *ITEMS* IS THE NUMBER OF ELEMENTS IN *S*.
         DO 10 I=1,ITEMS
10       ITAG(I)=I
         M=ITEMS
20       M=M/2
         IF(M.EQ.0) GO TO 60
         K=ITEMS-M
         J=1
30       I=J
40       IM=I+M
         L=ITAG(I)
         LM=ITAG(IM)
C     FOR DECREASING SEQUENCE CHANGE FROM .LE. TO .GE.
         IF(S(L).LE.S(LM)) GO TO 50
         ITAG(I)=LM
         ITAG(IM)=L
         I=I-M
         IF(I.GE.1) GO TO 40
50       J=J+1
         IF(J.GT.K) GO TO 20
         GO TO 30
C     THE ORDINARY VSORT ROUTINE WOULD HAVE A *RETURN* FOR STATEMENT 60.
C     FOR USE WITH *CORKX* AND *CORKK* REPLACE *S* VALUES BY THEIR RANKS.
C     IGNORE *S* VALUES WHICH ARE ZERO.
60       J=0
         L=ITAG(1)
         L1=L
         IF(S(L).EQ.0.) GO TO 70
         J=J+1
         ITAG(ITEMS+L)=J
         GO TO 80
70       ITAG(ITEMS+L)=0
80       DO 110 I=2,ITEMS
         L=ITAG(I)
         IF(S(L).EQ.0.) GO TO 100
         IF(ABS((S(L)-S(L1))/S(L)).LT.0.0001) GO TO 90
         J=J+1
90       ITAG(ITEMS+L)=J
         L1=L
         GO TO 110
100      ITAG(ITEMS+L)=0
110      CONTINUE
         DO 120 I=1,ITEMS
120      S(I)=ITAG(ITEMS+I)
         RETURN
         END
```

Function CORXX

```
      FUNCTION CORXX(X,XMEAN,VAR,NS,NVX,IDX1,IDX2)
C     THIS SUBROUTINE CALCULATES THE SQUARED CORRELATION BETWEEN TWO
C     INTERVAL VARIABLES WITH INDICES IDX1 AND IDX2
      DIMENSION X(1),XMEAN(1),VAR(1)
```

```
C  CALCULATE SUM OF CROSS PRODUCTS
      NOW1=IDX1
      NOW2=IDX2
      XY=0.
      DO 10 J=1,NS
      XY=XY+X(NOW1)*X(NOW2)
      NOW1=NOW1+NVX
10    NOW2=NOW2+NVX
      CORXX=((XY/NS-XMEAN(IDX1)*XMEAN(IDX2))**2)/(VAR(IDX1)*VAR(IDX2))
      RETURN
      END
```

Function CORKX

VERSION 1

```
      FUNCTION CORKX(X,XMEAN,VAR,K,KAT,XKAT,NKAT,SCORE,NPRNT,NS,NVX,NVK,
     AIDK,IDX)
C  THIS SUBROUTINE CALCULATES THE SQUARED CORRELATION BETWEEN A
C  CATEGORICAL VARIABLE WITH INDEX IDK AND AN INTERVAL VARIABLE WITH
C  INDEX IDX.
      DIMENSION X(1),XMEAN(1),VAR(1),K(1),KAT(1),XKAT(1),NKAT(1)
      DIMENSION SCORE(1)
C  KAT(IDK)=NUMBER OF CATEGORIES FOR VARIABLE IDK
C  XKAT(I)=SUM OF X VALUES FOR SUBJECT IN CATEGORY I
C  NKAT(I)=NUMBER OF SUBJECTS IN CATEGORY I
C
C  BUILD UP XKAT AND NKAT ARRAYS
      L1=KAT(IDK)
      DO 10 I=1,L1
      XKAT(I)=0.
10    NKAT(I)=0
      NOWX=IDX
      NOWK=IDK
      DO 20 J=1,NS
      L2=K(NOWK)
      XKAT(L2)=XKAT(L2)+X(NOWX)
      NKAT(L2)=NKAT(L2)+1
      NOWX=NOWX+NVX
20    NOWK=NOWK+NVK
C  CALCULATE WITHIN GROUP SUM OF SQUARES
      SS=0.
      DO 30 I=1,L1
      IF(NKAT(I).EQ.0) GO TO 30
      SS=SS+(XKAT(I)**2)/NKAT(I)
30    CONTINUE
      SS=SS/NS-XMEAN(IDX)**2
      CORKX=SS/VAR(IDX)
      IF(NPRNT.EQ.1) RETURN
C  CALCULATE OPTIMAL SCORES
      DO 50 I=1,L1
      IF(NKAT(I).EQ.0) GO TO 40
      SCORE(I)=XKAT(I)/NKAT(I)
      GO TO 50
40    SCORE(I)=0.
50    CONTINUE
      WRITE(6,2000) IDK,IDX,(SCORE(I),I=1,L1)
      RETURN
2000  FORMAT(32H SCORES FOR CATEGORICAL VARIABLE,I3,39H WHEN CORRELATED
     AWITH INTERVAL VARIABLE,I3,/,(1X,10E12.4))
      END
```

```
      FUNCTION CORKX(X,XMEAN,VAR,K,KAT,XKAT,NKAT,SCORE,NPRNT,NS,NVX,NVK,
     AIDK,IDX)
C THIS SUBROUTINE CALCULATES THE SQUARED CORRELATION BETWEEN A
C CATEGORICAL VARIABLE WITH INDEX IDK AND AN INTERVAL VARIABLE WITH
C INDEX IDX, SCORES FOR THE NOMINAL VARIABLE ARE DETERMINED BY
C USING THE RANKS OF GROUP MEANS.
      DIMENSION X(1),XMEAN(1),VAR(1),K(1),KAT(1),XKAT(1),NKAT(1)
      DIMENSION SCORE(1)
C KAT(IDK)=NUMBER OF CATEGORIES FOR VARIABLE IDK
C XKAT(I)=SUM OF X VALUES FOR SUBJECT IN CATEGORY I
C NKAT(I)=NUMBER OF SUBJECTS IN CATEGORY I
C
C BUILD UP XKAT AND NKAT ARRAYS
      L1=KAT(IDK)
      DO 10 I=1,L1
      XKAT(I)=0.
10    NKAT(I)=0
      NOWX=IDX
      NOWK=IDK
      DO 20 J=1,NS
      L2=K(NOWK)
      XKAT(L2)=XKAT(L2)+X(NOWX)
      NKAT(L2)=NKAT(L2)+1
      NOWX=NOWX+NVX
20    NOWK=NOWK+NVK
C CALCULATE OPTIMAL SCORES
      DO 50 I=1,L1
      IF(NKAT(I).EQ.0) GO TO 40
      SCORE(I)=XKAT(I)/NKAT(I)
      GO TO 50
40    SCORE(I)=0.
50    CONTINUE
      CALL VSORT(SCORE(1),SCORE(L1+1),L1)
      CP=0.
      SS=0.
      SUM=0.
      DO 60 I=1,L1
      CP=CP+SCORE(I)*XKAT(I)
      SS=SS+NKAT(I)*(SCORE(I)**2)
      SUM=SUM+NKAT(I)*SCORE(I)
60    CONTINUE
      CP=CP-XMEAN(IDX)*SUM
      SS=SS-SUM*SUM/NS
      CORKX=CP*CP/(SS*NS*VAR(IDX))
      IF(NPRNT.EQ.1) RETURN
      WRITE(6,2000) IDK,IDX,(SCORE(I),I=1,L1)
      RETURN
2000  FORMAT(32H SCORES FOR CATEGORICAL VARIABLE,I3,39H WHEN CORRELATED
     AWITH INTERVAL VARIABLE,I3,/,(1X,10E12.4))
      END
```

Function CORKK

```
      FUNCTION CORKK(K,KAT,CTABL,THAT,QMT,PMT,SCORE,NPRNT,NS,NVK,IDK1,
     AIDK2)
      DIMENSION K(1),KAT(1),CTABL(1),THAT(1),QMT(1),PMT(1),SCORE(1)
      INTEGER FIRST
      COMMON/AB/ KSCOR,TITLE(20)
```

```
      DATA KSCOR,(TITLE(I),I=1,20)/2,4HCANO,4HNICA,4HL CO,4HRREL,4HATIO,
     A4HNS U,4HSED ,4HFOR ,4HNOMI,4HNAL-,4HNOMI,4HNAL ,4HASSO,4HCIAT,
     B4HIONS,5*4H    /
C
C     THIS SUBPROGRAM CALCULATES THE MAXIMUM POSSIBLE SQUARED CORRELATION
C     BETWEEN CATEGORICAL VARIABLES IDK1 AND IDK2.  THE METHOD IS ADOPTED
C     FROM
C     1.  WILLIAMS, E.J., USE OF SCORES FOR THE ANALYSIS OF ASSOCIATION
C             IN CONTINGENCY TABLES, BIOMETRIKA, VOLUME 39, NUMBER 3/4,
C             DECEMBER 1952, PP 274-289.
C
C     2.  KENDALL, M.G. AND A. STUART, THE ADVANCED THEORY OF STATISTICS,
C             VOLUME 2, INFERENCE AND RELATIONSHIP, HAFNER PUBLISHING
C             COMPANY, NEW YORK, 1961, SECTIONS 33.44 TO 33.49.
C
C     VARIABLES UNIQUE TO THIS PROGRAM
C     KAT(I)=NUMBER OF CATEGORIES FOR I-TH CATEGORICAL VARIABLE
C     CTABL=CONTINGENCY TABLE STORED BY ROWS
C     THAT=MATRIX WHOSE LARGEST ROOT IS THE SQUARED CORRELATION
C     QMT(J)=MARGINAL TOTAL FOR J-TH COLUMN IN CONTINGENCY TABLE
C     PMT(I)=MARGINAL TOTAL FOR I-TH ROW IN CONTINGENCY TABLE.
C     SCORE(1) TO SCORE(KP)=SCORES FOR ROW CATEGORIES
C     SCORE(KP+1) TO SCORE(KP+KQ)=SCORES FOR COLUMN CATEGORIES
C         SCORES FOR NULL CATEGORIES AND ISOLATED CELLS ARE SET TO ZERO.
C     IDP=VARIABLE WITH SMALLER NUMBER OF CATEGORIES (ROW VARIABLE)
C     IDQ=COLUMN VARIABLE
C     TOT=TOTAL NUMBER OF OBSERVATIONS IN THE CONTINGENCY TABLE LESS THE
C         ISOLATED CELLS (IF ANY)
C     TISOL=TOTAL NUMBER OF OBSERVATIONS IN ISOLATED CELLS
C
C     DETERMINE ROW AND COLUMN VARIABLES
      IF(KAT(IDK1).GT.KAT(IDK2)) GO TO 10
      IDQ=IDK2
      IDP=IDK1
      GO TO 20
10    IDQ=IDK1
      IDP=IDK2
20    KQ=KAT(IDQ)
      KP=KAT(IDP)
C     CONCEPTUALLY THE CONTINGENCY TABLE IS KP ROWS BY KQ COLUMNS, KP.GE.KQ
C     INITIALIZE CONTINGENCY TABLE AND MARGINAL TOTALS
      IJ=0
      DO 30 I=1,KP
      PMT(I)=0.
      DO 30 J=1,KQ
      IJ=IJ+1
30    CTABL(IJ)=0.
      DO 40 J=1,KQ
40    QMT(J)=0.
C     BUILD CONTINGENCY TABLE
      NOWP=IDP
      NOWQ=IDQ
      DO 50 L=1,NS
      IJ=(K(NOWP)-1)*KQ+K(NOWQ)
      CTABL(IJ)=CTABL(IJ)+1
      NOWP=NOWP+NVK
50    NOWQ=NOWQ+NVK
C     THE FOLLOWING ENTRY POINT PERMITS CALCULATION OF THE CORRELATION
C     STARTING FROM A CONTINGENCY TABLE
C     FOR THE CDC 6600 ONLY THE ENTRY NAME IS REQUIRED
C     FOR THE IBM 360 THE ENTRY NAME AND COMPLETE ARGUMENT LIST IS REQUIRED
          ENTRY ACORKK
C     COMPUTE MARGINAL TOTALS
      IJ=0
```

```
      DO 60 I=1,KP
      DO 60 J=1,KQ
      IJ=IJ+1
      PMT(I)=PMT(I)+CTABL(IJ)
60    QMT(J)=QMT(J)+CTABL(IJ)
C PRINT CONTINGENCY TABLE AND MARGINALS IF DESIRED
      IF(NPRNT.NE.3) GO TO 90
      WRITE(6,2000) IDP,IDQ
      FIRST=1
      LAST=KQ
      DO 80 I=1,KP
      WRITE(6,2100) (CTABL(IJ),IJ=FIRST,LAST),PMT(I)
      FIRST=LAST+1
      LAST=LAST+KQ
80    CONTINUE
      WRITE(6,2100) (QMT(J),J=1,KQ)
C SEARCH FOR ISOLATED CELLS AND ZERO OUT ANY THAT ARE FOUND
90    TISOL=0.
      DO 110 I=1,KP
      II=(I-1)*KQ
      DO 100 J=1,KQ
      IJ=II+J
      IF(CTABL(IJ).EQ.0.) GO TO 100
      IF(CTABL(IJ).NE.PMT(I)) GO TO 110
      IF(CTABL(IJ).NE.QMT(J)) GO TO 110
      WRITE(6,2200) IDP,IDQ,I,J,CTABL(IJ)
      TISOL=TISOL+CTABL(IJ)
      CTABL(IJ)=0.
      PMT(I)=0.
      QMT(J)=0.
100   CONTINUE
110   CONTINUE
      TOT=NS-TISOL
C REPLACE MARGINAL TOTALS BY THEIR SQUARE ROOTS
      DO 120 I=1,KP
      IF(PMT(I).EQ.0.) GO TO 120
      PMT(I)=SQRT(PMT(I))
120   CONTINUE
      DO 130 J=1,KQ
      IF(QMT(J).EQ.0.) GO TO 130
      QMT(J)=SQRT(QMT(J))
130   CONTINUE
C MODIFIY CONTINGENCY TABLE
      IJ=0
      DO 140 I=1,KP
      DO 140 J=1,KQ
      IJ=IJ+1
      IF(CTABL(IJ).EQ.0.) GO TO 140
      CTABL(IJ)=CTABL(IJ)/(PMT(I)*QMT(J))
140   CONTINUE
C BUILD LOWER TRIANGULAR PART OF THE THAT MATRIX
      IJ=0
      DO 160 I=1,KP
      DO 160 J=1,I
      IJ=IJ+1
      II=(I-1)*KQ
      JJ=(J-1)*KQ
      THAT(IJ)=0.
      DO 150 L=1,KQ
      II=II+1
      JJ=JJ+1
150   THAT(IJ)=THAT(IJ)+CTABL(II)*CTABL(JJ)
```

```
C   DEFLATE THE THAT MATRIX BY REMOVING THE LARGEST EIGENVECTOR
        THAT(IJ)=THAT(IJ)-PMT(I)*PMT(J)/TOT
160     CONTINUE
C   FIND SQUARED CORRELATION AND SCORES
        CALL EIGEN (THAT,SCORE,KP,NPRNT)
        CORKK=THAT(1)
        IF(TISOL.EQ.0.) GO TO 170
C   ADJUST SQUARED CORRELATION FOR ISOLATED CELLS
        CORKK=(THAT(1)*TOT+TISOL)/NS
        WRITE(6,2300) IDP,IDQ,THAT(1),TISOL,CORKK
170     IF(NPRNT.EQ.1) RETURN
C   FIND LINEAR TRANSFORM FOR EIGENVALUES
        C=0.
        D=0.
        DO 180 I=1,KP
        D=D+PMT(I)
180     C=C+PMT(I)*SCORE(I)
        C=C/D
        D=0.
        DO 190 I=1,KP
        IF(PMT(I).EQ.0.) GO TO 190
        D=D+(SCORE(I)-C)**2
190     CONTINUE
        D=SQRT(D)
C   CALCULATE ROW SCORES
        SQRTOT=SQRT(TOT)
        DO 210  I=1,KP
        IF(PMT(I).EQ.0.) GO TO 200
        SCORE(I)=((SCORE(I)-C)/D)*SQRTOT/PMT(I)
        GO TO 210
200     SCORE(I)=0.
210     CONTINUE
        WRITE(6,2400) IDP,IDQ
        WRITE(6,2500) (SCORE(I),I=1,KP)
C   CALCULATE COLUMN SCORES AS CLASS MEANS
        DO 230 J=1,KQ
        KPJ=KP+J
        SCORE(KPJ)=0.
        IF(QMT(J).EQ.0.) GO TO 230
        IJ=J
        DO 220 I=1,KP
        SCORE(KPJ)=SCORE(KPJ)+CTABL(IJ)*PMT(I)*QMT(J)*SCORE(I)
        IJ=IJ+KQ
220     CONTINUE
        SCORE(KPJ)=SCORE(KPJ)/(QMT(J)**2)
230     CONTINUE
        WRITE(6,2400) IDQ,IDP
        WRITE(6,2500)(SCORE(KP+J),J=1,KQ)
        RETURN
2000    FORMAT(36H CONTINGENCY TABLE.  ROW VARIABLE IS,I3,21H.  COLUMN VAR
       AIABLE IS,I3,1H.)
2100    FORMAT(1X,25F5.0)
2200    FORMAT(26H ISOLATED CELL.  VARIABLES,I3,4H AND,I3,6H.  ROW,I3,8H,
       ACOLUMN,I3,5X,F5.0,14H OBSERVATIONS.)
2300    FORMAT(10H VARIABLES,I3,4H AND,I3,39H.  REDUCED TABLE SQUARED CORR
       AELATION IS,F12.7,//,33H SQUARED CORRELATION ADJUSTED FOR,F4.0,25H I
       BSOLATED OBSERVATIONS IS,F12.7)
2400    FORMAT(20H SCORES FOR VARIABLE,I3,30H WHEN CORRELATED WITH VARIABL
       AE,I3)
2500    FORMAT(1X,10E12.4)
        END
```

VERSION 2

```
      FUNCTION CORKK(K,KAT,CTABL,THAT,QMT,PMT,SCORE,NPRNT,NS,NVK,IDK1,
     AIDK2)
      DIMENSION K(1),KAT(1),CTABL(1),THAT(1),QMT(1),PMT(1),SCORE(1)
      INTEGER FIRST
      COMMON/AB/KSCOR,TITLE(20)
      DATA KSCOR,(TITLE(I),I=1,20)/1,4HCORR,4HELAT,4HIONS,4H CAL,4HCULA,
     A4HTED ,4HFROM,4H RAN,4HKS 0,4HF OP,4HTIMA,4HL  SC,4HORES,7*4H      /
C
C     THIS SUBPROGRAM CALCULATES THE SQUARED CORRELATION BETWEEN TWO
C     NOMINAL VARIABLES BY TAKING THE OPTIMAL SCORES DEFINED IN THE
C     FOLLOWING REFERENCES, REPLACING THESE SCORES BY RANKS AND THEN
C     USING THE RANKS IN AN ORDINARY PRODUCT MOMENT CORRELATION CALCULATION.
C     1.  WILLIAMS, E.J., USE OF SCORES FOR THE ANALYSIS OF ASSOCIATION
C         IN CONTINGENCY TABLES, BIOMETRIKA, VOLUME 39, NUMBER 3/4,
C         DECEMBER 1952, PP 274-289.
C
C     2.  KENDALL, M.G. AND A. STUART, THE ADVANCED THEORY OF STATISTICS,
C         VOLUME 2, INFERENCE AND RELATIONSHIP, HAFNER PUBLISHING
C         COMPANY, NEW YORK, 1961, SECTIONS 33.44 TO 33.49.
C
C     VARIABLES UNIQUE TO THIS PROGRAM
C     KAT(I)=NUMBER OF CATEGORIES FOR I-TH CATEGORICAL VARIABLE
C     CTABL=CONTINGENCY TABLE STORED BY ROWS
C     THAT=MATRIX WHOSE LARGEST ROOT IS THE SQUARED CORRELATION
C     QMT(J)=MARGINAL TOTAL FOR J-TH COLUMN IN CONTINGENCY TABLE
C     PMT(I)=MARGINAL TOTAL FOR I-TH ROW IN CONTINGENCY TABLE.
C     SCORE(1) TO SCORE(KP)=SCORES FOR ROW CATEGORIES
C     SCORE(KP+1) TO SCORE(KP+KQ)=SCORES FOR COLUMN CATEGORIES
C        SCORES FOR NULL CATEGORIES AND ISOLATED CELLS ARE SET TO ZERO.
C     IDP=VARIABLE WITH SMALLER NUMBER OF CATEGORIES (ROW VARIABLE)
C     IDQ=COLUMN VARIABLE
C     TOT=TOTAL NUMBER OF OBSERVATIONS IN THE CONTINGENCY TABLE LESS THE
C         ISOLATED CELLS (IF ANY)
C     TISOL=TOTAL NUMBER OF OBSERVATIONS IN ISOLATED CELLS
C
C     DETERMINE ROW AND COLUMN VARIABLES
      IF(KAT(IDK1).GT.KAT(IDK2)) GO TO 10
      IDQ=IDK2
      IDP=IDK1
      GO TO 20
10    IDQ=IDK1
      IDP=IDK2
20    KQ=KAT(IDQ)
      KP=KAT(IDP)
C  CONCEPTUALLY THE CONTINGENCY TABLE IS KP ROWS BY KQ COLUMNS, KP.GE.KQ
C  INITIALIZE CONTINGENCY TABLE AND MARGINAL TOTALS
      IJ=0
      DO 30 I=1,KP
      PMT(I)=0.
      DO 30 J=1,KQ
      IJ=IJ+1
30    CTABL(IJ)=0.
      DO 40 J=1,KQ
40    QMT(J)=0.
C  BUILD CONTINGENCY TABLE
      NOWP=IDP
      NOWQ=IDQ
      DO 50 L=1,NS
      IJ=(K(NOWP)-1)*KQ+K(NOWQ)
      CTABL(IJ)=CTABL(IJ)+1
      NOWP=NOWP+NVK
```

```
50     NOWQ=NOWQ+NVK
C   THE FOLLOWING ENTRY POINT PERMITS CALCULATION OF THE CORRELATION
C   STARTING FROM A CONTINGENCY TABLE
C   FOR THE CDC 6600 ONLY THE ENTRY NAME IS REQUIRED
C   FOR THE IBM 360 THE ENTRY NAME AND COMPLETE ARGUMENT LIST IS REQUIRED
       ENTRY ACORKK
C   COMPUTE MARGINAL TOTALS
       IJ=0
       DO 60 I=1,KP
       DO 60 J=1,KQ
       IJ=IJ+1
       PMT(I)=PMT(I)+CTABL(IJ)
60     QMT(J)=QMT(J)+CTABL(IJ)
C   PRINT CONTINGENCY TABLE AND MARGINALS IF DESIRED
       IF(NPRNT.NE.3) GO TO 90
       WRITE(6,2000) IDP,IDQ
       FIRST=1
       LAST=KQ
       DO 80 I=1,KP
       WRITE(6,2100) (CTABL(IJ),IJ=FIRST,LAST),PMT(I)
       FIRST=LAST+1
       LAST=LAST+KQ
80     CONTINUE
       WRITE(6,2100) (QMT(J),J=1,KQ)
C   SEARCH FOR ISOLATED CELLS AND ZERO OUT ANY THAT ARE FOUND
90     TISOL=0.
       DO 110 I=1,KP
       II=(I-1)*KQ
       DO 100 J=1,KQ
       IJ=II+J
       IF(CTABL(IJ).EQ.0.) GO TO 100
       IF(CTABL(IJ).NE.PMT(I)) GO TO 110
       IF(CTABL(IJ).NE.QMT(J)) GO TO 110
       WRITE(6,2200) IDP,IDQ,I,J,CTABL(IJ)
       TISOL=TISOL+CTABL(IJ)
       CTABL(IJ)=0.
       PMT(I)=0.
       QMT(J)=0.
100    CONTINUE
110    CONTINUE
       TOT=NS-TISOL
C   REPLACE MARGINAL TOTALS BY THEIR SQUARE ROOTS
       DO 120 I=1,KP
       IF(PMT(I).EQ.0.) GO TO 120
       PMT(I)=SQRT(PMT(I))
120    CONTINUE
       DO 130 J=1,KQ
       IF(QMT(J).EQ.0.) GO TO 130
       QMT(J)=SQRT(QMT(J))
130    CONTINUE
C   MODIFIY CONTINGENCY TABLE
       IJ=0
       DO 140 I=1,KP
       DO 140 J=1,KQ
       IJ=IJ+1
       IF(CTABL(IJ).EQ.0.) GO TO 140
       CTABL(IJ)=CTABL(IJ)/(PMT(I)*QMT(J))
140    CONTINUE
C   BUILD LOWER TRIANGULAR PART OF THE THAT MATRIX
       IJ=0
       DO 160 I=1,KP
       DO 160 J=1,I
       IJ=IJ+1
       II=(I-1)*KQ
```

```
         JJ=(J-1)*KQ
         THAT(IJ)=0.
         DO 150 L=1,KQ
         II=II+1
         JJ=JJ+1
150      THAT(IJ)=THAT(IJ)+CTABL(II)*CTABL(JJ)
C  DEFLATE THE THAT MATRIX BY REMOVING THE LARGEST EIGENVECTOR
         THAT(IJ)=THAT(IJ)-PMT(I)*PMT(J)/TOT
160      CONTINUE
C  FIND EIGENVALUES
         CALL EIGEN(THAT,SCORE,KP,0)
C  FIND LINEAR TRANSFORM FOR EIGENVALUES
         C=0.
         D=0.
         DO 180 I=1,KP
         D=D+PMT(I)
180      C=C+PMT(I)*SCORE(I)
         C=C/D
         D=0.
         DO 190 I=1,KP
         IF(PMT(I).EQ.0.) GO TO 190
         D=D+(SCORE(I)-C)**2
190      CONTINUE
         D=SQRT(D)
C  CALCULATE ROW SCORES
         SQRTOT=SQRT(TOT)
         DO 210  I=1,KP
         IF(PMT(I).EQ.0.) GO TO 200
         SCORE(I)=((SCORE(I)-C)/D)*SQRTOT/PMT(I)
         GO TO 210
200      SCORE(I)=0.
210      CONTINUE
         CALL VSORT(SCORE(1),THAT,KP)
C  CALCULATE COLUMN SCORES AS CLASS MEANS
         DO 230 J=1,KQ
         KPJ=KP+J
         SCORE(KPJ)=0.
         IF(QMT(J).EQ.0.) GO TO 230
         IJ=J
         DO 220 I=1,KP
         SCORE(KPJ)=SCORE(KPJ)+CTABL(IJ)*PMT(I)*QMT(J)*SCORE(I)
         IJ=IJ+KQ
220      CONTINUE
         SCORE(KPJ)=SCORE(KPJ)/(QMT(J)**2)
230      CONTINUE
         CALL VSORT(SCORE(KP+1),THAT,KQ)
         IJ=0
         SSA=0.
         SSB=0.
         CP=0.
         SUMA=0.
         SUMB=0.
         DO 240 I=1,KP
         SSA=SSA+(SCORE(I)*PMT(I))**2
         SUMA=SUMA+SCORE(I)*(PMT(I)**2)
         DO 240 J=1,KQ
         IJ=IJ+1
         CP=CP+CTABL(IJ)*PMT(I)*QMT(J)*SCORE(I)*SCORE(KP+J)
240      CONTINUE
         DO 250 J=1,KQ
         SSB=SSB+(SCORE(KP+J)*QMT(J))**2
         SUMB=SUMB+SCORE(KP+J)*(QMT(J)**2)
250      CONTINUE
         CP=CP-SUMA*SUMB/TOT
```

```
        SSA=SSA-SUMA*SUMA/TOT
        SSB=SSB-SUMB*SUMB/TOT
        THAT(1)=CP*CP/(SSA*SSB)
        CORKK=THAT(1)
        IF(TISOL.EQ.0.) GO TO 260
C   ADJUST SQUARED CORRELATION FOR ISOLATED CELLS
        CORKK=(THAT(1)*TOT+TISOL)/NS
        WRITE(6,2300) IDP,IDQ,THAT(1),TISOL,CORKK
260     IF(NPRNT.EQ.1) RETURN
        WRITE(6,2400) IDP,IDQ
        WRITE(6,2100) (SCORE(I),I=1,KP)
        WRITE(6,2400) IDQ,IDP
        WRITE(6,2100)(SCORE(KP+J),J=1,KQ)
        RETURN
2000    FORMAT(36H CONTINGENCY TABLE.  ROW VARIABLE IS,I3,21H.  COLUMN VAR
       AIABLE IS,I3,1H.)
2100    FORMAT(1X,25F5.0)
2200    FORMAT(26H ISOLATED CELL.  VARIABLES,I3,4H AND,I3,6H.  ROW,I3,8H,
       ACOLUMN,I3,5X,F5.0,14H OBSERVATIONS.)
2300    FORMAT(10H VARIABLES,I3,4H AND,I3,39H.  REDUCED TABLE SQUARED CORR
       AELATION IS,F12.7,/,33H SQUARED CORRELATION ADJUSTED FOR,F4.0,25H I
       BSOLATED OBSERVATIONS IS,F12.7)
2400    FORMAT(20H SCORES FOR VARIABLE,I3,30H WHEN CORRELATED WITH VARIABL
       AE,I3)
        END
```

Version 3

```
        FUNCTION CORKK(K,KAT,CTABL,ROWMAX,QMT,PMT,SCORE,NPRNT,NS,NVK,IDK1,
       AIDK2)
        DIMENSION K(1),KAT(1),CTABL(1),ROWMAX(1),QMT(1),PMT(1),SCORE(1)
        INTEGER FIRST
        COMMON/AB/KSCOR,TITLE(20)
        DATA KSCOR,(TITLE(I),I=1,20)/3,4HGOOD,4HMAN ,4HAND ,4HKRUS,4HKAL ,
       A4HLAMB,4HDA U,4HSED ,4HFOR ,4HASSO,4HCIAT,4HION ,4HBETW,4HEEN ,
       B4HNOMI,4HNAL ,4HVARI,4HABLE,4HS   ,4H   /
C
C   THIS SUBROUTINE FINDS THE LAMBDA ASSOCIATION MEASURE DEFINED IN
C      GOODMAN, L.A. AND W.H. KRUSKAL, MEASURES OF ASSOCIATION FOR CROSS
C      CLASSIFICATIONS, JOURNAL AMERICAN STATISTICAL ASSOCIATION,  VOLUME
C      49, NUMBER 268, DECEMBER 1954, PP 732-764.  SEE P 743.
C
C   VARIABLES UNIQUE TO THIS PROGRAM
C   KAT(I)=NUMBER OF CATEGORIES FOR I-TH CATEGORICAL VARIABLE
C   CTABL=CONTINGENCY TABLE STORED BY ROWS
C   ROWMAX(I)=MAXIMUM ENTRY IN ROW I OF THE CONTINGENCY TABLE
C   QMT(J)=MARGINAL TOTAL FOR J-TH COLUMN IN CONTINGENCY TABLE
C   PMT(I)=MARGINAL TOTAL FOR I-TH ROW IN CONTINGENCY TABLE.
C   IDP=VARIABLE WITH SMALLER NUMBER OF CATEGORIES (ROW VARIABLE)
C   IDQ=COLUMN VARIABLE
C
C   DETERMINE ROW AND COLUMN VARIABLES
        IF(KAT(IDK1).GT.KAT(IDK2)) GO TO 10
        IDQ=IDK2
        IDP=IDK1
        GO TO 20
10      IDQ=IDK1
        IDP=IDK2
20      KQ=KAT(IDQ)
        KP=KAT(IDP)
C   CONCEPTUALLY THE CONTINGENCY TABLE IS KP ROWS BY KQ COLUMNS, KP.GE.KQ
C   INITIALIZE CONTINGENCY TABLE AND MARGINAL TOTALS
        IJ=0
        DO 30 I=1,KP
```

```
          ROWMAX(I)=0.
          PMT(I)=0.
          DO 30 J=1,KQ
          IJ=IJ+1
30        CTABL(IJ)=0.
          DO 40 J=1,KQ
40        QMT(J)=0.
C     BUILD CONTINGENCY TABLE
          NOWP=IDP
          NOWQ=IDQ
          DO 50 L=1,NS
          IJ=(K(NOWP)-1)*KQ+K(NOWQ)
          CTABL(IJ)=CTABL(IJ)+1
          NOWP=NOWP+NVK
50        NOWQ=NOWQ+NVK
C     THE FOLLOWING ENTRY POINT PERMITS CALCULATION OF THE ASSOCIATION
C     MEASURE STARTING FROM A CONTINGENCY TABLE
C     STARTING FROM A CONTINGENCY TABLE
C     FOR THE CDC 6600 ONLY THE ENTRY NAME IS REQUIRED
C     FOR THE IBM 360 THE ENTRY NAME AND COMPLETE ARGUMENT LIST IS REQUIRED
          ENTRY ACORKK
C     COMPUTE MARGINAL TOTALS
          IJ=0
          DO 60 I=1,KP
          DO 60 J=1,KQ
          IJ=IJ+1
          PMT(I)=PMT(I)+CTABL(IJ)
60        QMT(J)=QMT(J)+CTABL(IJ)
C     PRINT CONTINGENCY TABLE AND MARGINALS IF DESIRED
          IF(NPRNT.NE.3) GO TO 90
          WRITE(6,2000) IDP,IDQ
          FIRST=1
          LAST=KQ
          DO 80 I=1,KP
          WRITE(6,2100) (CTABL(IJ),IJ=FIRST,LAST),PMT(I)
          FIRST=LAST+1
          LAST=LAST+KQ
80        CONTINUE
          WRITE(6,2100) (QMT(J),J=1,KQ)
C     FIND SUM OF ROW MAXIMA AND COLUMN MAXIMA
90        COLSUM=0.
          QMTMAX=0.
          DO 110 J=1,KQ
          COLMAX=0.
          IJ=J-KQ
          DO 100 I=1,KP
          IJ=IJ+KQ
          IF(CTABL(IJ).GT.COLMAX) COLMAX=CTABL(IJ)
          IF(CTABL(IJ).GT.ROWMAX(I)) ROWMAX(I)=CTABL(IJ)
100       CONTINUE
          COLSUM=COLSUM+COLMAX
          IF(QMT(J).GT.QMTMAX) QMTMAX=QMT(J)
110       CONTINUE
          ROWSUM=0.
          PMTMAX=0.
          DO 120 I=1,KP
          ROWSUM=ROWSUM+ROWMAX(I)
          IF(PMT(I).GT.PMTMAX) PMTMAX=PMT(I)
120       CONTINUE
          CORKK=(COLSUM+ROWSUM-PMTMAX-QMTMAX)/(2*NS-PMTMAX-QMTMAX)
          RETURN
2000      FORMAT(36H CONTINGENCY TABLE.   ROW VARIABLE IS,I3,21H   COLUMN VA
         AIABLE IS,I3,1H.)
2100      FORMAT(1X,25F5.0)
          END
```

Version 4

```
       FUNCTION CORKK(K,KAT,CTABL,ROWMAX,QMT,PMT,SCORE,NPRNT,NS,NVK,IDK1,
      AIDK2)
       DIMENSION K(1),KAT(1),CTABL(1),ROWMAX(1),QMT(1),PMT(1),SCORE(1)
       INTEGER FIRST
       COMMON/AB/KSCOR,TITLE(20)
C
C THIS SUBROUTINE COMPUTES THE D MEASURE OF EQUATION (4-11).
C
       DATA KSCOR/3/
       DATA (TITLE(I),I=1,20)/
      A4HANDE,4HRBER,4HG D ,4HMEAS,4HURE ,15*4H       /
C
C VARIABLES UNIQUE TO THIS PROGRAM
C KAT(I)=NUMBER OF CATEGORIES FOR I-TH CATEGORICAL VARIABLE
C CTABL=CONTINGENCY TABLE STORED BY ROWS
C ROWMAX(I)=MAXIMUM ENTRY IN ROW I OF THE CONTINGENCY TABLE
C QMT(J)=MARGINAL TOTAL FOR J-TH COLUMN IN CONTINGENCY TABLE
C PMT(I)=MARGINAL TOTAL FOR I-TH ROW IN CONTINGENCY TABLE.
C IDP=VARIABLE WITH SMALLER NUMBER OF CATEGORIES (ROW VARIABLE)
C IDQ=COLUMN VARIABLE
C
C DETERMINE ROW AND COLUMN VARIABLES
       IF(KAT(IDK1).GT.KAT(IDK2)) GO TO 10
       IDQ=IDK2
       IDP=IDK1
       GO TO 20
10     IDQ=IDK1
       IDP=IDK2
20     KQ=KAT(IDQ)
       KP=KAT(IDP)
C CONCEPTUALLY THE CONTINGENCY TABLE IS KP ROWS BY KQ COLUMNS, KP.GE.KQ
C INITIALIZE CONTINGENCY TABLE AND MARGINAL TOTALS
       IJ=0
       DO 30 I=1,KP
       ROWMAX(I)=0.
       PMT(I)=0.
       DO 30 J=1,KQ
       IJ=IJ+1
30     CTABL(IJ)=0.
       DO 40 J=1,KQ
40     QMT(J)=0.
C BUILD CONTINGENCY TABLE
       NOWP=IDP
       NOWQ=IDQ
       DO 50 L=1,NS
       IJ=(K(NOWP)-1)*KQ+K(NOWQ)
       CTABL(IJ)=CTABL(IJ)+1
       NOWP=NOWP+NVK
50     NOWQ=NOWQ+NVK
C THE FOLLOWING ENTRY POINT PERMITS CALCULATION OF THE ASSOCIATION
C MEASURE STARTING FROM A CONTINGENCY TABLE
C STARTING FROM A CONTINGENCY TABLE
C FOR THE CDC 6600 ONLY THE ENTRY NAME IS REQUIRED
C FOR THE IBM 360 THE ENTRY NAME AND COMPLETE ARGUMENT LIST IS REQUIRED
       ENTRY ACORKK
C COMPUTE MARGINAL TOTALS
       IJ=0
       DO 60 I=1,KP
       DO 60 J=1,KQ
       IJ=IJ+1
       PMT(I)=PMT(I)+CTABL(IJ)
60     QMT(J)=QMT(J)+CTABL(IJ)
```

```
C  PRINT CONTINGENCY TABLE AND MARGINALS IF DESIRED
      IF(NPRNT.NE.3) GO TO 90
      WRITE(6,2000) IDP,IDQ
      FIRST=1
      LAST=KQ
      DO 80 I=1,KP
      WRITE(6,2100) (CTABL(IJ),IJ=FIRST,LAST),PMT(I)
      FIRST=LAST+1
      LAST=LAST+KQ
80    CONTINUE
      WRITE(6,2100) (QMT(J),J=1,KQ)
C  FIND SUM OF ROW MAXIMA AND COLUMN MAXIMA
90    COLSUM=0.
      QMTMAX=0.
      DO 110 J=1,KQ
      COLMAX=0.
      IJ=J-KQ
      DO 100 I=1,KP
      IJ=IJ+KQ
      IF(CTABL(IJ).GT.COLMAX) COLMAX=CTABL(IJ)
      IF(CTABL(IJ).GT.ROWMAX(I)) ROWMAX(I)=CTABL(IJ)
100   CONTINUE
      COLSUM=COLSUM+COLMAX
      IF(QMT(J).GT.QMTMAX) QMTMAX=QMT(J)
110   CONTINUE
      ROWSUM=0.
      PMTMAX=0.
      DO 120 I=1,KP
      ROWSUM=ROWSUM+ROWMAX(I)
      IF(PMT(I).GT.PMTMAX) PMTMAX=PMT(I)
120   CONTINUE
      CORKK=(COLSUM+ROWSUM-PMTMAX-QMTMAX)/(2*NS)
      RETURN
2000  FORMAT(36H CONTINGENCY TABLE.  ROW VARIABLE IS,I3,21H.  COLUMN VAR
     AIABLE IS,I3,1H.)
2100  FORMAT(1X,25F5.0)
      END
```

D

PROGRAMS FOR ASSOCIATION MEASURES INVOLVING BINARY VARIABLES

Section 4.3 includes a wide variety of association measures among binary variables. Section 5.4 reconsiders these measures from the viewpoint of association among data units. This appendix provides a computer program for computing such measures. The description of the program is written in terms of association among variables but can be used for association among data units simply by reversing the two roles; that is, instead of thinking of variables described by sets of scores across all data units, think of data units described by sets of scores across all variables.

D.1 Bit-Level Storage

The concept of bit-level storage was introduced in Section 4.3.4 and is a central feature of this program. Only one bit is used to store the score of one data unit on one variable rather than a full word of NBIT† bits as in ordinary FORTRAN. This form of storage requires the setting of individual bits to establish in core the desired sequence of zeroes and ones. Three approaches are available as alternative versions of subroutine BDATA and are described in the following sections.

---

† The number of bits in a machine word is taken as a parameter, NBIT, to promote machine independence.

D.1.1   OCTAL CODING TECHNIQUES

A sequence of zeroes and ones cannot be read from cards as a binary number in FORTRAN. However, octal numbers can be read directly. The following correspondence exists between octal and binary numbers:

Binary	Octal	Binary	Octal
000	0	100	4
001	1	101	5
010	2	110	6
011	3	111	7

By coding each three digit group of binary numbers as one octal digit and reading the coded data in the Ø(octal) format, the data can be stored directly in core. To illustrate the procedure, suppose it is desired to store the ten-digit sequence 1101000111. Counting off three-digit groups from right to left the octal equivalent is 1507; counting from left to right and adding two dummy zeroes for filler, the octal equivalent is 6434. If 1507 is read in Ø4 format the original sequence is reproduced except for the addition of two zeroes for filler on the left; if 6434 is read in Ø4 format the sequence is reproduced again but this time with two zeroes for filler on the right. It makes no difference in this program whether zero fillers are added to the right or left as long as the same procedure is used throughout the data set.

D.1.2   SPARSE DATA TECHNIQUE

Depending on which half of each dichotomy is chosen to be scored as a one, the data set might be quite sparse (few ones). In such a case it may be simplest to list as input the sequence numbers of the data units which score one on each variable. For example in the ten-digit sequence 1001000010 only data units 1, 4, and 9 score one. A word reproducing this sequence of zeroes and ones may be set up using the .OR. masking operator and a set of NBIT masks; the ith mask is all zero except for the ith bit which is one. To illustrate, suppose that NBIT = 12 and it is desired to store the ten-digit sequence in a word labelled as IWRD. Let

$$\text{IWRD} = 000000000000 \qquad \text{MASK}(4) = 000100000000$$
$$\text{MASK}(1) = 100000000000 \qquad \text{MASK}(9) = 000000001000.$$

The sequence of zeroes and ones is stored by the following successive operations:

$$\text{IWRD} = \text{IWRD}.\text{OR}.\text{MASK}(1) = 100000000000$$
$$\text{IWRD} = \text{IWRD}.\text{OR}.\text{MASK}(4) = 100100000000$$
$$\text{IWRD} = \text{IWRD}.\text{OR}.\text{MASK}(9) = 100100001000.$$

The .OR. masking operator takes the logical sum of two words so that a bit set to one in the resulting word corresponds to a bit that was set to one in either or both of the original words.

D.1.3   DIRECT INPUT OF ZERO–ONE CODING

In spite of the advantages or appeal of the two preceding methods, it still may be most convenient to provide input as a sequence of zeroes and ones. The .OR. masking operator and a set of masks are used just as in Section D.1.2 but with a different procedure for deciding which masks to use. The zeroes and ones are read in I (integer) format and each digit is stored in a different word. Then a block of NBIT digits is compressed to one word. The word of compressed storage is built up by using the .OR. operator between this word and the masks corresponding to digits which are set to one. That is, in the sequence 100100001000 the first, fourth, and ninth masks are used because the corresponding digits are one and those remaining are zero. Note that both the method in this section and in Section D.1.2 begin storage with the leftmost bit and put zero fillers on the right. Also, the subroutines for these latter two methods include an option to save the data in octal coded form for subsequent processing.

D.2   Computing Association Measures

To compute an association measure it is necessary to determine the parameters of Table 4.3. There are three essential elements involved in finding these parameters when using bit level storage. The first is the .OR. masking operator which has been discussed already. The second is the .AND. masking operator which computes the logical product of two words; that is, a bit is set to one if and only if it was one in both of the original words. The third element is a function subprogram named KOUNT which counts the number of bits set to one in a word. KOUNT is written in assembly language because methods for counting the number of one-bits in FORTRAN would be excessively time consuming. The version listed here is written in COMPASS, the assembly language for CDC 6000-series computers. The subprogram is very simple and it should not be difficult to write an equivalent subprogram in other assembly languages.

To illustrate use of these elements for computing the entries in Table 4.3, suppose the scores for variables A and B are stored in IWRD and JWRD, respectively. Then various useful quantities are computed as follows:

$$a = \text{KOUNT (IWRD.AND.JWRD)}$$
$$a + b = \text{KOUNT (IWRD)}$$
$$a + c = \text{KOUNT (JWRD)}$$
$$a + b + c = \text{KOUNT (IWRD.OR.JWRD)}.$$

Used in conjunction with n (number of data units) these quantities can be used to determine any entry in the table. Computation of the association measures is performed in the function subprogram BASSN. This appendix includes fourteen versions of BASSN which accounts for all the interpretable association measures in Sections 4.3 and 5.4.

D.3   Use of the Program

The main program is named BINARY. The input specifications are given on comment cards at the beginning of the program. All problem dependent storage is accommodated by partitioning the X array; be sure to set the value of LIMIT as the dimension of this array. The choice of input mode for the data is effected through the selection of one of the three interchangeable versions of subroutine BDATA. The choice of association measure is made by selecting one of the several versions of function BASSN, also all interchangeable. If the user wishes to use the program merely for converting the data to the octal coded form, insert a STOP card in BINARY after the call to subroutine BDATA.

The program is written for CDC 6000-series computers but it should not be a formidable proposition to make conversions for other machines. Function KOUNT will have to be written in the applicable assembly language; subroutine BDATA should require no more than a change in the masks to correspond to the machine's word length. In any case the programs listed here can serve as examples if more extensive modifications are required.

Program BINARY

```
      PROGRAM BINARY(INPUT,OUTPUT,PUNCH,TAPE5=INPUT,TAPE6=OUTPUT,
     ATAPE7=PUNCH,TAPE1,TAPE2,TAPE3)
C
C     THIS PROGRAM IS WRITTEN TO COMPUTE ASSOCIATION MEASURES AMONG
C     BINARY VARIABLES.  DATA IS STORED AT THE BIT LEVEL.  SUBROUTINE
C     *BDATA* READS THE DATA AND PACKS IT INTO CORE.  THREE DIFFERENT
C     VERSIONS ARE AVAILABLE.  THE LOWER TRIANGULAR PORTION OF THE
C     SIMILARITY MATRIX IS COMPUTED AND WRITTEN ON BOTH UNITS 6 AND
C     *NTMAM*.  THE ASSOCIATION MEASURE IS COMPUTED IN ANY OF THE
C     SEVERAL VERSIONS OF SUBROUTINE *BASSN*.
C
C     THE PROGRAM AND ALL COMMENTS ARE WRITTEN IN TERMS OF ASSOCIATION
C     AMONG VARIABLES.  HOWEVER, ASSOCIATIONS AMONG DATA UNITS MAY BE
C     COMPUTED BY SWAPPING THE ROLES OF VARIABLES AND DATA UNITS.
C
C     ANY PRE-POSITIONING OF UNITS *NTIN*, *NTOUT*, AND *NTMAM* MUST BE
C     ACCOMPLISHED EXTERNALLY TO THIS PROGRAM WITH CONTROL CARDS.
C
C-------------------------------------------------------------------
C     INPUT SPECIFICATIONS
C     CARD 1    TITLE CARD
```

```
C   CARD 2   RUN PARAMETERS
C      COLS  1- 5   NV=NUMBER OF VARIABLES
C      COLS  6-10   NE=NUMBER OF ENTITIES (DATA UNITS)
C      COLS 11-15   NTIN=INPUT UNIT FOR DATA
C                   NTIN=5, CARD READER
C                   NTIN.NE.5, TAPE OR DISK
C      COLS 16-20   NTOUT=OUTPUT UNIT FOR SAVING CODED DATA
C                   NTOUT.LE.0, DO NOT SAVE
C                   NTOUT=7, CARD PUNCH
C      COLS 21-25   NTMAM=OUTPUT UNIT FOR SAVING MATRIX OF ASSOCIATION
C                         MEASURES
C
C   CARD(S) 3  INPUT DATA--EACH VERSION OF *BDATA* EMPLOYS A DIFFERENT
C                          INPUT PROCEDURE
C-----------------------------------------------------------------------
C
C   NBIT=NUMBER OF BITS IN A MACHINE WORD
C   NW=NUMBER OF WORDS NEEDED TO STORE AT THE BIT LEVEL ALL DATA FOR ONE
C      VARIABLE
C   LIMIT=DIMENSION OF THE *X* ARRAY
C   X=STORAGE ARRAY FOR PROBLEM DEPENDENT STORAGE SPACE
C
C   STORAGE ALLOCATIONS
C   X(N1) TO X(N2-1)  NV*NW WORDS--STORAGE OF DATA AT BIT LEVEL
C   X(N2) TO X(N3)    NE WORDS--STORAGE OF *KTEMP* ARRAY IN ONE VERSION
C                        OF *BDATA*, 0 WORDS FOR OTHER VERSIONS
C   X(N3+1) TO X(N4)  NV-1 WORDS--AREA FOR BUILDUP OF THE MATRIX ONE ROW
C                        AT A TIME
C
C-----------------------------------------------------------------------
        COMMON/ONE/KSCOR,TITLEA(20)
        DIMENSION TITLE(20)
        DIMENSION X(500)
        LIMIT=500
        NBIT=60
        READ(5,1000) TITLE
        READ(5,1100) NV,NE,NTIN,NTOUT,NTMAM
        WRITE(6,2000) TITLE
        WRITE(6,2100) NV,NE,NTIN,NTOUT,NTMAM
C   DETERMINE *NW* AND ALLOCATE STORAGE
        NW=NE/NBIT
        IF(NW*NBIT.LT.NE) NW=NW+1
        N1=1
        N2=N1+NV*NW
        N3=N2+KSCOR*(NE-1)
        N4=N3+NV-1
        WRITE(6,2200) N4,LIMIT
        IF(N4.GT.LIMIT) STOP
C   STORE DATA
        CALL BDATA(NV,NE,NW,NTIN,NTOUT,X(N1),X(N2))
C   BUILD LOWER TRIANGULAR MATRIX
        IWRD=0
        WRITE(6,2000) TITLEA
        IF(NTMAM.LE.0) GO TO 5
        WRITE(NTMAM,1000) TITLE
        WRITE(NTMAM,1000) TITLEA
5       DO 20 I=2,NV
        I1=I-1
        IWRD=IWRD+NW
        JWRD=-NW
        DO 10 J=1,I1
        JWRD=JWRD+NW
C   X(IWRD+K)=K-TH WORD STORING DATA ON VARIABLE I
C   X(JWRD+K)=K-TH WORD STORING DATA ON VARIABLE J
```

```
10      X(N3+J)=BASSN(NE,NW,X(IWRD),X(JWRD))
        WRITE(6,2300) I
        WRITE(6,2400) (J,X(N3+J),J=1,I1)
        IF(NTMAM.LE.0) GO TO 20
        WRITE(NTMAM,2500) (X(N3+J),J=1,I1)
20      CONTINUE
        WRITE(6,2600)
        IF(NTMAM.GT.0) ENDFILE NTMAM
1000    FORMAT(20A4)
1100    FORMAT(5I5)
2000    FORMAT(1H1,20A4)
2100    FORMAT(5HONV =,I8,/,5H NE =,I8,/,7H NTIN =,I6,/,8H NTOUT =,I5,/,
       A8H NTMAM =,I5)
2200    FORMAT(19HOREQUIRED STORAGE =,I10,6H WORDS,/,
       A       19HOALLOTTED STORAGE =,I10,6H WORDS)
2300    FORMAT(7HOENTITY,I5)
2400    FORMAT(1X,8(I5,F10.7))
2500    FORMAT(8F10.7)
2600    FORMAT(10HOFINISHED.)
        END
```

Subroutine BDATA

VERSION 1

```
        SUBROUTINE BDATA(NV,NE,NW,NTIN,NTOUT,KDATA,KTEMP)
C
C       THIS VERSION READS THE VECTOR OF DATA UNIT SCORES FOR EACH VARIABLE
C       AS A SEQUENCE OF ZEROES AND ONES AND PACKS THE INFORMATION AT THE
C       BIT LEVEL THROUGH USE OF MASKING OPERATIONS.
C
C       THE INPUT FORMAT IS 80I1.  EACH VARIABLE SHOULD START ON A NEW
C       CARD AND BE CONTINUED ON ADDITIONAL CARDS AS NECESSARY UNTIL ALL
C       DATA UNITS ARE SCORED ON THE VARIABLE.
C
C       NTIN=INPUT UNIT FOR DATA
C       NE=NUMBER OF ENTITIES (DATA UNITS)
C       NV=NUMBER OF VARIABLES
C       NW=NUMBER OF WORDS TO STORE DATA AT THE BIT LEVEL FOR ONE VARIABLE
C       KDATA=ONE-DIMENSIONAL ARRAY OF DATA STORED AT THE BIT LEVEL
C       NTOUT=OUTPUT UNIT FOR SAVING DATA IN OCTAL CODED FORM
C       KTEMP=TEMPORARY STORAGE FOR THE *NE* SCORES ON ONE VARIABLE
C
        COMMON/ONE/KSCOR,TITLEA(20)
        INTEGER FIRST
        DIMENSION KDATA(1),KTEMP(1),MASK(60)
        DATA KSCOR/1/
        DATA (MASK(I), I= 1,30)/
       A04000000000000000000000,02000000000000000000000,01000000000000000000
       B00400000000000000000000,00200000000000000000000,00100000000000000000
       C00040000000000000000000,00020000000000000000000,00010000000000000000
       D00004000000000000000000,00002000000000000000000,00001000000000000000
       E00000400000000000000000,00000200000000000000000,00000100000000000000
       F00000040000000000000000,00000020000000000000000,00000010000000000000
       G00000004000000000000000,00000002000000000000000,00000001000000000000
       H00000000400000000000000,00000000200000000000000,00000000100000000000
       I00000000040000000000000,00000000020000000000000,00000000010000000000
       J00000000004000000000000,00000000002000000000000,00000000001000000C
        DATA (MASK(I),I=31,60)/
       K00000000000400000000000,00000000000200000000000,00000000000100000C
       L00000000000040000000000,00000000000020000000000,00000000000010000C
       M00000000000004000000000,00000000000002000000000,00000000000001000C
```

```
      N0000000000000004000000,00000000000000002000000,00000000000000001000000,
      00000000000000000400000,00000000000000000200000,00000000000000000100000,
      P0000000000000000040000,00000000000000000020000,00000000000000000010000,
      Q0000000000000000004000,00000000000000000002000,00000000000000000001000,
      R0000000000000000000400,00000000000000000000200,00000000000000000000100,
      S0000000000000000000040,00000000000000000000020,00000000000000000000010,
      T0000000000000000000004,00000000000000000000002,00000000000000000000001/
      NBIT=60
      ISTOP=0
      IWRD=0
      DO 20 I=1,NV
      READ(NTIN,1000) (KTEMP(L),L=1,NE)
      IF(EOF,NTIN) 50,10
10    LDU=0
      DO 20 J=1,NW
      IWRD=IWRD+1
      KDATA(IWRD)=0
      DO 20 K=1,NBIT
      LDU=LDU+1
      IF(KTEMP(LDU).NE.1) GO TO 20
      KDATA(IWRD)=KDATA(IWRD).OR.MASK(K)
20    CONTINUE
C  FINISHED WITH INPUT. SAVE OCTAL CODED DATA
30    IF(NTOUT.LE.0) RETURN
      WRITE(6,3100) NTOUT
      LAST=0
      DO 40 I=1,NV
      FIRST=LAST+1
      LAST=LAST+NW
      WRITE(NTOUT,2000) (KDATA(J),J=FIRST,LAST)
40    CONTINUE
      IF(ISTOP.EQ.1) STOP
      RETURN
C  UNEXPECTEDLY READ THE END OF FILE
50    WRITE(6,3000) NTIN,I
C  SAVE PARTIAL RESULTS IF SAVE WAS DESIRED FOR FULL RESULTS
      IF(NTOUT.LE.0) STOP
      NV=I-1
      ISTOP=1
      GO TO 30
1000  FORMAT(80I1)
2000  FORMAT(4O20)
3000  FORMAT(24H0EOF ENCOUNTERED ON UNIT,I3, 6H.  I =,I5)
3100  FORMAT(32H0OCTAL CODED DATA STORED ON UNIT,I3)
      END
```

VERSION 2

```
      SUBROUTINE BDATA(NV,NE,NW,NTIN,NTOUT,KDATA,KDUM)
C
C  THIS VERSION READS THE ID NUMBERS OF THE DATA UNITS WHICH SCORE 1 ON
C  THE VARIABLE AND PACKS THE INFORMATION AT THE BIT LEVEL THROUGH
C  THE USE OF MASKING OPERATIONS
C
C  THE INPUT FORMAT IS 20I4.  EACH VARIABLE SHOULD START ON A NEW
C  CARD AND BE CONTINUED ON ADDITIONAL CARDS AS NECESSARY UNTIL ALL
C  DATA UNITS ARE SCORED ON THE VARIABLE.  TO TERMINATE THE LIST FOR A
C  VARIABLE USE A ZERO OR NEGATIVE DUMMY ID NUMBER.
C
C  NTIN=INPUT UNIT FOR DATA
C  NE=NUMBER OF ENTITIES (DATA UNITS)
C  NV=NUMBER OF VARIABLES
C  NW=NUMBER OF WORDS TO STORE DATA AT THE BIT LEVEL FOR ONE VARIABLE
```

```
C   KDATA=ONE-DIMENSIONAL ARRAY OF DATA STORED AT THE BIT LEVEL
C   NTOUT=OUTPUT UNIT FOR SAVING DATA IN OCTAL CODED FORM
C
      COMMON/ONE/KSCOR,TITLEA(20)
      INTEGER FIRST
      DIMENSION KDATA(1),KTEMP(20),MASK(60)
      DATA KSCOR/0/
      DATA (MASK(I), I= 1,30)/
     A0400000000000000000000,0200000000000000000000,0100000000000000000000,
     B0040000000000000000000,0020000000000000000000,0010000000000000000000,
     C0004000000000000000000,0002000000000000000000,0001000000000000000000,
     D0000400000000000000000,0000200000000000000000,0000100000000000000000,
     E0000040000000000000000,0000020000000000000000,0000010000000000000000,
     F0000004000000000000000,0000002000000000000000,0000001000000000000000,
     G0000000400000000000000,0000000200000000000000,0000000100000000000000,
     H0000000040000000000000,0000000020000000000000,0000000010000000000000,
     I0000000004000000000000,0000000002000000000000,0000000001000000000000,
     J0000000000400000000000,0000000000200000000000,0000000000100000000000/
      DATA (MASK(I),I=31,60)/
     K0000000000040000000000,0000000000020000000000,0000000000010000000000,
     L0000000000004000000000,0000000000002000000000,0000000000001000000000,
     M0000000000000400000000,0000000000000200000000,0000000000000100000000,
     N0000000000000040000000,0000000000000020000000,0000000000000010000000,
     O0000000000000004000000,0000000000000002000000,0000000000000001000000,
     P0000000000000000400000,0000000000000000200000,0000000000000000100000,
     Q0000000000000000040000,0000000000000000020000,0000000000000000010000,
     R0000000000000000004000,0000000000000000002000,0000000000000000001000,
     S0000000000000000000400,0000000000000000000200,0000000000000000000100,
     S0000000000000000000040,0000000000000000000020,0000000000000000000010,
     T0000000000000000000004,0000000000000000000002,0000000000000000000001
      NBIT=60
      LAST=0
      ISTOP=0
      NVNW=NV*NW
      DO 5 I=1,NVNW
5     KDATA(I)=0
      DO 50 I=1,NV
      FIRST=LAST+1
      LAST=LAST+NW
10    READ(NTIN,1000) KTEMP
      IF(EOF,NTIN) 80,20
20    DO 40 J=1,20
      IF(KTEMP(J).LE.0) GO TO 50
      IWRD=KTEMP(J)/NBIT
      IBIT=KTEMP(J)-IWRD*NBIT
      IF(IBIT.NE.0) GO TO 30
C   THE DATA UNIT CORRESPONDS TO THE LAST BIT IN A WORD
      IWRD=IWRD-1
      IBIT=NBIT
30    IWRD=IWRD+FIRST
      KDATA(IWRD)=KDATA(IWRD).OR.MASK(IBIT)
40    CONTINUE
      GO TO 10
50    CONTINUE
C   FINISHED WITH INPUT.  SAVE OCTAL CODED DATA
60    IF(NTOUT.LE.0) RETURN
      WRITE(6,3100) NTOUT
      LAST=0
      DO 70 I=1,NV
      FIRST=LAST+1
      LAST=LAST+NW
      WRITE(NTOUT,2000) (KDATA(J),J=FIRST,LAST)
70    CONTINUE
      IF(ISTOP.EQ.1) STOP
      RETURN
```

```
C  UNEXPECTEDLY READ THE END OF FILE
80    WRITE(6,3000) NTIN,I
C  SAVE PARTIAL RESULTS IF SAVE WAS DESIRED FOR FULL RESULTS
      IF(NTOUT.LE.0) STOP
      NV=I-1
      ISTOP=1
      GO TO 60
1000  FORMAT(20I4)
2000  FORMAT(4020)
3000  FORMAT(24H0EOF ENCOUNTERED ON UNIT,I3, 6H.  I =,I5)
3100  FORMAT(32H0OCTAL CODED DATA STORED ON UNIT,I3)
      END
```

VERSION 3

```
      SUBROUTINE BDATA(NV,NE,NW,NTIN,NTOUT,KDATA,KDUM)
C
C  THIS VERSION READS PACKED DATA IN OCTAL CODE AS PREPARED BY HAND
C  OR BY THE ALTERNATIVE VERSIONS OF *BDATA*.
C
C  THE INPUT FORMAT IS 4020.  EACH VARIABLE SHOULD START ON A NEW CARD
C  AND BE CONTINUED ON ADDITIONAL CARDS AS NECESSARY UNTIL ALL DATA
C  UNITS ARE SCORED ON THE VARIABLE.
C
C  NE=NUMBER OF ENTITIES (DATA UNITS)
C  NV=NUMBER OF VARIABLES
C  NW=NUMBER OF WORDS TO STORE DATA AT THE BIT LEVEL FOR ONE VARIABLE
C  KDATA=ONE-DIMENSIONAL ARRAY OF DATA STORED AT THE BIT LEVEL
      COMMON/ONE/KSCOR,TITLEA(20)
      DIMENSION KDATA(1)
      DATA KSCOR/0/
      INTEGER FIRST
      LAST=0
      DO 10 I=1,NV
      FIRST=LAST+1
      LAST=LAST+NW
      READ(NTIN,1000) (KDATA(J),J=FIRST,LAST)
      IF(EOF,NTIN) 20,10
10    CONTINUE
      RETURN
C  UNEXPECTEDLY READ THE END OF FILE
20    WRITE(6,3000) NTIN,I
      STOP
1000  FORMAT(4020)
3000  FORMAT(24H0EOF ENCOUNTERED ON UNIT,I3, 6H.  I =,I5)
      END
```

Function Subprogram KOUNT

```
      IDENT KOUNT
      ENTRY KOUNT
      VFD 30/5LKOUNT,30/0
KOUNT DATA 0
      SA1 B1
      CX6 X1
      EQ KOUNT
      END
```

Function BASSN

VERSION 1

```
      FUNCTION BASSN(NE,NW,IWRD,JWRD)
C   AVERAGE CONDITIONAL PROBABILITY OF A 1-1 MATCH
C   (A/(A+B)+A/(A+C))/2
      COMMON/ONE/KSCOR,TITLEA(20)
      DIMENSION IWRD(1),JWRD(1)
      DATA (TITLEA(I),I=1,20)/
     A4HAVER,4HAGE ,4HCOND,4HITIO,4HNAL ,4HPROB,4HABIL,4HITY ,4HOF A,
     B4H 1-1,4H MAT,4HCH =,4H  (A/,4H(A+B,4H)+A/,4H(A+C,4H))/2,3*4H      /
      A=0.
      AB=0.
      AC=0.
      DO 10 K=1,NW
      NA=IWRD(K).AND.JWRD(K)
      A=A+KOUNT(NA)
      AB=AB+KOUNT(IWRD(K))
      AC=AC+KOUNT(JWRD(K))
10    CONTINUE
      BASSN=(A/AB+A/AC)/2
      RETURN
      END
```

VERSION 2

```
      FUNCTION BASSN(NE,NW,IWRD,JWRD)
C   COSINE OF THE ANGLE BETWEEN VECTORS
C   SQRT(A*A/((A+B)(A+C)))
      COMMON/ONE/KSCOR,TITLEA(20)
      DIMENSION IWRD(1),JWRD(1)
      DATA (TITLEA(I),I=1,20)/
     A4HCOSI,4HNE O,4HF TH,4HE AN,4HGLE ,4HBETW,4HEEN ,4HVECT,4HORS ,
     B4H= SQ,4HRT(A,4H*A/(,4H(A+B,4H)(A+,4HC))),5*4H      /
      A=0.
      AB=0.
      AC=0.
      DO 10 K=1,NW
      NA=IWRD(K).AND.JWRD(K)
      A=A+KOUNT(NA)
      AB=AB+KOUNT(IWRD(K))
      AC=AC+KOUNT(JWRD(K))
10    CONTINUE
      BASSN=SQRT(A*A/(AC*AB))
      RETURN
      END
```

VERSION 3

```
      FUNCTION BASSN(NE,NW,IWRD,JWRD)
C   AVERAGE CONDITIONAL PROBABILITY OF A MATCH=
C   (A/(A+B)+A/(A+C)+D/(B+D)+D/(C+D))/4
      COMMON/ONE/KSCOR,TITLEA(20)
      DIMENSION IWRD(1),JWRD(1)
      DATA (TITLEA(I),I=1,20)/
     A4HAVER,4HAGE ,4HCOND,4HITIO,4HNAL ,4HPROB,4HABIL,4HITY ,4HOF A,
     B4H MAT,4HCH =,4H  (A/,4H(A+B,4H)+A/,4H(A+C,4H)+D/,4H(B+D,4H)+D/,
     C4H(C+D,4H))/4/
      A=0.
      AB=0.
```

```
      AC=0.
      DO 10 K=1,NW
      NA=IWRD(K).AND.JWRD(K)
      A=A+KOUNT(NA)
      AB=AB+KOUNT(IWRD(K))
      AC=AC+KOUNT(JWRD(K))
10    CONTINUE
      CD=NE-AB
      BD=NE-AC
      C=AC-A
      D=NE-AB-C
      BASSN=(A/AB+A/AC+D/BD+D/CD)/4.
      RETURN
      END
```

VERSION 4

```
      FUNCTION BASSN(NE,NW,IWRD,JWRD)
C GEOMETRIC AVERAGE OF COSINES
C SQRT(A*A*D*D/((A+B)(A+C)(B+D)(C+D))
      COMMON/ONE/KSCOR,TITLEA(20)
      DIMENSION IWRD(1),JWRD(1)
      DATA (TITLEA(I),I=1,20)/
     A4HGEOM,4HETRI,4HC AV,4HERAG,4HE OF,4H COS,4HINES,4H = S,4HQRT(,
     B4HA*A*,4HD*D/,4H((A+,4HB)(A,4H+C)(,4HB+D),4H(C+D,4H))  ,3*4H     /
      A=0.
      AB=0.
      AC=0.
      DO 10 K=1,NW
      NA=IWRD(K).AND.JWRD(K)
      A=A+KOUNT(NA)
      AB=AB+KOUNT(IWRD(K))
      AC=AC+KOUNT(JWRD(K))
10    CONTINUE
      CD=NE-AB
      BD=NE-AC
      C=AC-A
      D=NE-AB-C
      BASSN=SQRT(A*A*D*D/(AB*AC*BD*CD))
      RETURN
      END
```

VERSION 5

```
      FUNCTION BASSN(NE,NW,IWRD,JWRD)
C PHI-SQUARED (CORRELATION SQUARED)
C (AD-BC)**2/((A+B)(C+D)(A+C)(B+D))
      COMMON/ONE/KSCOR,TITLEA(20)
      DIMENSION IWRD(1),JWRD(1)
      DATA (TITLEA(I),I=1,20)/
     A4HPHI-,4HSQUA,4HRED ,4H= (A,4HD-BC,4H)**2,4H/((A,4H+B)(,4HC+D),
     B4H(A+C,4H)(B+,4HD))  ,8*4H     /
      A=0.
      AB=0.
      AC=0.
      DO 10 K=1,NW
      NA=IWRD(K).AND.JWRD(K)
      A=A+KOUNT(NA)
      AB=AB+KOUNT(IWRD(K))
      AC=AC+KOUNT(JWRD(K))
10    CONTINUE
      B=AB-A
      C=AC-A
```

```
      D=NE-AB-C
      BASSN=(A*D-B*C)**2
      BASSN=BASSN/(AB*(C+D)*AC*(B+D))
      RETURN
      END
```

Version 6

```
      FUNCTION BASSN(NE,NW,IWRD,JWRD)
C  ORDINARY GOODMAN AND KRUSKAL LAMBDA
      COMMON/ONE/KSCOR,TITLEA(20)
      DIMENSION IWRD(1),JWRD(1)
      DATA (TITLEA(I),I=1,20)/
     A4HORDI,4HNARY,4H GOO,4HDMAN,4H AND,4H KRU,4HSKAL,4H LAM,4HBDA ,
     B11*4H     /
      A=0.
      AB=0.
      AC=0.
      DO 10 K=1,NW
      NA=IWRD(K).AND.JWRD(K)
      A=A+KOUNT(NA)
      AB=AB+KOUNT(IWRD(K))
      AC=AC+KOUNT(JWRD(K))
10    CONTINUE
      B=AB-A
      C=AC-A
      D=NE-AB-C
      T1=AMAX1(A,B)+AMAX1(C,D)+AMAX1(A,C)+AMAX1(B,D)
      BD=NE-AC
      CD=NE-AB
      T2=AMAX1(AC,BD)+AMAX1(AB,CD)
      BASSN=(T1-T2)/(2.*NE-T2)
      RETURN
      END
```

Version 7

```
      FUNCTION BASSN(NE,NW,IWRD,JWRD)
C  GOODMAN AND KRUSKAL LAMBDA FOR FIXED MARGINALS.  YULE Y COEFFICIENT
      COMMON/ONE/KSCOR,TITLEA(20)
      DIMENSION IWRD(1),JWRD(1)
      DATA (TITLEA(I),I=1,20)/
     A4HGOOD,4HMAN ,4HAND ,4HKRUS,4HKAL ,4HLAMB,4HDA F,4HOR F,4HIXED,
     B4H MAR,4HGINA,4HLS. ,4HYULE,4H Y C,4HOEFF,4HICIE,4HNT  , 3*4H
      A=0.
      AB=0.
      AC=0.
      DO 10 K=1,NW
      NA=IWRD(K).AND.JWRD(K)
      A=A+KOUNT(NA)
      AB=AB+KOUNT(IWRD(K))
      AC=AC+KOUNT(JWRD(K))
10    CONTINUE
      B=AB-A
      C=AC-A
      D=NE-AB-C
      T1=SQRT(A*D)
      T2=SQRT(B*C)
      BASSN=(T1-T2)/(T1+T2)
      RETURN
      END
```

VERSION 8

```
      FUNCTION BASSN(NE,NW,IWRD,JWRD)
C  RUSSELL AND RAO COEFFICIENT,  PROBABILITY OF A 1-1 MATCH
C  A/(A+B+C+D)
      COMMON/ONE/KSCOR,TITLEA(20)
      DIMENSION IWRD(1),JWRD(1)
      DATA (TITLEA(I),I=1,20)/
     A4HRUSS,4HELL ,4HAND ,4HRAO ,4HCOEF,4HFICI,4HENT ,4H= A/,4HN
     B11*4H   /
      A=0.
      DO 10 K=1,NW
      NA=IWRD(K).AND.JWRD(K)
      A=A+KOUNT(NA)
10    CONTINUE
      BASSN=A/NE
      RETURN
      END
```

VERSION 9

```
      FUNCTION BASSN(NE,NW,IWRD,JWRD)
C  ANDERBERG D MEASURE
      COMMON/ONE/KSCOR,TITLEA(20)
      DIMENSION IWRD(1),JWRD(1)
      DATA (TITLEA(I),I=1,20)/
     A4HANDE,4HRBER,4HG D ,4HMEAS,4HURE ,15*4H     /
      A=0.
      AB=0.
      AC=0.
      DO 10 K=1,NW
      NA=IWRD(K).AND.JWRD(K)
      A=A+KOUNT(NA)
      AB=AB+KOUNT(IWRD(K))
      AC=AC+KOUNT(JWRD(K))
10    CONTINUE
      B=AB-A
      C=AC-A
      D=NE-AB-C
      T1=AMAX1(A,B)+AMAX1(C,D)+AMAX1(A,C)+AMAX1(B,D)
      BD=NE-AC
      CD=NE-AB
      T2=AMAX1(AC,BD)+AMAX1(AB,CD)
      BASSN=(T1-T2)/(2.*NE)
      RETURN
      END
```

VERSION 10

```
      FUNCTION BASSN(NE,NW,IWRD,JWRD)
C  SIMPLE MATCHING COEFFICIENT,  PROBABILITY OF A MATCH
C  (A+D)/(A+B+C+D)
      COMMON/ONE/KSCOR,TITLEA(20)
      DIMENSION IWRD(1),JWRD(1)
      DATA (TITLEA(I),I=1,20)/
     A4HSIMP,4HLE M,4HATCH,4HING ,4HCOEF,4HFICI,4HENT ,4H= (A,4H+B)/,
     B4HN   ,10*4H   /
      A=0.
      ABC=0.
      DO 10 K=1,NW
```

```
      NA=IWRD(K).AND.JWRD(K)
      NABC=IWRD(K).OR.JWRD(K)
      A=A+KOUNT(NA)
      ABC=ABC+KOUNT(NABC)
10    CONTINUE
      D=NE-ABC
      BASSN=(A+D)/NE
      RETURN
      END
```

Version 11

```
      FUNCTION BASSN(NE,NW,IWRD,JWRD)
C  JACCARD COEFFICIENT
C  A/(A+B+C)
      COMMON/ONE/KSCOR,TITLEA(20)
      DIMENSION IWRD(1),JWRD(1)
      DATA (TITLEA(I),I=1,20)/
     A4HJACC,4HARD ,4HCOEF,4HFICI,4HENT ,4H= A/,4H(A+B,4H+C) ,12*4H      /
      A=0.
      ABC=0.
      DO 10 K=1,NW
      NA=IWRD(K).AND.JWRD(K)
      NABC=IWRD(K).OR.JWRD(K)
      A=A+KOUNT(NA)
      ABC=ABC+KOUNT(NABC)
10    CONTINUE
      BASSN=A/ABC
      RETURN
      END
```

Version 12

```
      FUNCTION BASSN(NE,NW,IWRD,JWRD)
C  ROGERS AND TANIMOTO COEFFICIENT
C  (A+D)/(A+D+2(B+C))
      COMMON/ONE/KSCOR,TITLEA(20)
      DIMENSION IWRD(1),JWRD(1)
      DATA (TITLEA(I),I=1,20)/
     A4HROGE,4HRS A,4HND T,4HANIM,4HOTO ,4HCOEF,4HFICI,4HENT ,4H= (A,
     B4H+D)/,4H(A+D,4H+2(B,4H+C)),7*4H      /
      A=0.
      ABC=0.
      DO 10 K=1,NW
      NA=IWRD(K).AND.JWRD(K)
      NABC=IWRD(K).OR.JWRD(K)
      A=A+KOUNT(NA)
      ABC=ABC+KOUNT(NABC)
10    CONTINUE
      D=NE-ABC
      BASSN=(A+D)/(A+NE)
      RETURN
      END
```

Version 13

```
      FUNCTION BASSN(NE,NW,IWRD,JWRD)
C  DICE COEFFICIENT
C  2A/(2A+B+C)
      COMMON/ONE/KSCOR,TITLEA(20)
      DIMENSION IWRD(1),JWRD(1)
      DATA (TITLEA(I),I=1,20)/
```

```
      A4HDICE,4H COE,4HFFIC,4HIENT,4H = 2,4HA/(2,4HA+B+,4HC)   ,12*4H
      A=0.
      ABC=0.
      DO 10 K=1,NW
      NA=IWRD(K).AND.JWRD(K)
      NABC=IWRD(K).OR.JWRD(K)
      A=A+KOUNT(NA)
      ABC=ABC+KOUNT(NABC)
10    CONTINUE
      BASSN=2.*A/(A+ABC)
      RETURN
      END
```

Version 14

```
      FUNCTION BASSN(NE,NW,IWRD,JWRD)
C  METRIC DISTANCE
C  (B+C)/(A+B+C+D)
      COMMON/ONE/KSCOR,TITLEA(20)
      DIMENSION IWRD(1),JWRD(1)
      DATA (TITLEA(I),I=1,20)/
      A4HMETR,4HIC D,4HISTA,4HNCE ,4H= (B,4H+C)/,4HN   ,13*4H
      A=0.
      ABC=0.
      DO 10 K=1,NW
      NA=IWRD(K).AND.JWRD(K)
      NABC=IWRD(K).OR.JWRD(K)
      A=A+KOUNT(NA)
      ABC=ABC+KOUNT(NABC)
10    CONTINUE
      D=NE-ABC
      BASSN=1.-(A+D)/NE
      RETURN
      END
```

Version 15

```
      FUNCTION BASSN(NE,NW,IWRD,JWRD)
C
C  THIS SUBROUTINE IS AN EXAMPLE OF HOW TO USE VARIABLE WEIGHTS FOR
C  BINARY VARIABLES WHILE USING THE BIT LEVEL STORAGE SCHEME.  UNLIKE
C  OTHER VERSIONS OF *BASSN*, FUNCTION *KOUNT* IS NOT REQUIRED.
C  THIS VERSION IS WRITTEN IN TERMS OF ASSOCIATION BETWEEN DATA UNITS.
C
C  THE EXAMPLE PROBLEM HAS 100 VARIABLES.  THE WEIGHTS FOR EACH VARIABLE
C  ARE STORED IN ARRAY *W* AND SPECIFIED IN A DATA STATEMENT.  THE
C  STORED DATA IS CHECKED BIT BY BIT WITH MASKING OPERATIONS AND THE
C  APPPOPRIATE TABLE ENTRY IS INCREMENTED BY THE VARIABLE WEIGHT RATHER
C  THAN BY 1.  THIS VERSION IS COMPLETELY COMPATIBLE WITH  PROGRAM
C  *BINARY*.
C
C  SIMPLE MATCHING COEFFICIENT,  PROBABILITY OF A MATCH
C  (A+D)/(A+B+C+D)
C
      COMMON/ONE/KSCOR,TITLEA(20)
      DIMENSION IWRD(1),JWRD(1)
      DATA (TITLEA(I),I=1,20)/
      A4HSIMP,4HLE M,4HATCH,4HING ,4HCOEF,4HFICI,4HENT ,4H= (A,4H+B)/,
      B4HN   ,10*4H   /
```

```
C
      DIMENSION MASK(60)
      DATA (MASK(I), I= 1,30)/
     A040000000000000000000,02000000000000000000000,010000000000000000000,
     B004000000000000000000,00200000000000000000000,001000000000000000000,
     C000400000000000000000,00020000000000000000000,000100000000000000000,
     D000040000000000000000,00002000000000000000000,000010000000000000000,
     E000004000000000000000,00000200000000000000000,000001000000000000000,
     F000000400000000000000,00000020000000000000000,000000100000000000000,
     G000000040000000000000,00000002000000000000000,000000010000000000000,
     H000000004000000000000,00000000200000000000000,000000001000000000000,
     I000000000400000000000,00000000020000000000000,000000000100000000000,
     J000000000040000000000,00000000002000000000000,000000000010000000000/
      DATA (MASK(I),I=31,60)/
     K000000000004000000000,00000000000200000000000,000000000001000000000,
     L000000000000400000000,00000000000020000000000,000000000000100000000,
     M000000000000040000000,00000000000002000000000,000000000000010000000,
     N000000000000004000000,00000000000000200000000,000000000000001000000,
     O000000000000000400000,00000000000000020000000,000000000000000100000,
     P000000000000000040000,00000000000000002000000,000000000000000010000,
     Q000000000000000004000,00000000000000000200000,000000000000000001000,
     R000000000000000000400,00000000000000000020000,000000000000000000100,
     S000000000000000000040,00000000000000000002000,000000000000000000010,
     T000000000000000000004,00000000000000000000200,000000000000000000001/
C
      DIMENSION W(100)
      DATA (W(I),I=1,100)/ 7.2,1.4,5*2.0,6.1,.75,14*1.,23.,17.5,27*1.,
     A4.3,5.2,36*2.0/
C THE FOLLOWING SEQUENCE OF STATEMENTS WILL GENERATE ALL OF THE TABLE
C ENTRIES NEEDED FOR COMPUTING ANY MEASURE ON A 2X2 TABLE.
      NBIT=60
      A=0.
      B=0.
      C=0.
      D=0.
      DO 50 I=1,NW
      KBIT=NBIT
      INBIT=(I-1)*NBIT
      LBIT=NV-INBIT
      IF(LBIT.LT.NBIT) KBIT=LBIT
      DO 40 J=1,KBIT
C TEST BIT *J* IN IWRD(I) AND JWRD(I)
      KAB=IWRD(I).AND.MASK(J)
      KCD=JWRD(I).AND.MASK(J)
      WJ=W(INBIT+J)
      IF(KAB.EQ.0) GO TO 20
      IF(KCD.EQ.0) GO TO 10
      A=A+WJ
      GO TO 40
10    B=B+WJ
      GO TO 40
20    IF(KCD.EQ.0) GO TO 30
      C=C+WJ
      GO TO 40
30    D=D+WJ
40    CONTINUE
50    CONTINUE
C COMPUTE ASSOCIATION MEASURE
      BASSN=(A+D)/(A+B+C+D)
      RETURN
      END
```

APPENDIX

E

PROGRAMS FOR HIERARCHICAL CLUSTER
ANALYSIS

Three different computational approaches for hierarchical cluster analysis
are given detailed discussion in Chapter 6. This appendix presents computer
programs suitable for implementing these approaches. The three sets of
programs in this appendix have the following characteristics in common.

1. The overall program design philosophy is the same as that used in
Appendix C. The main program DRIVER is a dummy program which establishes input/output units, sets the dimension of one all-encompassing array,
and calls a subroutine which is the principal deck segment. The principal
subroutine called from DRIVER includes complete input specifications, deck
setup specifications, and an explanation as to the computed allocation of
storage for various subarrays within the single overall storage array. The use
of a single storage array permits all deck segments except program DRIVER
to be independent of problem size; the user should never have to alter any
of the subroutines in these program sets unless making a substantive change
to the clustering method.

2. All three sets of programs utilize subroutine TREE to convert the
sequence of cluster merges into a tree diagram which is drawn on the line
printer. Subroutine TREE is documented and explained in Appendix G.

3. In all three sets of programs the last set of cards in the job deck contains
labels for the clustered entities. If desired, the user can identify each entity
with up to 20 characters of information which will be printed on the tree
diagram. If labeling is not desired, then the user need only include a card with
the characters NOLB as the last card in the deck (before the end of file card).

275

E.1   Stored Similarity Matrix Approach

In Section 6.2 a computational approach is described in which a similarity matrix is provided externally and stored in the computer's central memory. A complete job deck for this approach includes the following segments.

1. Program DRIVER is a dummy main program. An example is listed in subroutine CNTRL.

2. Subroutine CNTRL is the principal deck segment. This subroutine describes the input specifications, deck setup, and storage allocation; it also reads input cards, computes storage allocations, and calls the other deck segments.

3. Subroutine MTXIN is used to read the similarity matrix into the central memory. The matrix may be read either in constant length records (set INOPT as the number of similarity measures in a record) or in variable length records where each record is one row of the lower triangular matrix (set INOPT less than or equal to 0). The subroutine listing is given in Appendix G.

4. Subroutine CLSTR performs the steps of the clustering algorithm described in Section 6.2.

5. Subroutine LFIND computes the location of the similarity measure s_{ij} within the similarity matrix which is stored as a one-dimensional array.

6. Subroutine METHOD computes the similarity measured between clusters and updates the similarity matrix after each merger. Seven different versions of subroutine METHOD are provided to implement all the methods described in Section 6.2; all versions are interchangeable so that the user can obtain alternative analyses by simply removing one subroutine and inserting another in the deck. The first call to subroutine METHOD causes a message to be printed which identifies the particular version being used.

7. Subroutine TREE draws a tree diagram on the line printer. See Appendix G for documentation and description.

E.2   Stored Data Approach

In Section 6.3 a computational approach is described in which each data unit is described by a vector of scores and this information is stored in the central memory. A complete job deck for this approach includes the following segments.

1. Program DRIVER is a dummy main program. An example is listed in subroutine MANAGE.

2. Subroutine MANAGE is the principal deck segment. This subroutine describes the input specifications, deck setup, and storage allocation; it also computes storage allocations and calls the other deck segments.

3. Subroutine USER is a job-dependent subprogram which the user provides to read in a full set of data for one data unit. This subroutine is called once for each data unit. An example is included in the comments at the beginning of subroutine MANAGE.

4. Subroutine GROUP performs the steps of the clustering algorithm described in Section 6.3.

5. Subroutine PROC computes the similarity measure between clusters and updates the cluster description after each merger. Seven different versions of subroutine PROC are provided. Four versions implement the four methods described in Section 6.3; the other three versions repeat the methods of Sections 6.3.2–6.3.4 but report the error sum of squares (the Ward criterion) at each merger; the centroid method will not show reversals against this latter criterion because the error sum of squares is a nondecreasing function. The first call to subroutine PROC causes printing of a message which identifies the clustering method being used.

6. Subroutine TREE draws a tree diagram on the line printer. See Appendix G for documentation and discussion.

E.3 Sorted Matrix Approach

In Section 6.4 a computational approach is described in which the similarity matrix is computed and sorted into ascending (for distance-like measures) or descending (for correlation-like measures) sequence as a preliminary step before beginning to cluster. A complete job deck for this approach includes the following segments.

1. Program DRIVER is a dummy main program. An example is listed in subroutine ALLIN1.

2. Subroutine ALLIN1 is the principal deck segment. This subroutine describes the input specifications, deck setup, and storage allocation; it also reads input cards, computes storage allocations, and calls the other deck segments.

3. Subroutine PREP performs the steps of the clustering algorithm described in Section 6.4.1. The sorted similarity triples (s_{ij}, i, j) are read in blocks of NREC triples using either binary (IOPT $\neq$ 1) or formatted BCD (IOPT = 1) READ statements. The greatest efficiency is realized when using binary READ statements and the largest possible value of NREC consistent with the maximum number of words permitted in an input record.

4. Subroutine TREE draws a tree diagram on the line printer. See Appendix G for documentation and description.

This program uses no more of the list of sorted similarity triples than necessary; all required links may be found in only a small portion of the list and therefore it is not necessary to examine the entire list. If the sorted similarity triples are read from cards (which is feasible for small problems) the list of labels for the clustered entities may not be reached because some of the similarity triples are not read; in such a case, steps are taken in subroutine PREP to permit printing of the tree diagram without labels.

This set of programs also can be used to draw the tree diagram depicting the results of *any* hierarchical clustering. For a data set of n entities, prepare a list of $n - 1$ similarity triples (s_{ij}, i, j), where i and j are the identification or sequence numbers of *any two entities in the two clusters* merged at a similarity value s_{ij}. Referring to Fig. 6.1 in Chapter 6, suppose the four merges occur with distance-like similarities 1, 3, 4, and 7; then this set of programs could draw the tree diagram in Fig. 6.1 from the sequence of similarity triples

$$(1, 1, 2) \qquad (3, 3, 4) \qquad (4, 3, 5) \qquad (7, 1, 3).$$

Hierarchical clustering results from other programs or from published reports can be transformed to a tree diagram through use of this program.

Subroutine CNTRL

```
      SUBROUTINE CNTRL(X,LIMIT)
C
C  THIS SUBROUTINE ALLOCATES STORAGE, READS INPUT AND CONTROLS
C  EXECUTION FOR A HIERARCHICAL CLUSTERING JOB BASED ON A PROVIDED
C  SIMILARITY MATRIX.
C
C------------------------------------------------------------------
C  INPUT SPECIFICATIONS
C
C  CARD 1   TITLE CARD
C  CARD 2   INFORMATION FOR SUBROUTINES CLSTR AND TREE
C     COLS  1- 3   NE=NUMBER OF ENTITIES (DATA UNITS OR VARIABLES) TO BE
C                     CLUSTERED
C     COLS  4- 5   ISIGN=OPTION FOR SIMILARITY FUNCTION
C                     ISIGN=+1, DISTANCE MEASURE
C                     ISIGN=-1, CORRELATION MEASURE
C     COLS  6- 7   NTSV=TAPE UNIT ON WHICH CLSTR RESULTS ARE SAVED
C                     NTSV=7, PUNCH RESULTS ON CARDS
C                     NTSV.LE.0, DO NOT SAVE RESULTS
C     COLS  8- 9   NTIN=UNIT FROM WHICH SIMILARITY MATRIX IS READ
C                     NTIN=5, CARD READER
C                     NTIN.NE.5, DISK OR TAPE
C     COLS 10-12   INOPT=INPUT OPTION FOR SIMILARITY MATRIX
C                     INOPT.LE.0, EACH RECORD IS ONE ROW OF A LOWER TRIANG-
C                        ULAR MATRIX
C                     INOPT.GT.0, THE LOWER TRIANGULAR MATRIX IS CONSIDERED
C                        TO BE STORED BY ROWS IN ONE LONG LINEAR
C                        ARRAY AND IS READ IN BLOCKS *INOPT* LONG.
```

```
C     COLS 13-14  KOUT=OUTPUT OPTION
C                 KOUT=+2, STANDARD OUTPUT
C                 KOUT=-2, STANDARD OUTPUT PLUS PUNCHED SEQUENCE LIST
C                         FROM SUBROUTINE *TREE*
C
C***ANY PREPOSITIONING OF THE I/O UNITS NTSV AND NTIN MUST BE
C   ACCOMPLISHED IN PROGRAM DRIVER OR THROUGH USE OF CONTROL CARDS.
C
C   CARD 3  INPUT FORMAT FOR SIMILARITY MATRIX (20A4 FORMAT)
C   CARD(S) 4  SIMILARITY MATRIX
C   CARD 5  END OF RECORD CARD (7/8/9)
C
C***INCLUDE CARDS 4 AND 5 ONLY IF THE SIMILARITY MATRIX IS ON CARDS***
C
C   CARD(S) 6  LABEL CARDS FOR ENTITIES.  THERE ARE TWO OPTIONS
C     1.  INCLUDE 1 CARD WITH THE 4 CHARACTERS *NOLB* IN COLUMNS 1-4.
C         UNDER THIS OPTION LABELS ARE NOT PRINTED ON THE TREE OUTPUT.
C
C     2.  INCLUDE NE CARDS, COLUMNS 1 TO 20 CONTAINING A LABEL FOR ONE
C         ENTITY.  ORDER THE LABEL CARDS IN THE SAME SEQUENCE AS THE
C         ENTITIES ARE REPRESENTED IN THE SIMILARITY MATRIX.
C-----------------------------------------------------------------------
C
C   DECK SETUP SPECIFICATIONS
C
C   THE USER PROVIDES PROGRAM DRIVER WHICH PERFORMS THE FOLLOWING TASKS.
C     1.  ASSIGNS INPUT/OUTPUT UNITS
C     2.  ESTABLISHES THE DIMENSION OF THE X ARRAY AND SETS THIS
C         DIMENSION EQUAL TO LIMIT.
C     3.  CALLS SUBROUTINE CNTRL.
C   THE FOLLOWING EXAMPLE WILL SUFFICE IN MOST CASES.
C
C     PROGRAM DRIVER (INPUT,OUTPUT,PUNCH,TAPE5=INPUT,TAPE6=OUTPUT,
C    ATAPE7=PUNCH,TAPE1,TAPE2)
C     DIMENSION X(7000)
C     LIMIT=7000
C     CALL CNTRL(X,LIMIT)
C     END
C
C   A SECOND JOB DEPENDENT SEGMENT IS SUBROUTINE METHOD.  THE USER
C   SELECTS AMONG THE SEVERAL ALTERNATIVE VERSIONS OF THIS SUBROUTINE TO
C   IMPLEMENT THE DESIRED CLUSTERING TECHNIQUE.
C
C   THE SUBPROGRAMS CNTRL, CLSTR, MTXIN, LFIND AND TREE GO IN EVERY JOB.
C-----------------------------------------------------------------------
C
C   THE X ARRAY IS PARTITIONED FOR STORAGE AS FOLLOWS
C   STORAGE FOR ARRAYS NEEDED AT ALL STAGES OF THE JOB
C   X(N1) TO X(N2-1)  NE WORDS--STORAGE OF THE II ARRAY
C   X(N2) TO X(N3-1)  NE WORDS--STORAGE OF THE JJ ARRAY
C   X(N3) TO X(N4-1)  NE WORDS--STORAGE OF THE SS ARRAY
C   X(N4) TO X(N5-1)  NE WORDS--STORAGE OF THE IL ARRAY
C   X(N5) TO X(N6-1)  NE WORDS--STORAGE OF THE JL ARRAY
C   X(N6) TO X(N7-1)  NE WORDS--STORAGE OF THE NEXT ARRAY
C   STORAGE FOR ARRAYS NEEDED IN SUBROUTINE CLSTR
C   M1=N7
C   X(M1) TO X(M2-1)  (NE*(NE-1))/2 WORDS--STORAGE OF THE S ARRAY
C   X(M2) TO X(M3-1)  NE WORDS--STORAGE OF THE LAST ARRAY
C   X(M3) TO X(M4-1)  NE WORDS--STORAGE OF THE NEAR ARRAY
C   X(M4) TO X(M5-1)  NE WORDS--STORAGE OF THE SREF ARRAY
C   X(M5) TO X(M6-1)  NE WORDS--STORAGE OF THE LIST ARRAY
C   X(M6) TO X(M7-1)  NE WORDS--STORAGE OF THE A ARRAY
C   X(M7) TO X(M8)    NE WORDS--STORAGE OF THE B ARRAY
C   STORAGE FOR ARRAYS NEEDED IN SUBROUTINE TREE (OVERLAY ARRAYS NEEDED
C   IN SUBROUTINE CLSTR)
```

```
C   L1=N7
C   X(L1) TO X(L2-1)   25*NE WORDS--STORAGE OF THE A ARRAY
C   X(L2) TO X(L3-1)   5*NE WORDS--STORAGE OF THE LABEL ARRAY
C   X(L3) TO X(L4-1)   NE WORDS--STORAGE OF THE LCLNO ARRAY
C   X(L4) TO X(L5-1)   NE WORDS--STORAGE OF THE LINE ARRAY
C   X(L5) TO X(L6-1)   NE WORDS--STORAGE OF THE IS ARRAY
C   X(L6) TO X(L7)     NE WORDS--STORAGE OF THE LAST ARRAY
C------------------------------------------------------------------
C
      INTEGER FIRST
      DIMENSION X(1),FMT(20),TITLE(20),EPS(25)
      READ(5,1000) TITLE
      READ(5,1100) NE,ISIGN,NTSV,NTIN,INOPT,KOUT
      WRITE(6,2500) TITLE
      WRITE(6,2200) NE,ISIGN,NTSV,NTIN,INOPT,KOUT
C   PARTITION THE STORAGE ARRAY
      N1=1
      N2=N1+NE
      N3=N2+NE
      N4=N3+NE
      N5=N4+NE
      N6=N5+NE
      N7=N6+NE
      M2=N7+(NE*(NE-1))/2
      M3=M2+NE
      M4=M3+NE
      M5=M4+NE
      M6=M5+NE
      M7=M6+NE
      M8=M7+NE-1
      L2=N7+25*NE
      L3=L2+5*NE
      L4=L3+NE
      L5=L4+NE
      L6=L5+NE
      L7=L6+NE-1
C   CHECK FOR SUFFICIENT STORAGE
      MAX=M8
      IF(L7.GT.MAX) MAX=L7
      WRITE(6,2300) MAX,LIMIT
      IF(MAX.GT.LIMIT) STOP
C   READ THE SIMILARITY MATRIX
      READ(5,1000) FMT
      WRITE(6,2100) FMT
      CALL MTXIN(X(N7),INOPT,NE,NTIN,FMT)
C   READY TO CLUSTER
60    CALL CLSTR(X(N1),X(N2),X(N3),X(N4),X(N5),X(N6),X(N7),X(M2),X(M3),
     AX(M4),X(M5),X(M6),X(M7),TITLE,NE,ISIGN,NTSV)
C   READ LABEL CARD(S)
      FIRST=L2
      LAST=L2+4
      READ(5,1000) (X(I),I=FIRST,LAST)
      IF(X(FIRST).EQ.4HNOLB) GO TO 80
C   READ REMAINING LABELS
      DO 70 K=2,NE
      FIRST=LAST+1
      LAST=LAST+5
70    READ(5,1000) (X(I),I=FIRST,LAST)
C   DRAW THE TREE CORRESPONDING TO THE CLUSTERING
80    MERGES=NE-1
      CALL TREE(X(N1),X(N2),X(N3),X(N4),X(N5),X(N6),X(N7),X(L2),X(L3),
     AX(L4),X(L5),X(L6),EPS,TITLE,MERGES,1,6,1,KOUT,NE)
      RETURN
1000  FORMAT(20A4)
```

```
1100   FORMAT(I3,3I2,I3,I2)
2100   FORMAT(7H0FORMAT,20A4)
2200   FORMAT(5HONE =,I8,/,8H ISIGN =,I5,/,7H NTSV =,I6,/,7H NTIN =,I6,
       A/,8H INOPT =,I5,/,7H KOUT =,I6)
2300   FORMAT(19H0REQUIRED STORAGE =,I5,6H WORDS,/,
       A       19H0ALLOTTED STORAGE =,I5,6H WORDS,/)
2500   FORMAT(1H0,20A4)
       END
```

Subroutine CLSTR

```
       SUBROUTINE CLSTR(II,JJ,SS,IL,JL,NEXT,S,LAST,NEAR,SREF,LIST,A,B,
       ATITLE,N,ISIGN,NT)
C   IN THIS VERSION THE LOWER TRIANGULAR PORTION OF THE SIMILARITY MATRIX
C   IS STORED BY ROWS IN THE ONE-DIMENSIONAL ARRAY S.
C
C   THE FOLLOWING VARIABLES ARE SPECIFIED IN THE CALLING PROGRAM AND
C   ARE PASSED THROUGH THE ARGUMENT LIST
C     N=NUMBER OF OBJECTS TO BE CLUSTERED
C     S(J)=J-TH ELEMENT IN LOWER TRIANGULAR SIMILARITY MATRIX
C     ISIGN=OPTION SPECIFYING TYPE OF SIMILARITY FUNCTION USED
C     ISIGN=+1=DISTANCE MEASURE (DECREASING FUNCTION OF SIMILARITY)
C     ISIGN=-1=CORRELATION MEASURE (INCREASING FUNCTION OF SIMILARITY)
C     NT=TAPE UNIT ON WHICH THE RESULTS ARE SAVED
C     NT.LE.0=DO NOT SAVE RESULTS ON TAPE
C     NT=7=SAVE RESULTS ON PUNCHED CARDS
C     TITLE=IDENTIFYING TITLE FOR THIS RUN
C
C   THE FOLLOWING VARIABLES REPRESENT THE OUTPUT OF THE PROGRAM AND ARE
C   PASSED BACK THROUGH THE ARGUMENT LIST.  THESE RESULTS ARE READY FOR
C   SUBROUTINE TREE.
C     K=STAGE OF CLUSTERING
C     II(K)=LOWER NUMBERED CLUSTER MERGED AT STAGE K
C     JJ(K)=UPPER NUMBERED CLUSTER MERGED AT STAGE K
C     SS(K)=VALUE OF SIMILARITY FUNCTION ASSOCIATED WITH MERGE AT STAGE K
C     IL(K)=PRECEDING STAGE AT WHICH II(K) WAS LAST IN A MERGE
C     JL(K)=PRECEDING STAGE AT WHICH JJ(K) WAS LAST IN A MERGE
C     NEXT(K)=NEXT STAGE AT WHICH II(K) IS IN A MERGE
C
C   IN ADDITION, THE FOLLOWING VARIABLES PLAY IMPORTANT ROLES IN THE PROGRAM
C     NEAR(I)=ID NUMBER OF EXTREME ELEMENT IN ROW I OF THE LOWER
C             TRIANGULAR SIMILARITY MATRIX.
C     SREF(I)=SIMILARITY MEASURE FOR THE PAIR (I,NEAR(I))
C     LIST(I)=I-TH CLUSTER ID NUMBER IN SEQUENTIAL LIST OF CURRENT CLUSTERS
C     NCL=NUMBER OF CLUSTERS AT CURRENT STAGE
C     LAST(I)=STAGE NUMBER AT WHICH CLUSTER I WAS LAST IN A MERGE
C     A=WORKING AREA FOR SUBROUTINE METHOD
C     B=WORKING AREA FOR SUBROUTINE METHOD
C
C   THIS SUBROUTINE USES FUNCTION LFIND(I,J) TO FIND THE ADDRESS IN S
C   FOR THE SIMILARITY MEASURE BETWEEN CLUSTERS I AND J
       DIMENSION S(1),II(1),JJ(1),SS(1),IL(1),JL(1),NEXT(1),NEAR(1),
       ASREF(1),LIST(1),LAST(1),A(1),B(1)
       DIMENSION TITLE(20)
C   INITIALIZE VARIABLES AND SET CONSTANTS
       NCL=N
       K=1
       SIGN=ISIGN
       BIG=SIGN*1.E50
       CALL METHOD(S,NEAR,SREF,LIST,A,B,SREFX,SIGN,N,NCL,LREF,NREF,1)
```

```
C   INITIALIZE ARRAYS
        DO 10 J=1,N
        LAST(J)=0
        NEXT(J)=0
        LIST(J)=J
        SREF(J)=BIG
10      CONTINUE
C   FIND EXTREME ENTRY IN EACH ROW
        L=0
        DO 30 I=2,N
        I1=I-1
        DO 30 J=1,I1
        L=L+1
C   IN EFFECT S(L)=S(I,J)
        IF((((S(L)-SREF(I))*SIGN).GT.0.) GO TO 30
        NEAR(I)=J
        SREF(I)=S(L)
30      CONTINUE
C   MAIN LOOP.  FIND EXTREME VALUE IN SREF ARRAY
40      SREFX=BIG
        DO 50 I=2,NCL
        LISTI=LIST(I)
        IF((((SREF(LISTI)-SREFX)*SIGN).GT.0) GO TO 50
        IREF=I
        LREF=LISTI
        SREFX=SREF(LISTI)
50      CONTINUE
C   LREF IS THE ROW NUMBER CONTAINING THE EXTREME ENTRY IN THE S ARRAY.
C   IF THERE ARE TIES, THEN LREF IS THE HIGHEST NUMBERED ROW WITH THIS
C   EXTREME VALUE.  HENCE LREF.GT.NEAR(LREF).  IREF IDENTIFIES THE
C   PLACEMENT OF LREF IN THE LIST ARRAY.
        NREF=NEAR(LREF)
        CALL METHOD(S,NEAR,SREF,LIST,A,B,SREFX,SIGN,N,NCL,LREF,NREF,2)
C   GENERATE MERGE DATA NEEDED FOR SUBROUTINE TREE
        II(K)=NREF
        JJ(K)=LREF
        SS(K)=SREFX
        IL(K)=LAST(NREF)
        JL(K)=LAST(LREF)
        LAST(NREF)=K
        IF(IL(K).EQ.0) GO TO 60
        ILK=IL(K)
        NEXT(ILK)=K
60      IF(JL(K).EQ.0) GO TO 70
        JLK=JL(K)
        NEXT(JLK)=K
70      K=K+1
C   TERMINATE IF N-1 MERGES HAVE OCCURED
        IF(K.EQ.N) GO TO 140
C   UPDATE FOR THE NEXT CYCLE
        NCL=NCL-1
        IF(IREF.GT.NCL) GO TO 90
C   UPDATE LIST ARRAY BY REMOVING LREF AND PUSHING DOWN THE LIST
        DO 80 I=IREF,NCL
80      LIST(I)=LIST(I+1)
C   UPDATE FOR NEXT CYCLE
90      CALL METHOD(S,NEAR,SREF,LIST,A,B,SREFX,SIGN,N,NCL,LREF,NREF,3)
        GO TO 40
C   CLUSTERING FINISHED AND ALL ANCILLARY INFORMATION GENERATED.
C   SAVE RESULTS AS DESIRED.
140     K=K-1
160     IF(NT.LE.0) RETURN
        WRITE(NT,2300) TITLE
        DO 170 I=1,K
```

```
170     WRITE(NT,2200) I,II(I),JJ(I),SS(I),IL(I),JL(I),NEXT(I)
        RETURN
2200    FORMAT(3I10,E16.8,3I10)
2300    FORMAT(20A4)
        END
```

Function LFIND

```
        FUNCTION LFIND(I,J)
C   IF THE LOWER TRIANGULAR PORTION OF A SYMMETRIC MATRIX IS STORED BY
C   ROWS IN A ONE-DIMENSIONAL ARRAY, THEN THE ELEMENT (I,J) IN THE FULL
C   MATRIX IS ELEMENT LFIND(I,J) IN THE LINEAR ARRAY
        IF(I.GT.J) GO TO 10
C   ROW J, COLUMN I
        LFIND=((J-1)*(J-2))/2+I
        RETURN
C   ROW I, COLUMN J
10      LFIND=((I-1)*(I-2))/2+J
        RETURN
        END
```

Subroutine METHOD

Version 1

```
        SUBROUTINE METHOD(S,NEAR,SREF,LIST,NUMBR,SUM,SREFX,SIGN,N,NCL,
        ALREF,NREF,JOB)
C
C   HIERARCHICAL CLUSTERING BY CENTROID SORTING
C
C   THE PARTICULAR ALGORITHM USED HERE IS DESCRIBED IN
C       LANCE, G.N. AND W.T. WILLIAMS, A GENERAL THEORY OF CLASSIFICATORY
C       SORTING STRATEGIES, 1. HIERARCHICAL SYSTEMS, THE COMPUTER JOURNAL,
C       VOLUME 9, NUMBER 4, FEBRUARY 1967, PP373-380.
C
        DIMENSION S(1),NEAR(1),SREF(1),LIST(1),NUMBR(1),SUM(1)
        GO TO (10,25,30),JOB
C   JOB=1, INITIALIZE.
C   NUMBR(I)=NUMBER OF ENTITIES CURRENTLY IN THE I-TH CLUSTER
C           CLUSTER
10      WRITE(6,2000)
2000    FORMAT(42H0CENTROID CLUSTERING.  BEWARE OF REVERSALS)
        DO 20 J=1,N
20      NUMBR(J)=1
        BIG=SIGN*1.E50
        RETURN
C   JOB=2, DUMMY ENTRY.
25      RETURN
C   JOB=3, UPDATE FOR NEXT ROUND.
C   UPDATE THE NEW CLUSTER
30      NTOT=NUMBR(NREF)+NUMBR(LREF)
        TOT=NTOT
        ALL=NUMBR(LREF)/TOT
        ALN=NUMBR(LREF)/TOT
        PROD=ALN*ALL
        LBET=LFIND(LREF,NREF)
        DO 40 J=1,NCL
        I=LIST(J)
        IF(I.EQ.NREF) GO TO 40
```

```
C   RECALL THAT LREF HAS BEEN REMOVED FROM LIST AND THEREFORE I NEED NOT
C   BE TESTED FOR EQUALITY WITH LREF.
        LL=LFIND(I,LREF)
        LN=LFIND(I,NREF)
        S(LN)=ALL*S(LL)+ALN*S(LN)-PROD*S(LBET)
40      CONTINUE
C   UPDATE THE NEAR AND SREF ARRAYS.  IF THE EXTREME ELEMENT IN ROW I
C   WAS EITHER LREF OR NREF, THEN IT IS NECESSARY TO FIND A NEW EXTREME
C   ELEMENT.  ROWS PRIOR TO NREF NEED NOT BE CONSIDERED.
        DO 50 J=1,NCL
        I=LIST(J)
        IF(I.EQ.NREF) GO TO 55
50      CONTINUE
55      IF(J.EQ.1) GO TO 80
60      SREF(I)=BIG
        J1=J-1
        DO 70 L=1,J1
        LISTL=LIST(L)
        LL=LFIND(I,LISTL)
        IF(((S(LL)-SREF(I))*SIGN).GE.0.) GO TO 70
        NEAR(I)=LISTL
        SREF(I)=S(LL)
70      CONTINUE
80      J=J+1
        IF(J.GT.NCL) RETURN
        I=LIST(J)
        IF(NEAR(I).EQ.LREF.OR.NEAR(I).EQ.NREF) GO TO 60
        GO TO 80
        END
```

Version 2

```
        SUBROUTINE METHOD(S,NEAR,SREF,LIST,A,B,SREFX,SIGN,N,NCL,LREF,NREF,
       AJOB)
C
C   HIERARCHICAL CLUSTERING BY COMPLETE LINKAGE.  THE ALGORITHM IS
C   DERIVED FROM
C   JOHNSON, S.C., HIERARCHICAL CLUSTERING SCHEMES, PSYCHOMETRIKA,
C   VOLUME 32, NUMBER 3, SEPTEMBER 1967, PP 241-254.
C
        DIMENSION S(1),NEAR(1),SREF(1),LIST(1),A(1),B(1)
        GO TO (10,15,20),JOB
C   JOB=1. INITIALIZATION
10      WRITE(6,2000)
2000    FORMAT(28H0COMPLETE LINKAGE CLUSTERING)
        BIG=SIGN*1.E50
        RETURN
C   JOB=2, DUMMY ENTRY.
15      RETURN
C   JOB=3, UPDATE FOR NEXT ROUND.
20      DO 30 J=1,NCL
        I=LIST(J)
        IF(I.EQ.NREF) GO TO 30
C   RECALL THAT LREF HAS BEEN REMOVED FROM LIST SO I NEED NOT BE
C   TESTED FOR EQUALITY WITH LREF.
        LL=LFIND(I,LREF)
        LN=LFIND(I,NREF)
        IF(((S(LL)-S(LN))*SIGN).LE.0) GO TO 30
        S(LN)=S(LL)
30      CONTINUE
C   UPDATE THE NEAR AND SREF ARRAYS.  IF THE EXTREME ELEMENT IN ROW I
C   WAS EITHER LREF OR NREF, THEN IT IS NECESSARY TO FIND A NEW EXTREME
C   ELEMENT.  ROWS PRIOR TO NREF NEED NOT BE CONSIDERED.
```

```
40      DO 50 J=1,NCL
        I=LIST(J)
        IF(I.EQ.NREF) GO TO 55
50      CONTINUE
55      IF(J.EQ.1) GO TO 80
60      SREF(I)=BIG
        J1=J-1
        DO 70 L=1,J1
        LISTL=LIST(L)
        LL=LFIND(I,LISTL)
        IF(((S(LL)-SREF(I))*SIGN).GE.0.) GO TO 70
        NEAR(I)=LISTL
        SREF(I)=S(LL)
70      CONTINUE
80      J=J+1
        IF(J.GT.NCL) RETURN
        I=LIST(J)
        IF(NEAR(I).EQ.LREF.OR.NEAR(I).EQ.NREF) GO TO 60
        GO TO 80
        END
```

VERSION 3

```
        SUBROUTINE METHOD(S,NEAR,SREF,LIST,NUMBR,SUM,SREFX,SIGN,N,NCL,
        ALREF,NREF,JOB)
C
C HIERARCHICAL CLUSTERING BY MINIMIZING THE AVERAGE DISTANCE OR
C MAXIMIZING THE AVERAGE CORRELATION WITHIN THE NEW GROUP.  THAT IS,
C FOR EACH POTENTIAL MERGE THE AVERAGE OF ALL LINKAGES WITHIN THE
C NEW GROUP IS CALCULATED.
        DIMENSION S(1),NEAR(1),SREF(1),LIST(1),NUMBR(1),SUM(1)
        GO TO (10,25,30),JOB
C   JOB=1,  INITIALIZE.
C NUMBR(I)=NUMBER OF ENTITIES CURRENTLY IN THE I-TH CLUSTER
C SUM(I)=SUM OF ALL PAIRWISE SIMILARITIES AMONG ENTITIES IN THE I-TH
C         CLUSTER
10      WRITE(6,2000)
2000    FORMAT(37H0AVERAGE LINKAGE WITHIN THE NEW GROUP)
        DO 20 J=1,N
        NUMBR(J)=1
20      SUM(J)=0.
        BIG=SIGN*1.E50
        RETURN
C   JOB=2,  DUMMY ENTRY.
25      RETURN
C   JOB=3,  UPDATE FOR NEXT ROUND.
C UPDATE THE NEW CLUSTER
30      NUMBR(NREF)=NUMBR(NREF)+NUMBR(LREF)
        LN=LFIND(LREF,NREF)
        SUM(NREF)=SUM(NREF)+SUM(LREF)+S(LN)
C UPDATE ENTRIES IN THE REDUCED SIMILARITY MATRIX.  THE ENTRIES ARE
C THE SUM TOTAL OF SIMILARITY VALUES ASSOCIATED WITH ALL
C PAIRWISE LINKS BETWEEN THE ELEMENTS OF THE TWO CLUSTERS.
        DO 40 J=1,NCL
        I=LIST(J)
        IF(I.EQ.NREF) GO TO 40
C RECALL THAT LREF HAS BEEN REMOVED FROM LIST AND THEREFORE I NEED NOT
C BE TESTED FOR EQUALITY WITH LREF.
        LL=LFIND(I,LREF)
        LN=LFIND(I,NREF)
        S(LN)=S(LN)+S(LL)
40      CONTINUE
```

```
C  UPDATE THE NEAR AND SREF ARRAYS.  IF THE EXTREME ELEMENT IN ROW I
C  WAS EITHER LREF OR NREF, THEN IT IS NECESSARY TO FIND A NEW EXTREME
C  ELEMENT.  ROWS PRIOR TO NREF NEED NOT BE CONSIDERED.
          DO 50 J=1,NCL
          I=LIST(J)
          IF(I.EQ.NREF) GO TO 55
50        CONTINUE
55        IF(J.EQ.1) GO TO 80
60        SREF(I)=BIG
          J1=J-1
          DO 70 L=1,J1
          LISTL=LIST(L)
          LL=LFIND(I,LISTL)
          NTOT=NUMBR(I)+NUMBR(LISTL)
          NTOT=(NTOT*(NTOT-1))/2
          SREFX=(SUM(I)+SUM(LISTL)+S(LL))/NTOT
          IF(((SREFX-SREF(I))*SIGN).GE.0.) GO TO 70
          NEAR(I)=LISTL
          SREF(I)=SREFX
70        CONTINUE
80        J=J+1
          IF(J.GT.NCL) RETURN
          I=LIST(J)
          IF(NEAR(I).EQ.LREF.OR.NEAR(I).EQ.NREF) GO TO 60
          GO TO 80
          END
```

Version 4

```
      SUBROUTINE METHOD(S,NEAR,SREF,LIST,A,B,SREFX,SIGN,N,NCL,LREF,NREF,
     AJOB)
C
C  HIERARCHICAL CLUSTERING BY SINGLE LINKAGE.  THE ALGORITHM IS DERIVED
C  FROM
C  JOHNSON, S.C., HIERARCHICAL CLUSTERING SCHEMES, PSYCHOMETRIKA,
C  VOLUME 32, NUMBER 3, SEPTEMBER 1967, PP 241-254.
C
      DIMENSION S(1),NEAR(1),SREF(1),LIST(1),A(1),B(1)
      GO TO (10,15,20),JOB
C  JOB=1. INITIALIZATION
10    WRITE(6,3000)
3000  FORMAT(26H0SINGLE LINKAGE CLUSTERING)
      BIG=SIGN*1.E50
      RETURN
C  JOB=2, DUMMY ENTRY.
15    RETURN
C  JOB=3, UPDATE FOR NEXT ROUND.
20    CONTINUE
      DO 50 J=1,NCL
C  UPDATE ENTRIES IN S ARRAY ASSOCIATED WITH NREF
      I=LIST(J)
      IF(I.EQ.NREF) GO TO 50
C  RECALL THAT LREF HAS BEEN REMOVED FROM LIST SO I NEED NOT BE TESTED
C  FOR EQUALITY WITH LREF
      LL=LFIND(I,LREF)
      LN=LFIND(I,NREF)
      IF(((S(LL)-S(LN))*SIGN).GE.0.) GO TO 35
      S(LN)=S(LL)
      IF(I.GT.NREF) GO TO 30
C  I.LT.NREF
C  CHECK WHETHER S(LN) HAS A BETTER VALUE THAN SREF(NREF)
      IF(((S(LN)-SREF(NREF))*SIGN).GT.0.) GO TO 50
```

```
      NEAR(NREF)=I
      SREF(NREF)=S(LN)
      GO TO 50
30    IF(I.GT.LREF) GO TO 40
C     I.GT.NREF.AND.I.LT.LREF
C     CHECK WHETHER S(LN) HAS A BETTER VALUE THAN SREF(I)
      IF(((S(LN)-SREF(I))*SIGN).GE.0.) GO TO 50
      SREF(I)=S(LN)
      NEAR(I)=NREF
      GO TO 50
35    IF(I.LT.LREF) GO TO 50
C     I.GT.LREF
C     UPDATE NEAR ARRAY FOR THOSE ROWS WHOSE EXTREME ELEMENT WAS LREF
40    IF(NEAR(I).NE.LREF) GO TO 50
      NEAR(I)=NREF
      SREF(I)=S(LN)
50    CONTINUE
      RETURN
      END
```

VERSION 5

```
      SUBROUTINE METHOD(S,NEAR,SREF,LIST,NUMBR,SUM,SREFX,SIGN,N,NCL,
     ALREF,NREF,JOB)
C
C     HIERARCHICAL CLUSTERING BY MINIMIZING THE AVERAGE DISTANCE OR
C     MAXIMIZING THE AVERAGE CORRELATION BETWEEN THE MERGED GROUPS.
C
C     THE ALGORITHM IS DERIVED FROM THE *GROUP AVERAGE* METHOD DESCRIBED IN
C     LANCE, G.N. AND W.T. WILLIAMS, A GENERAL THEORY OF CLASSIFICATORY
C     SORTING STRATEGIES, 1. HIERARCHICAL SYSTEMS, THE COMPUTER JOURNAL,
C     VOLUME 9, NUMBER 4, FEBRUARY 1967, PP373-380.
C
      DIMENSION S(1),NEAR(1),SREF(1),LIST(1),NUMBR(1),SUM(1)
      GO TO (10,25,30),JOB
C     JOB=1, INITIALIZE.
C     NUMBR(I)=NUMBER OF ENTITIES CURRENTLY IN THE I-TH CLUSTER
10    WRITE(6,2000)
2000  FORMAT(42HOAVERAGE LINKAGE BETWEEN THE MERGED GROUPS)
      DO 20 J=1,N
20    NUMBR(J)=1
      BIG=SIGN*1.E50
      RETURN
C     JOB=2, DUMMY ENTRY.
25    RETURN
C     JOB=3, UPDATE FOR NEXT ROUND.
C     UPDATE THE NEW CLUSTER
30    NUMBR(NREF)=NUMBR(NREF)+NUMBR(LREF)
C     UPDATE ENTRIES IN THE REDUCED SIMILARITY MATRIX. THE ENTRIES ARE
C     THE SUM TOTAL OF SIMILARITY VALUES ASSOCIATED WITH ALL
C     PAIRWISE LINKS BETWEEN THE ELEMENTS OF THE TWO CLUSTERS.
      DO 40 J=1,NCL
      I=LIST(J)
      IF(I.EQ.NREF) GO TO 40
C     RECALL THAT LREF HAS BEEN REMOVED FROM LIST AND THEREFORE I NEED NOT
C     BE TESTED FOR EQUALITY WITH LREF.
      LL=LFIND(I,LREF)
      LN=LFIND(I,NREF)
      S(LN)=S(LN)+S(LL)
40    CONTINUE
C     UPDATE THE NEAR AND SREF ARRAYS. IF THE EXTREME ELEMENT IN ROW I
```

```
C  WAS EITHER LREF OR NREF, THEN IT IS NECESSARY TO FIND A NEW EXTREME
C  ELEMENT.  ROWS PRIOR TO NREF NEED NOT BE CONSIDERED.
          DO 50 J=1,NCL
          I=LIST(J)
          IF(I.EQ.NREF) GO TO 55
50        CONTINUE
55        IF(J.EQ.1) GO TO 80
60        SREF(I)=BIG
          J1=J-1
          DO 70 L=1,J1
          LISTL=LIST(L)
          LL=LFIND(I,LISTL)
          SREFX=S(LL)/(NUMBR(I)*NUMBR(LISTL))
          IF(((SREFX-SREF(I))*SIGN).GE.0.) GO TO 70
          NEAR(I)=LISTL
          SREF(I)=SREFX
70        CONTINUE
80        J=J+1
          IF(J.GT.NCL) RETURN
          I=LIST(J)
          IF(NEAR(I).EQ.LREF.OR.NEAR(I).EQ.NREF) GO TO 60
          GO TO 80
          END
```

VERSION 6

```
          SUBROUTINE METHOD(S,NEAR,SREF,LIST,A,B,SREFX,SIGN,N,NCL,LREF,NREF,
         AJOB)
C
C  HIERARCHICAL CLUSTERING BY THE MEDIAN METHOD OF
C     GOWER, J.C., A COMPARISON OF SOME METHODS OF CLUSTER ANALYSIS,
C     BIOMETRICS, VOLUME 23, NUMBER 4, DECEMBER 1967, PP 623-637.
C
C
          DIMENSION S(1),NEAR(1),SREF(1),LIST(1),A(1),B(1)
          GO TO (10,15,20),JOB
C  JOB=1. INITIALIZATION
10        WRITE(6,2000)
2000      FORMAT(44H0MEDIAN METHOD OF GOWER. BEWARE OF REVERSALS)
          BIG=SIGN*1.E50
          RETURN
C  JOB=2. DUMMY ENTRY.
15        RETURN
C  JOB=3. UPDATE FOR NEXT ROUND.
20        LBET=LFIND(LREF,NREF)
          DO 30 J=1,NCL
          I=LIST(J)
          IF(I.EQ.NREF) GO TO 30
C  RECALL THAT LREF HAS BEEN REMOVED FROM LIST SO I NEED NOT BE
C  TESTED FOR EQUALITY WITH LREF.
          LL=LFIND(I,LREF)
          LN=LFIND(I,NREF)
C  IF S IS A DECREASING FUNCTION OF SIMILARITY (E.G. DISTANCE) THEN
C     S(LN)=(S(LN)+S(LL))/2.-S(LBET)/4.
C  IF S IS AN INCREASING FUNCTION OF SIMILARITY (E.G. CORRELATION) THEN
C     S(LN)=(S(LN)+S(LL))/2.+(1.-S(LBET))/4.
          S(LN)=(S(LN)+S(LL))/2.-S(LBET)/4.
30        CONTINUE
C  UPDATE THE NEAR AND SREF ARRAYS.  IF THE EXTREME ELEMENT IN ROW I
C  WAS EITHER LREF OR NREF, THEN IT IS NECESSARY TO FIND A NEW EXTREME
C  ELEMENT.  ROWS PRIOR TO NREF NEED NOT BE CONSIDERED.
40        DO 50 J=1,NCL
          I=LIST(J)
```

```
          IF(I.EQ.NREF) GO TO 55
50        CONTINUE
55        IF(J.EQ.1) GO TO 80
60        SREF(I)=BIG
          J1=J-1
          DO 70 L=1,J1
          LISTL=LIST(L)
          LL=LFIND(I,LISTL)
          IF(((S(LL)-SREF(I))*SIGN).GE.0.) GO TO 70
          NEAR(I)=LISTL
          SREF(I)=S(LL)
70        CONTINUE
80        J=J+1
          IF(J.GT.NCL) RETURN
          I=LIST(J)
          IF(NEAR(I).EQ.LREF.OR.NEAR(I).EQ.NREF) GO TO 60
          GO TO 80
          END
```

VERSION 7

```
          SUBROUTINE METHOD(S,NEAR,SREF,LIST,NUMBR,SUM,SREFX,SIGN,N,NCL,
          ALREF,NREF,JOB)
C
C    HIERARCHICAL CLUSTERING BY THE METHOD OF
C       WARD, J.H.,JR. HIERARCHICAL GROUPING TO OPTIMISE AN OBJECTIVE
C       FUNCTION, JOURNAL OF THE AMERICAN STATISTICAL ASSOCIATION, VOLUME
C       58, 1963, PP 236-244.
C
C    THE PARTICULAR ALGORITHM USED HERE IS DESCRIBED IN
C       WISHART, D., AN ALGORITHM FOR HIERARCHICAL CLASSIFICATIONS,
C       BIOMETRICS, VOLUME 22, NUMBER 1, MARCH 1969, PP 165-170.
C
          DIMENSION S(1),NEAR(1),SREF(1),LIST(1),NUMBR(1),SUM(1)
          GO TO (10,25,30),JOB
C    JOB=1, INITIALIZE.
C    NUMBR(I)=NUMBER OF ENTITIES CURRENTLY IN THE I-TH CLUSTER
10        WRITE(6,2000)
2000      FORMAT(44H0HIERARCHICAL GROUPING BY THE METHOD OF WARD)
          DO 20 J=1,N
20        NUMBR(J)=1
          W=0.
          BIG=SIGN*1.E50
          RETURN
C    JOB=2, CALCULATE OBJECTIVE FUNCTION VALUE
25        W=W+SREFX/2.
          SREFX=W
          RETURN
C    JOB=3, UPDATE FOR NEXT ROUND.
30        LBET=LFIND(LREF,NREF)
          NTOT=NUMBR(LREF)+NUMBR(NREF)
          DO 40 J=1,NCL
          I=LIST(J)
          IF(I.EQ.NREF) GO TO 40
C    RECALL THAT LREF HAS BEEN REMOVED FROM LIST SO I NEED NOT BE
C    TESTED FOR EQUALITY WITH LREF.
          LL=LFIND(I,LREF)
          LN=LFIND(I,NREF)
          S(LN)=(S(LN)*(NUMBR(I)+NUMBR(NREF))+S(LL)*(NUMBR(I)+NUMBR(LREF))-
          AS(LBET)*NUMBR(I))/(NTOT+NUMBR(I))
40        CONTINUE
          NUMBR(NREF)=NTOT
```

```
C   UPDATE THE NEAR AND SREF ARRAYS.  IF THE EXTREME ELEMENT IN ROW I
C   WAS EITHER LREF OR NREF, THEN IT IS NECESSARY TO FIND A NEW EXTREME
C   ELEMENT.  ROWS PRIOR TO NREF NEED NOT BE CONSIDERED.
          DO 50 J=1,NCL
          I=LIST(J)
          IF(I.EQ.NREF) GO TO 55
50        CONTINUE
55        IF(J.EQ.1) GO TO 80
60        SREF(I)=BIG
          J1=J-1
          DO 70 L=1,J1
          LISTL=LIST(L)
          LL=LFIND(I,LISTL)
          IF(((S(LL)-SREF(I))*SIGN).GE.0.) GO TO 70
          NEAR(I)=LISTL
          SREF(I)=S(LL)
70        CONTINUE
80        J=J+1
          IF(J.GT.NCL) RETURN
          I=LIST(J)
          IF(NEAR(I).EQ.LREF.OR.NEAR(I).EQ.NREF) GO TO 60
          GO TO 80
          END
```

Subroutine MANAGE

```
      SUBROUTINE MANAGE(X,LIMIT)
C
C   THIS SUBROUTINE ALLOCATES STORAGE, READS INPUT AND CONTROLS
C   EXECUTION FOR A HIERARCHICAL CLUSTERING JOB BASED ON AN INPUT DATA
C   VECTOR FOR EACH DATA UNIT.  ALL CLUSTERING CRITERIA USED WITH THIS
C   PROGRAM ARE OF THE MINIMUM VARIANCE TYPE BECAUSE THE APPROACH
C   DEPENDS ON CALCULATING THE SIMILARITY BETWEEN TWO CLUSTERS USING
C   ONLY THE MEAN (OR TOTAL) VECTOR AND SUM OF SQUARES VECTOR FOR THE
C   TWO CLUSTERS.  THIS TECHNIQUE PERMITS TRADING OFF THE NUMBER OF
C   DATA UNITS WITH THE NUMBER OF VARIABLES.
C
C-------------------------------------------------------------------
C   INPUT SPECIFICATIONS
C
C   CARD 1   TITLE CARD
C   CARD 2   INFORMATION FOR SUBROUTINES *GROUP* AND *TREE*
C     COLS   1- 4   NE=NUMBER OF ENTITIES (DATA UNITS) TO BE CLUSTERED
C     COLS   5- 8   NV=NUMBER OF VARIABLES
C     COLS   9-10   NTSV=TAPE UNIT ON WHICH *GROUP* RESULTS ARE SAVED
C                   NTSV=7, PUNCH RESULTS ON CARDS
C                   NTSV.LE.0, DO NOT SAVE RESULTS
C     COLS  11-12   KOUT=OUTPUT OPTION
C                   KOUT=+2, STANDARD OUTPUT
C                   KOUT=-2, STANDARD OUTPUT PLUS PUNCHED SEQUENCE LIST
C                       FROM SUBROUTINE *TREE*
C
C***ANY PREPOSITIONING OF THE OUTPUT UNIT *NTSV* MUST BE
C   ACCOMPLISHED IN PROGRAM *DRIVER* OR THROUGH USE OF CONTROL CARDS.
C
C   CARD(S) 3  DATA CARDS FOR THIS RUN (NO CARDS IF DATA IS INPUT
C              FROM TAPE OR DISK FILES)
C   CARD 4   END OF RECORD CARD (7/8/9)
C            INCLUDE CARD 4 IN ALL RUNS.  INSURES ALL DATA CARDS HAVE
C            BEEN PASSED SO LABEL CARDS CAN BE READ.
C   CARD(S) 5  LABEL CARDS FOR DATA UNITS.  THERE ARE TWO OPTIONS
```

```
C   1.  INCLUDE 1 CARD WITH THE 4 CHARACTERS *NOLB* IN COLUMNS 1-4.
C       UNDER THIS OPTION LABELS ARE NOT PRINTED ON THE *TREE* OUTPUT.
C
C   2.  INCLUDE *NE* CARDS, COLUMNS 1 TO 20 CONTAINING A LABEL FOR ONE
C       DATA UNIT.   ORDER THE LABEL CARDS IN THE SAME SEQUENCE AS THE
C       DATA UNITS ARE RETURNED FROM SUBROUTINE *USER*.
C-------------------------------------------------------------------------
C
C   DECK SETUP SPECIFICATIONS
C
C   THE USER PROVIDES PROGRAM *DRIVER* WHICH PERFORMS THE FOLLOWING TASKS
C   1.  ASSIGNS INPUT/OUTPUT UNITS
C   2.  ESTABLISHES THE DIMENSION OF THE *X* ARRAY AND SETS THIS
C       DIMENSION EQUAL TO *LIMIT*
C   3.  CALLS SUBROUTINE *MANAGE*.
C   THE FOLLOWING EXAMPLE WILL SUFFICE IN MOST CASES.
C
C       PROGRAM DRIVER (INPUT,OUTPUT,PUNCH,TAPE5=INPUT,TAPE6=OUTPUT,
C      ATAPE7=PUNCH,TAPE1,TAPE2)
C       DIMENSION X(7000)
C       LIMIT=7000
C       CALL MANAGE(X,LIMIT)
C       END
C
C   A SECOND JOB DEPENDENT SEGMENT IS SUBROUTINE *PROC*.  THE USER
C   SELECTS AMONG THE SEVERAL ALTERNATIVE VERSIONS OF THIS SUBROUTINE TO
C   IMPLEMENT THE DESIRED CLUSTERING TECHNIQUE.
C
C   THE USER ALSO PROVIDES SUBROUTINE *USER* WHICH IS THE INPUT ROUTINE
C   FOR THE DATA TO BE USED IN THE RUN.  EACH CALL TO SUBROUTINE *USER*
C   SHOULD RESULT IN THE READING AND/OR GENERATION OF DATA FOR ONE DATA
C   UNIT.  THE FOLLOWING OPERATIONS ARE AMONG THE MANY POSSIBILITIES THAT
C   MAY BE INCLUDED IN SUBROUTINE *USER*.
C   1.  MERGE RECORDS FROM SEVERAL FILES TO GENERATE A SINGLE DATA
C       VECTOR FOR EACH DATA UNIT.
C   2.  APPLY TRANSFORMATIONS TO ANY VARIABLE.
C   3.  GENERATE NEW VARIABLES AS FUNCTIONS OF THOSE THAT WERE READ IN.
C
C   ALL DATA RETURNED FROM SUBROUTINE *USER* MUST BE IN FLOATING POINT
C   MODE FOR THE CLUSTERING PROGRAM TO OPERATE PROPERLY.
C
C   THE FOLLOWING EXAMPLE ILLUSTRATES SOME USES OF SUBROUTINE *USER*.
C       SUBROUTINE USER(X)
C       DIMENSION X(1)
C       READ(5,100) (X(I),I=1,3)
C       READ(1,200) X(4)
C       X(3)=X(4)/X(3)
C       X(5)=2.*X(4)+X(1)*X(2)
C       RETURN
C 100   FORMAT(F10.6,2F9.3)
C 200   FORMAT(12X,F5.3)
C       END
C
C-------------------------------------------------------------------------
C
C   THE *X* ARRAY IS PARTITIONED FOR STORAGE AS FOLLOWS
C   STORAGE FOR ARRAYS NEEDED AT ALL STAGES OF THE JOB
C   X(N1) TO X(N2-1)  NE WORDS--STORAGE OF THE II ARRAY
C   X(N2) TO X(N3-1)  NE WORDS--STORAGE OF THE JJ ARRAY
C   X(N3) TO X(N4-1)  NE WORDS--STORAGE OF THE SS ARRAY
C   X(N4) TO X(N5-1)  NE WORDS--STORAGE OF THE IL ARRAY
C   X(N5) TO X(N6-1)  NE WORDS--STORAGE OF THE JL ARRAY
C   X(N6) TO X(N7-1)  NE WORDS--STORAGE OF THE NEXT ARRAY
C   STORAGE FOR ARRAYS NEEDED IN SUBROUTINE *GROUP*
```

```
C    M1=N7
C    X(M1) TO X(M2-1)   NE*NV WORDS-- STORAGE OF THE DATA VECTORS
C    X(M2) TO X(M3-1)   NE WORDS--STORAGE OF THE SUMSQ ARRAY
C    X(M3) TO X(M4-1)   NE WORDS--STORAGE OF THE SQDEV ARRAY
C    X(M4) TO X(M5-1)   NE WORDS--STORAGE OF THE NUMBR ARRAY
C    X(M5) TO X(M6-1)   NE WORDS--STORAGE OF THE LIST ARRAY
C    X(M6) TO X(M7-1)   NE WORDS--STORAGE OF THE LAST ARRAY
C    X(M7) TO X(M8-1)   NE WORDS--STORAGE OF THE NEAR ARRAY
C    X(M8) TO X(M9)     NE WORDS--STORAGE OF THE SREF ARRAY
C    STORAGE FOR ARRAYS NEEDED IN SUBROUTINE *TREE* (OVERLAY ARRAYS NEEDED
C    IN SUBROUTINE *GROUP*)
C    L1=N7
C    X(L1) TO X(L2-1)   25*NE WORDS--STORAGE OF THE A ARRAY
C    X(L2) TO X(L3-1)   5*NE WORDS--STORAGE OF THE LABEL ARRAY
C    X(L3) TO X(L4-1)   NE WORDS--STORAGE OF THE LCLNO ARRAY
C    X(L4) TO X(L5-1)   NE WORDS--STORAGE OF THE LINE ARRAY
C    X(L5) TO X(L6-1)   NE WORDS--STORAGE OF THE IS ARRAY
C    X(L6) TO X(L7)     NE WORDS--STORAGE OF THE LAST ARRAY
C------------------------------------------------------------------------
C
      INTEGER FIRST
      DIMENSION X(1),TITLE(20),EPS(25)
      READ(5,1000) TITLE
      READ(5,1100) NE,NV,NTSV,KOUT
      WRITE(6,2500) TITLE
      WRITE(6,2200) NE,NV,NTSV,KOUT
C    PARTITION THE STORAGE ARRAY
      N1=1
      N2=N1+NE
      N3=N2+NE
      N4=N3+NE
      N5=N4+NE
      N6=N5+NE
      N7=N6+NE
      M2=N7+NE*NV
      M3=M2+NE
      M4=M3+NE
      M5=M4+NE
      M6=M5+NE
      M7=M6+NE
      M8=M7+NE
      M9=M8+NE-1
      L2=N7+25*NE
      L3=L2+5*NE
      L4=L3+NE
      L5=L4+NE
      L6=L5+NE
      L7=L6+NE-1
C    CHECK FOR SUFFICIENT STORAGE
      MAX=M9
      IF(L7.GT.MAX) MAX=L7
      WRITE(6,2300) MAX,LIMIT
      IF(MAX.GT.LIMIT) STOP
C    READ AND/OR GENERATE THE DATA VECTOR FOR EACH DATA UNIT
      FIRST=N7
      DO 40 I=1,NE
      CALL USER(X(FIRST))
40    FIRST=FIRST+NV
C    PASS THE EOF
      READ(5,1000) Z
      IF(EOF,5) 50,90
50    CONTINUE
C    READY TO CLUSTER
      CALL GROUP(X(N1),X(N2),X(N3),X(N4),X(N5),X(N6),X(N7),X(M2),X(M3),
     AX(M4),X(M5),X(M6),X(M7),X(M8),TITLE,NE,NV,NTSV)
```

```
C   READ LABEL CARD(S)
        FIRST=L2
        LAST=L2+4
        READ(5,1000) (X(I),I=FIRST,LAST)
        IF(X(FIRST).EQ.4HNOLB) GO TO 80
C   READ REMAINING LABELS
        DO 70 K=2,NE
        FIRST=LAST+1
        LAST=LAST+5
70      READ(5,1000) (X(I),I=FIRST,LAST)
C   DRAW THE TREE CORRESPONDING TO THE CLUSTERING
80      MERGES=NE-1
        CALL TREE(X(N1),X(N2),X(N3),X(N4),X(N5),X(N6),X(N7),X(L2),X(L3),
       AX(L4),X(L5),X(L6),EPS,TITLE,MERGES,1,6,1,KOUT,NE)
        RETURN
C   ERROR MESSAGES
90      WRITE(6,2700)
        LAST=FIRST-1
        FIRST=FIRST-NV
        WRITE(6,2800) I,FIRST,LAST,Z,(X(I),I=FIRST,LAST)
        RETURN
1000    FORMAT(20A4)
1100    FORMAT(2I4,2I2)
2200    FORMAT(5HONE =,I8,/,5H NV =,I8,/,7H NTSV =,I6,/,7H KOUT =,I6)
2300    FORMAT(19HOREQUIRED STORAGE =,I5,6H WORDS,/,
       A        19HOALLOTTED STORAGE =,I5,6H WORDS,/)
2500    FORMAT(1H0,20A4)
2700    FORMAT(30HONO EOR WHEN ONE WAS EXPECTED.)
2800    FORMAT(1X,3I10,E12.4,/,(1X,10E12.4))
        END
```

Subroutine GROUP

```
        SUBROUTINE GROUP(II,JJ,SS,IL,JL,NEXT,DATA,SUMSQ,SQDEV,NUMBR,LIST,
       ALAST,NEAR,SREF,TITLE,NS,NV,NT)
C
C   THIS SUBROUTINE IMPLEMENTS THE WARD HIERARCHICAL CLUSTERING METHOD
C   FOR A VARIETY OF CRITERION FUNCTIONS
C
C   THE SOURCE PAPER FOR THIS METHOD IS
C   WARD, J.H., JR., HIERARCHICAL GROUPING TO OPTIMIZE AN OBJECTIVE
C   FINCTION, JOURNAL OF THE AMERICAN STATISTICAL ASSOCIATION, VOLUME 58,
C   1963, PP 236-244.
C
C   THE FOLLOWING VARIABLES ARE SPECIFIED IN THE CALLING PROGRAM AND
C   ARE PASSED THROUGH THE ARGUMENT LIST
C     NS=NUMBER OF SUBJECTS (DATA UNITS) TO BE CLUSTERED
C     NV=NUMBER OF VARIABLES
C     DATA=DATA MATRIX STORED AS A ONE-DIMENSIONAL ARRAY.  EACH BLOCK OF
C       *NV* ELEMENTS REPRESENTS ONE SUBJFCT
C     NT=TAPE UNIT ON WHICH THE RESULTS ARE SAVED
C     TITLE=IDENTIFYING TITLE FOR THIS RUN
C
C   THE FOLLOWING VARIABLES REPRESENT THE OUTPUT OF THE PROGRAM AND ARE
C   PASSED BACK THROUGH THE ARGUMENT LIST.  THESE RESULTS ARE READY FOR
C   SUBROUTINE *TREE*.
C     K=STAGE OF CLUSTERING
C     II(K)=LOWER NUMBERED CLUSTER MERGED AT STAGE K
C     JJ(K)=UPPER NUMBERED CLUSTER MERGED AT STAGE K
C     SS(K)=VALUE OF SIMILARITY FUNCTION ASSOCIATED WITH MERGE AT STAGE K
C     IL(K)=PRECEDING STAGE AT WHICH II(K) WAS LAST IN A MERGE
C     JL(K)=PRECEDING STAGE AT WHICH JJ(K) WAS LAST IN A MERGE
```

```
C     NEXT(K)=NEXT STAGE AT WHICH II(K) IS IN A MERGE
C
C  IN ADDITION, THE FOLLOWING VARIABLES PLAY IMPORTANT ROLES IN THE PROGRA
C     NEAR(I)=ID NUMBER OF NEAREST NEIGHBOR FOR CLUSTER I
C     SREF(I)=SIMILARITY MEASURE FOR THE PAIR (I,NEAR(I))
C     LIST(I)=I TH CLUSTER ID NUMBER IN SEQUENTIAL LIST OF CURRENT CLUSTERS
C     NCL=NUMBER OF CLUSTERS AT CURRENT STAGE
C     LAST(I)=STAGE NUMBER AT WHICH CLUSTER I WAS LAST IN A MERGE
C     SUMSQ(I)=SUM OF SQUARED ATTRIBUTE VALUES IN CLUSTER I
C     NUMBR(I)=NUMBER OF SUBJECTS IN CLUSTER I
C     SQDEV(I)=SUM OF WITHIN GROUP SQUARED DEVIATIONS FOR CLUSTER I
C
      DIMENSION II(1),JJ(1),SS(1),IL(1),JL(1),NEXT(1),DATA(1),SUMSQ(1),
     ASQDEV(1),NUMBR(1),LIST(1),LAST(1),NEAR(1),SREF(1),TITLE(20)
C  INITIALIZE VARIABLES AND SET CONSTANTS
      NCL=NS
      K=1
      BIG=1.E50
C  INITIALIZE THE OBJECTIVE FUNCTION
      CALL PROC (DATA,SUMSQ,SQDEV,NUMBR,NV,NS,I,J,SREFX,1)
C  INITIALIZE ARRAYS
      DO 10 J=1,NS
      LAST(J)=0
      NEXT(J)=0
      LIST(J)=J
      SREF(J)=BIG
10    CONTINUE
C  FIND EXTREME ENTRY IN EACH ROW OF SIMULATED LOWER TRIANGULAR
C  SIMILARITY MATRIX
      DO 30 I=2,NS
      I1=I-1
      DO 30 J=1,I1
      CALL   PROC(DATA,SUMSQ,SQDEV,NUMBR,NV,NS,I,J,STEST,2)
      IF(STEST.GE.SREF(I)) GO TO 30
      NEAR(I)=J
      SREF(I)=STEST
30    CONTINUE
C  MAIN LOOP.  FIND EXTREME VALUE IN SREF ARRAY
40    SREFX=BIG
      DO 50 I=2,NCL
      LISTI=LIST(I)
      IF(SREF(LISTI).GT.SREFX) GO TO 50
      IREF=I
      LREF=LISTI
      SREFX=SREF(LISTI)
50    CONTINUE
C  LREF AND NEAR(LREF) FORM THE CLOSEST PAIR.  SINCE WE ACCEPTED EQUALIT
C  AS AN IMPROVEMENT IN FINDING EXTREME VALUES IN THE S ARRAY, WE INSURE
C  THAT LREF.GT.NEAR(LREF) SINCE THESE TWO ARE MUTUALLY CLOSEST TO
C  EACH OTHER.
C  UPDATE THE DATA REPRESENTATION FOR CLUSTER NREF
      NREF=NEAR(LREF)
      CALL   PROC(DATA,SUMSQ,SQDEV,NUMBR,NV,NS,NREF,LREF,SREFX,3)
C  GENERATE MERGE DATA NEEDED FOR SUBROUTINE TREE
      II(K)=NREF
      JJ(K)=LREF
      SS(K)=SREFX
      IL(K)=LAST(NREF)
      JL(K)=LAST(LREF)
      LAST(NREF)=K
      IF(IL(K).EQ.0) GO TO 60
      ILK=IL(K)
      NEXT(ILK)=K
60    IF(JL(K).EQ.0) GO TO 70
```

```
      JLK=JL(K)
      NEXT(JLK)=K
70    K=K+1
C   TERMINATE IF N-1 MERGES HAVE OCCURED
      IF(K.EQ.NS) GO TO 140
C   UPDATE FOR THE NEXT CYCLE
      NCL=NCL-1
      IF(IREF.GT.NCL) GO TO 90
C   UPDATE LIST ARRAY BY REMOVING LREF AND PUSHING DOWN THE LIST
      DO 80 I=IREF,NCL
80    LIST(I)=LIST(I+1)
C   UPDATE THE SREF AND NEAR LISTS
90    DO 130 J=1,NCL
      I=LIST(J)
      IF(I-NREF) 130,110,95
C   CHECK WHETHER THE NEW CLUSTER (NREF) MIGHT BE A MORE DESIREABLE
C   MATE FOR CLUSTER I
95    CALL   PROC(DATA,SUMSQ,SQDEV,NUMBR,NV,NS,I,NREF,STEST,2)
      IF(STEST.GT.SREF(I)) GO TO 100
      SREF(I)=STEST
      NEAR(I)=NREF
      GO TO 130
C   TEST WHETHER THE NEAREST NEIGHBOR FOR I WAS LREF OR NREF
100   IF(NEAR(I).NE.LREF.AND.NEAR(I).NE.NREF) GO TO 130
C   SEARCH FOR NEW NEAREST NEIGHBOR
110   SREF(I)=BIG
      DO 125 L=1,NCL
      LISTL=LIST(L)
      IF(LISTL.EQ.I) GO TO 130
      CALL   PROC(DATA,SUMSQ,SQDEV,NUMBR,NV,NS,I,LISTL,STEST,2)
      IF(STEST.GE.SREF(I)) GO TO 125
      NEAR(I)=LISTL
      SREF(I)=STEST
125   CONTINUE
130   CONTINUE
      GO TO 40
C   CLUSTERING FINISHED AND ALL ANCILLARY INFORMATION GENERATED.
C   SAVE RESULTS AS DESIRED.
140   K=K-1
160   IF(NT.LE.0) RETURN
      WRITE(NT,2300) TITLE
      DO 170 I=1,K
170   WRITE(NT,2200) I,II(I),JJ(I),SS(I),IL(I),JL(I),NEXT(I)
      RETURN
2200  FORMAT(3I10,E16.8,3I10)
2300  FORMAT(20A4)
      END
```

Subroutine PROC

VERSION 1

```
      SUBROUTINE   PROC(DATA,SUMSQ,SQDEV,NUMBR,NV,NS,I,J,SCOEF,JOB)
C   TWGSD=TOTAL WITHIN GROUP SUM OF SQUARED DEVIATIONS
      DIMENSION DATA(1),SUMSQ(1),SQDEV(1),NUMBR(1)
      GO TO (5,15,25),JOB
5     WRITE(6,2000)
2000  FORMAT(48HOMINIMUM INCREASE IN WITHIN GROUP SUM OF SQUARES,/,
     A16HOWARD CRITERION.)
C   INITIALIZE ARRAYS
      LWRD=0
      DO 10 L=1,NS
```

```
         NUMBR(L)=1
         SQDEV(L)=0.
         SUMSQ(L)=0.
         DO 10 K=1,NV
         LWRD=LWRD+1
10       SUMSQ(L)=SUMSQ(L)+DATA(LWRD)*DATA(LWRD)
         TWGSD=0.
         RETURN
C  CALCULATE THE WITHIN GROUP SQUARED DEVIATIONS
15       IWRD=(I-1)*NV
         JWRD=(J-1)*NV
         TSUM2=0.
         N=NUMBR(I)+NUMBR(J)
         DO 20 K=1,NV
         SUM=DATA(IWRD+K)+DATA(JWRD+K)
20       TSUM2=TSUM2+SUM*SUM
         SCOEF=SUMSQ(I)+SUMSQ(J)-TSUM2/N-SQDEV(I)-SQDEV(J)
         RETURN
C  UPDATE THE INFORMATION NEEDED FOR THIS CRITERION
25       IWRD=(I-1)*NV
         JWRD=(J-1)*NV
C  I IS TAKEN AS THE LABEL FOR THE NEW CLUSTER
         DO 30 K=1,NV
30       DATA(IWRD+K)=DATA(IWRD+K)+DATA(JWRD+K)
         SUMSQ(I)=SUMSQ(I)+SUMSQ(J)
         NUMBR(I)=NUMBR(I)+NUMBR(J)
         SQDEV(I)=SQDEV(I)+SQDEV(J)+SCOEF
         TWGSD=TWGSD+SCOEF
         SCOEF=TWGSD
         RETURN
         END
```

Version 2

```
         SUBROUTINE  PROC(DATA,SUMSQ,SQDEV,NUMBR,NV,NS,I,J,SCOEF,JOB)
         DIMENSION DATA(1),SUMSQ(1),SQDEV(1),NUMBR(1)
         GO TO (5,15,25),JOB
5        WRITE(6,2000)
2000     FORMAT(54H0CENTROID CLUSTERING.  REPORTED CRITERION VALUE IS THE,
        A/,59H SQUARED EUCLIDEAN DISTANCE BETWEEN THE CENTROIDS OF MERGED,
        B/,33H CLUSTERS.  BEWARED OF REVERSALS.)
C  INITIALIZE ARRAYS
         DO 10 L=1,NS
10       NUMBR(L)=1
         RETURN
C  CALCULATE SQUARED DISTANCE BETWEEN CLUSTER MEANS
15       IWRD=(I-1)*NV
         JWRD=(J-1)*NV
         SCOEF=0.
         DO 20 K=1,NV
20       SCOEF=SCOEF+(DATA(IWRD+K)/NUMBR(I)-DATA(JWRD+K)/NUMBR(K))**2
         RETURN
C  UPDATE THE INFORMATION NEEDED FOR THIS CRITERION
25       IWRD=(I-1)*NV
         JWRD=(J-1)*NV
C  I IS TAKEN AS THE LABEL FOR THE NEW CLUSTER
         DO 30 K=1,NV
30       DATA(IWRD+K)=DATA(IWRD+K)+DATA(JWRD+K)
         NUMBR(I)=NUMBR(I)+NUMBR(J)
         RETURN
         END
```

VERSION 3

```
      SUBROUTINE   PROC(DATA,SUMSQ,SQDEV,NUMBR,NV,NS,I,J,SCOEF,JOB)
C   TWGSD=TOTAL WITHIN GROUP SUM OF SQUARED DEVIATIONS
      DIMENSION DATA(1),SUMSQ(1),SQDEV(1),NUMBR(1)
      GO TO (5,15,25),JOB
5     WRITE(6,2000)
2000  FORMAT(66H0MEAN WITHIN GROUP SQUARED DEVIATION IN THE NEW CLUSTER
     AIS MINIMAL)
C   INITIALIZE ARRAYS
      LWRD=0
      DO 10 L=1,NS
      NUMBR(L)=1
      SUMSQ(L)=0.
      DO 10 K=1,NV
      LWRD=LWRD+1
10    SUMSQ(L)=SUMSQ(L)+DATA(LWRD)*DATA(LWRD)
      TWGSD=0.
      RETURN
C   CALCULATE THE VARIANCE IN THE GROUP FORMED BY MERGING I AND J
15    IWRD=(I-1)*NV
      JWRD=(J-1)*NV
      TSUM2=0.
      N=NUMBR(I)+NUMBR(J)
      DO 20 K=1,NV
      SUM=DATA(IWRD+K)+DATA(JWRD+K)
20    TSUM2=TSUM2+SUM*SUM
      SCOEF=(SUMSQ(I)+SUMSQ(J)-TSUM2/N)/N
      RETURN
C   UPDATE THE INFORMATION NEEDED FOR THIS CRITERION
25    IWRD=(I-1)*NV
      JWRD=(J-1)*NV
C   I IS TAKEN AS THE LABEL FOR THE NEW CLUSTER
      DO 30 K=1,NV
30    DATA(IWRD+K)=DATA(IWRD+K)+DATA(JWRD+K)
      SUMSQ(I)=SUMSQ(I)+SUMSQ(J)
      NUMBR(I)=NUMBR(I)+NUMBR(J)
      RETURN
      END
```

VERSION 4

```
      SUBROUTINE   PROC(DATA,SUMSQ,SQDEV,NUMBR,NV,NS,I,J,SCOEF,JOB)
C   TWGSD=TOTAL WITHIN GROUP SUM OF SQUARED DEVIATIONS
      DIMENSION DATA(1),SUMSQ(1),SQDEV(1),NUMBR(1)
      GO TO (5,15,25),JOB
5     WRITE(6,2000)
2000  FORMAT(58H0WITHIN GROUP SUM OF SQUARES IN THE NEW CLUSTER IS MINIM
     AAL)
C   INITIALIZE ARRAYS
      LWRD=0
      DO 10 L=1,NS
      NUMBR(L)=1
      SUMSQ(L)=0.
      DO 10 K=1,NV
      LWRD=LWRD+1
10    SUMSQ(L)=SUMSQ(L)+DATA(LWRD)*DATA(LWRD)
      TWGSD=0.
      RETURN
C   CALCULATE THE WITHIN GROUP SQUARED DEVIATION IN THE CLUSTER FORMED BY
C   MERGING I AND J.
```

```
15      IWRD=(I-1)*NV
        JWRD=(J-1)*NV
        TSUM2=0.
        N=NUMBR(I)+NUMBR(J)
        DO 20 K=1,NV
        SUM=DATA(IWRD+K)+DATA(JWRD+K)
20      TSUM2=TSUM2+SUM*SUM
        SCOEF=SUMSQ(I)+SUMSQ(J)-TSUM2/N
        RETURN
C  UPDATE THE INFORMATION NEEDED FOR THIS CRITERION
25      IWRD=(I-1)*NV
        JWRD=(J-1)*NV
C   I IS TAKEN AS THE LABEL FOR THE NEW CLUSTER
        DO 30 K=1,NV
30      DATA(IWRD+K)=DATA(IWRD+K)+DATA(JWRD+K)
        SUMSQ(I)=SUMSQ(I)+SUMSQ(J)
        NUMBR(I)=NUMBR(I)+NUMBR(J)
        RETURN
        END
```

VERSION 5

```
        SUBROUTINE  PROC(DATA,SUMSQ,SQDEV,NUMBR,NV,NS,I,J,SCOEF,JOB)
C  TWGSD=TOTAL WITHIN GROUP SUM OF SQUARED DEVIATIONS
        DIMENSION DATA(1),SUMSQ(1),SQDEV(1),NUMBR(1)
        GO TO (5,15,25),JOB
5       WRITE(6,2000)
2000    FORMAT(66H0MEAN WITHIN GROUP SQUARED DEVIATION IN THE NEW CLUSTER
       AIS MINIMAL,/,84H0PRINTED CRITERION IS TOTAL WITHIN GROUPS SUM OF S
       AQUARED DEVIATIONS (WARD CRITERION))
C  INITIALIZE ARRAYS
        LWRD=0
        DO 10 L=1,NS
        NUMBR(L)=1
        SQDEV(L)=0.
        SUMSQ(L)=0.
        DO 10 K=1,NV
        LWRD=LWRD+1
10      SUMSQ(L)=SUMSQ(L)+DATA(LWRD)*DATA(LWRD)
        TWGSD=0.
        RETURN
C  CALCULATE THE VARIANCE IN THE GROUP FORMED BY MERGING I AND J
15      IWRD=(I-1)*NV
        JWRD=(J-1)*NV
        TSUM2=0.
        N=NUMBR(I)+NUMBR(J)
        DO 20 K=1,NV
        SUM=DATA(IWRD+K)+DATA(JWRD+K)
20      TSUM2=TSUM2+SUM*SUM
        SCOEF=(SUMSQ(I)+SUMSQ(J)-TSUM2/N)/N
        RETURN
C  UPDATE THE INFORMATION NEEDED FOR THIS CRITERION
25      IWRD=(I-1)*NV
        JWRD=(J-1)*NV
C   I IS TAKEN AS THE LABEL FOR THE NEW CLUSTER
        DO 30 K=1,NV
30      DATA(IWRD+K)=DATA(IWRD+K)+DATA(JWRD+K)
        SUMSQ(I)=SUMSQ(I)+SUMSQ(J)
        NUMBR(I)=NUMBR(I)+NUMBR(J)
        SCOEF=SCOEF*NUMBR(I)
        TWGSD=TWGSD+SCOEF-SQDEV(I)-SQDEV(J)
        SQDEV(I)=SCOEF
        SCOEF=TWGSD
        RETURN
        END
```

VERSION 6

```
      SUBROUTINE  PROC(DATA,SUMSQ,SQDEV,NUMBR,NV,NS,I,J,SCOEF,JOB)
C   TWGSD=TOTAL WITHIN GROUP SUM OF SQUARED DEVIATIONS
      DIMENSION DATA(1),SUMSQ(1),SQDEV(1),NUMBR(1)
      GO TO (5,15,25),JOB
5     WRITE(6,2000)
2000  FORMAT(58HOWITHIN GROUP SUM OF SQUARES IN THE NEW CLUSTER IS MINIM
     AAL,       /,84HOPRINTED CRITERION IS TOTAL WITHIN GROUPS SUM OF S
     AQUARED DEVIATIONS (WARD CRITERION))
C   INITIALIZE ARRAYS
      LWRD=0
      DO 10 L=1,NS
      NUMBR(L)=1
      SQDEV(L)=0.
      SUMSQ(L)=0.
      DO 10 K=1,NV
      LWRD=LWRD+1
10    SUMSQ(L)=SUMSQ(L)+DATA(LWRD)*DATA(LWRD)
      TWGSD=0.
      RETURN
C   CALCULATE THE WITHIN GROUP SQUARED DEVIATION IN THE CLUSTER FORMED BY
C   MERGING I AND J.
15    IWRD=(I-1)*NV
      JWRD=(J-1)*NV
      TSUM2=0.
      N=NUMBR(I)+NUMBR(J)
      DO 20 K=1,NV
      SUM=DATA(IWRD+K)+DATA(JWRD+K)
20    TSUM2=TSUM2+SUM*SUM
      SCOEF=SUMSQ(I)+SUMSQ(J)-TSUM2/N
      RETURN
C   UPDATE THE INFORMATION NEEDED FOR THIS CRITERION
25    IWRD=(I-1)*NV
      JWRD=(J-1)*NV
C   I IS TAKEN AS THE LABEL FOR THE NEW CLUSTER
      DO 30 K=1,NV
30    DATA(IWRD+K)=DATA(IWRD+K)+DATA(JWRD+K)
      SUMSQ(I)=SUMSQ(I)+SUMSQ(J)
      NUMBR(I)=NUMBR(I)+NUMBR(J)
      TWGSD=TWGSD+SCOEF-SQDEV(I)-SQDEV(J)
      SQDEV(I)=SCOEF
      SCOEF=TWGSD
      RETURN
      END
```

VERSION 7

```
      SUBROUTINE  PROC(DATA,SUMSQ,SQDEV,NUMBR,NV,NS,I,J,SCOEF,JOB)
C   TWGSD=TOTAL WITHIN GROUP SUM OF SQUARED DEVIATIONS
      DIMENSION DATA(1),SUMSQ(1),SQDEV(1),NUMBR(1)
      GO TO (5,15,25),JOB
5     WRITE(6,2000)
2000  FORMAT(20HOCENTROID CLUSTERING,
     A          /,84HOPRINTED CRITERION IS TOTAL WITHIN GROUPS SUM OF S
     AQUARED DEVIATIONS (WARD CRITERION))
C   INITIALIZE ARRAYS
      LWRD=0
      DO 10 L=1,NS
      NUMBR(L)=1
      SQDEV(L)=0.
      SUMSQ(L)=0.
      DO 10 K=1,NV
      LWRD=LWRD+1
```

```
10      SUMSQ(L)=SUMSQ(L)+DATA(LWRD)*DATA(LWRD)
        TWGSD=0.
        RETURN
C  CALCULATE SQUARED DISTANCE BETWEEN CLUSTER MEANS
15      IWRD=(I-1)*NV
        JWRD=(J-1)*NV
        SCOEF=0.
        DO 20 K=1,NV
20      SCOEF=SCOEF+(DATA(IWRD+K)/NUMBR(I)-DATA(JWRD+K)/NUMBR(K))**2
        RETURN
C  UPDATE THE INFORMATION NEEDED FOR THIS CRITERION
25      IWRD=(I-1)*NV
        JWRD=(J-1)*NV
        TSUM2=0.
C  I IS TAKEN AS THE LABEL FOR THE NEW CLUSTER
        DO 30 K=1,NV
        DATA(IWRD+K)=DATA(IWRD+K)+DATA(JWRD+K)
30      TSUM2=TSUM2+DATA(IWRD+K)**2
        SUMSQ(I)=SUMSQ(I)+SUMSQ(J)
        NUMBR(I)=NUMBR(I)+NUMBR(J)
        SCOEF=SUMSQ(I)-TSUM2/NUMBR(I)-SQDEV(I)-SQDEV(J)
        SQDEV(I)=SQDEV(I)+SQDEV(J)+SCOEF
        TWGSD=TWGSD+SCOEF
        SCOEF=TWGSD
        RETURN
        END
```

Subroutine ALLIN1

```
        SUBROUTINE ALLIN1(X,LIMIT)
C
C  THIS SUBROUTINE ALLOCATES STORAGE, READS INPUT AND CONTROLS EXECUTION
C  FOR A HIERARCHICAL CLUSTERING JOB BASED ON AN INPUT SEQUENCE OF
C  SORTED SIMILARITY MEASURES.  TWO SITUATIONS ARE OF INTEREST.
C     1  S(I,J) DESCRIBES THE SIMILARITY (DISSIMILARITY) BETWEEN OBJECTS
C        I AND J.  THE ENTIRE LOWER TRIANGULAR SIMILARITY MATRIX OF
C        (S(I,J),I,J) TRIPLES HAS BEEN SORTED ON S(I,J) INTO A
C        NON-INCREASING (NON-DECREASING) SEQUENCE WHICH IS STORED ON
C        UNIT *NTIN*.  THIS PROGRAM WILL FIND THE MAXIMUM (MINIMUM)
C        SPANNING TREE AND THEREBY PERFORM A SINGLE LINKAGE CLUSTERING
C        OF THE DATA.
C
C     2. S(I,J) IS THE CRITERION VALUE ASSOCIATED WITH THE MERGER OF
C        CLUSTERS I AND J THROUGH SOME HIERARCHICAL CLUSTERING METHOD.
C        THE NE-1 TRIPLES (S(I,J),I,J) ARE ASSUMED TO BE SORTED IN EITHER
C        NON-INCREASING (CORRELATION) OR NON-DECREASING (DISTANCE)
C        SEQUENCE AND STORED ON UNIT *NTIN*.  THIS PROGRAM WILL GENERATE
C        THE NECESSARY INPUT TO SUBROUTINE *TREE* FOR DRAWING A TREE ON
C        THE LINE PRINTER.
C  IN EITHER CASE THE SIMILARITY TRIPLES MUST BE SORTED EXTERNALLY TO
C  THIS PROGRAM.  MOST LARGE COMPUTING CENTERS HAVE SORT-MERGE
C  PACKAGES WELL SUITED TO THIS PRELIMINARY SORTING.
C
C------------------------------------------------------------------------
C  INPUT SPECIFICATIONS
C
C  CARD 1   TITLE CARD
C  CARD 2   PARAMETER CARD
C     COLS  1- 4   NE=NUMBER OF ENTITIES (SUBJECTS OR ATTRIBUTES) TO BE
C                     CLUSTERED
C     COLS  5- 8   NREC=NUMBER OF (S(I,J),I,J) TRIPLES IN A RECORD ON
C                     UNIT *NTIN*
```

```
C     COLS  9-10   NTIN=UNIT STORING SORTED SIMILARITIES
C                  NTIN=5, CARD READER
C                  NTIN.NE.5, DISK OR TAPE
C     COLS 11-12   IOPT=OPTION FOR READING SIMILARITIES
C                  IOPT=1, READ ACCORDING TO FORMAT *FMT* ON CARD 3 BELOW
C                  IOPT.NE.1, BINARY READ, ORDINARILY FROM TAPE OR DISK
C     COLS 13-14   NTSV=UNIT ON WHICH CLUSTERING RESULTS ARE SAVED
C                  NTSV=7, PUNCH RESULTS ON CARDS
C                  NTSV.LE.0, DO NOT SAVE RESULTS
C     COLS 15-16   KOUT=OUTPUT OPTION
C                  KOUT=+2, STANDARD OUTPUT
C                  KOUT=-2, STANDARD OUTPUT PLUS PUNCHED SEQUENCE LIST
C                            FROM SUBROUTINE *TRFE*
C**ANY PREPOSITIONING OF THE I/O UNITS *NTSV* AND *NTIN* MUST BE
C**ACCOMPLISHED IN PROGRAM *DRIVER* OR THROUGH CONTROL CARDS.
C
C   CARD 3  INPUT FORMAT FOR SIMILARITIES
C   CARD(S) 4  SORTED SIMILARITY TRIPLES
C   CARD 5  END OF RECORD CARD (7/8/9)
C
C***INCLUDE CARDS 4 AND 5 ONLY IF THE SIMILARITIES ARE ON CARDS
C
C   CARD(S) 6  LABEL CARDS FOR ENTITIES.  THERE ARE TWO OPTIONS.
C   1.  INCLUDE 1 CARD WITH THE 4 CHARACTERS *NOLB* IN COLUMNS 1-4.
C       UNDER THIS OPTION LABELS ARE NOT PRINTED ON THE TREE OUTPUT.
C
C   2.  INCLUDE NE CARDS, COLUMNS 1 TO 20 CONTAINING A LABEL FOR ONE
C       ENTITY.  ORDER THE LABEL CARDS IN THE SAME SEQUENCE AS THE
C       ENTITIES ARE REPRESENTED IN THE SIMILARITY MATRIX.
C-------------------------------------------------------------------
C
C   DECK SETUP SPECIFICATIONS
C
C   THE USER PROVIDES PROGRAM DRIVER WHICH PERFORMS THE FOLLOWING TASKS.
C   1.  ASSIGNS INPUT/OUTPUT UNITS.
C   2.  ESTABLISHES THE DIMENSION OF THE *X* ARRAY AND SETS THIS
C       DIMENSION EQUAL TO *LIMIT*.
C   3.  CALLS SUBROUTINE ALLIN1.
C   THE FOLLOWING EXAMPLE WILL SUFFICE IN MOST CASES.
C
C       PROGRAM DRIVER(INPUT,OUTPUT,PUNCH,TAPE5=INPUT,TAPE6=OUTPUT,
C      ATAPE7=PUNCH,TAPE1,TAPE2)
C       DIMENSION X(7000)
C       LIMIT=7000
C       CALL ALLIN1(X,LIMIT)
C       END
C
C-------------------------------------------------------------------
C   THE *X* ARRAY IS PARTITIONED FOR STORAGE AS FOLLOWS
C
C   STORAGE FOR ARRAYS NEEDED AT ALL STAGES OF THE JOB
C   X(N1) TO X(N2-1)   NE WORDS--STORAGE OF THE II ARRAY
C   X(N2) TO X(N3-1)   NE WORDS--STORAGE OF THE JJ ARRAY
C   X(N3) TO X(N4-1)   NE WORDS--STORAGE OF THE SS ARRAY
C   X(N4) TO X(N5-1)   NE WORDS--STORAGE OF THE IL ARRAY
C   X(N5) TO X(N6-1)   NE WORDS--STORAGE OF THE JL ARRAY
C   X(N6) TO X(N7-1)   NE WORDS--STORAGE OF THE NEXT ARRAY
C   STORAGE FOR ARRAYS NEEDED IN SUBROUTINE *PREP*
C   M1=N7
C   X(M1) TO X(M2-1)   NE WORDS--STORAGE OF THE MEMBR ARRAY
C   X(M2) TO X(M3-1)   NE WORDS--STORAGE OF THE LAST ARRAY
C   X(M3) TO X(M4-1)   NE WORDS--STORAGE OF THE LINK ARRAY
C   X(M4) TO X(M5-1)   NE WORDS--STORAGE OF THE LOBJ ARRAY
C   X(M5) TO X(M6-1)   NREC WORDS--STORAGE OF THE IA ARRAY
```

```
C    X(M6) TO X(M7-1)   NREC WORDS--STORAGE OF THE JA ARRAY
C    X(M7) TO X(M8)      NREC WORDS--STORAGE OF THE SA ARRAY
C    STORAGE FOR ARRAYS NEEDED IN SUBROUTINE TREE (OVERLAY ARRAYS NEEDED
C    IN SUBROUTINE ALLIN1)
C    L1=N7
C    X(L1) TO X(L2-1)   25*NE WORDS--STORAGE OF THE A ARRAY
C    X(L2) TO X(L3-1)   5*NE WORDS--STORAGE OF THE LABEL ARRAY
C    X(L3) TO X(L4-1)   NE WORDS--STORAGE OF THE LCLNO ARRAY
C    X(L4) TO X(L5-1)   NE WORDS--STORAGE OF THE LINE ARRAY
C    X(L5) TO X(L6-1)   NE WORDS--STORAGE OF THE IS ARRAY
C    X(L6) TO X(L7)     NE WORDS--STORAGE OF THE LAST ARRAY
C
C---------------------------------------------------------------------
      DIMENSION X(1),TITLE(20),FMT(20),EPS(25)
      INTEGER FIRST
      READ(5,1000) TITLE
      READ(5,1100) NE,NREC,NTIN,IOPT,NTSV,KOUT
      WRITE(6,2000) TITLE
      WRITE(6,2200) NE,NREC,NTIN,IOPT,NTSV,KOUT
      IF(IOPT.NE.1) GO TO 10
      READ (5,1000) FMT
      WRITE(6,2100) FMT
10    N1=1
      N2=N1+NE
      N3=N2+NE
      N4=N3+NE
      N5=N4+NE
      N6=N5+NE
      N7=N6+NE
      M2=N7+NE
      M3=M2+NE
      M4=M3+NE
      M5=M4+NE
      M6=M5+NREC
      M7=M6+NREC
      M8=M7+NREC-1
      L2=N7+25*NE
      L3=L2+5*NE
      L4=L3+NE
      L5=L4+NE
      L6=L5+NE
      L7=L6+NE-1
C    CHECK FOR SUFFICIENT STORAGE
      MAX=M8
      IF(L7.GT.MAX) MAX=L7
      WRITE(6,2300) MAX,LIMIT
      IF(MAX.GT.LIMIT) STOP
C    READY TO CLUSTER
      CALL PREP(X(N1),X(N2),X(N3),X(N4),X(N5),X(N6),X(N7),X(M2),X(M3),
     AX(M4),X(M5),X(M6),X(M7),TITLE,FMT,NE,NREC,NTIN,IOPT,NTSV,X(L2))
      IF(X(L2).EQ.4HNOLB) GO TO 80
C    READ LABEL CARD(S)
      FIRST=L2
      LAST=L2+4
      READ(5,1000) (X(I),I=FIRST,LAST)
      IF(X(FIRST).EQ.4HNOLB) GO TO 80
C    READ REMAINING LABELS
      DO 70 K=2,NE
      FIRST=LAST+1
      LAST=LAST+5
70    READ(5,1000) (X(I),I=FIRST,LAST)
C    DRAW THE TREE CORRESPONDING TO THE CLUSTERING
80    MERGES=NE-1
      CALL TREE(X(N1),X(N2),X(N3),X(N4),X(N5),X(N6),X(N7),X(L2),X(L3),
```

```
       AX(L4),X(L5),X(L6),EPS,TITLE,MERGES,1,6,1,KOUT,NE)
       RETURN
 1000  FORMAT(20A4)
 1100  FORMAT(2I4,4I2)
 2000  FORMAT(1H1,20A4)
 2100  FORMAT(7H0FORMAT,20A4)
 2200  FORMAT(5H0NE =,I8,/,7H NREC =,I6,/,7H NTIN =,I6,/,7H IOPT =,I6,/,
      A7H NTSV =,I6,/,7H KOUT =,I6)
 2300  FORMAT(19H0REQUIRED STORAGE =,I5,6H WORDS,/,
      A       19H0ALLOTTED STORAGE =,I5,6H WORDS,/)
       END
```

Subroutine PREP

```
       SUBROUTINE PREP(II,JJ,SS,IL,JL,NEXT,MEMBR,LAST,LINK,LOBJ,IA,JA,
      ASA,TITLE,FMT,NE,NREC,NTIN,IOPT,NTSV,XL2)
C THIS SUBROUTINE FINDS THE MINIMUM SPANNING TREE IN THE GRAPH
C CORRESPONDING TO THE SORTED SEQUENCE OF SIMILARITY TRIPLES.
C
C THE FOLLOWING VARIABLES ARE SPECIFIED IN THE CALLING PROGRAM AND
C ARE PASSED THROUGH THE ARGUMENT LIST.
C   NE=NUMBER OF ENTITIES TO BE CLUSTERED
C   NREC=NUMBER OF (S(I,J),I,J) TRIPLES IN A RECORD ON UNIT *NTIN*
C   NTIN=INPUT UNIT FOR SIMILARITY TRIPLES
C   IOPT=INPUT OPTION FOR READING SIMILARITY TRIPLES
C   IOPT=1, BCD READ ACCORDING TO FORMAT *FMT*
C   IOPT.NE.1, BINARY READ
C   NTSV=UNIT ON WHICH CLUSTERING RESULTS ARE SAVED
C   NTSV.LE.0, RESULTS NOT SAVED
C   TITLE=IDENTIFYING TITLE FOR RUN
C   FMT=FORMAT FOR SIMILARITY TRIPLES
C IN ADDITION THE FOLLOWING VARIABLES PLAY IMPORTANT ROLES
C   MEMBR(I)=NUMBER OF CLUSTER TO WHICH ENTITY I CURRENTLY BELONGS
C   LAST(I)=ID NUMBER OF NEXT ENTITY AFTER I IN THE CHAIN DESCRIBING
C           THE CLUSTER TO WHICH I BELONGS (0 FOR LAST ENTITY IN CHAIN)
C   LOBJ(I)=ID NUMBER OF LAST ENTITY IN CHAIN DESCRIBING CLUSTER I
C
       DIMENSION II(1),JJ(1),SS(1),IL(1),JL(1),NEXT(1),MEMBR(1),LAST(1),
      ALINK(1),LOBJ(1),IA(1),JA(1),SA(1),TITLE(20),FMT(20)
       INTEGER HI
C INITIALIZE
       DO 10 I=1,NE
       LOBJ(I)=I
       MEMBR(I)=I
       LINK(I)=0
       LAST(I)=0
 10    NEXT(I)=0
       ISTOP=0
       K=1
       KEOF=0
       NREAD=0
C READ *NREC* TRIPLES
 20    IF(IOPT.EQ.1) GO TO 30
C BINARY READ
       READ(NTIN) (SA(I),IA(I),JA(I),I=1,NREC)
       IF(EOF,NTIN) 40,50
C BCD FORMATTED READ
 30    READ(NTIN,FMT) (SA(I),IA(I),JA(I),I=1,NREC)
       IF(EOF,NTIN) 40,50
C READ EOF.  DETERMINE HOW MANY GOOD RECORDS WERE READ
 40    KEOF=1
       NREC=NE-NREAD
       IF(NREC.LE.0) GO TO 170
```

```
C   PROCESS THE *NREC* TRIPLES
50    NREAD=NREAD+NREC
      DO 120 I=1,NREC
      IAI=IA(I)
      JAI=JA(I)
      IF(MEMBR(IAI)-MEMBR(JAI)) 60,120,70
C   IAI IS IN THE LOWER NUMBERED CLUSTER
60    LO=MEMBR(IAI)
      HI=MEMBR(JAI)
      GO TO 80
C   JAI IS IN THE LOWER NUMBERED CLUSTER
70    LO=MEMBR(JAI)
      HI=MEMBR(IAI)
C   STORE THE MERGE DATA NEEDED FOR SUBROUTINE *TREE*
80    II(K)=LO
      JJ(K)=HI
      SS(K)=SA(I)
      IL(K)=LAST(LO)
      JL(K)=LAST(HI)
      LAST(LO)=K
      IF(IL(K).EQ.0) GO TO 90
      ILK=IL(K)
      NEXT(ILK)=K
90    IF(JL(K).EQ.0) GO TO 100
      JLK=JL(K)
      NEXT(JLK)=K
C   ASSIGN MEMBERS OF THE *HI* CLUSTER TO THE *LO* CLUSTER
100   ID=HI
110   MEMBR(ID)=LO
      ID=LINK(ID)
      IF(ID.NE.0) GO TO 110
C   ATTACH THE TWO *LINK* LISTS
      ID=LOBJ(LO)
      LINK(ID)=HI
      LOBJ(LO)=LOBJ(HI)
      K=K+1
C   TERMINATE IF NE-1 MERGES HAVE OCCURRED
      IF(K.EQ.NE) GO TO 130
120   CONTINUE
C   TERMINATE IF EOF WAS READ
      IF(KEOF.EQ.1) GO TO 170
      GO TO 20
C   CLUSTERING FINISHED.  CHECK FOR ANY POSSIBLE HANGUPS IN OUTPUT OF
C   RESULTS
130   IF(NTIN.NE.5.OR.KEOF.EQ.1) GO TO 180
C   TRY TO READ THE EOF FOLLOWING THE SIMILARITY TRIPLES SO THE LABEL
C   CARDS CAN BE READ.
      IF(IOPT.EQ.1) GO TO 140
      READ(NTIN) (SA(I),IA(I),JA(I),I=1,NREC)
      IF(EOF,NTIN) 180,150
140   READ(NTIN,FMT)(SA(I),IA(I),JA(I),I=1,NREC)
      IF(EOF,NTIN) 180,150
C   WILL NOT BE ABLE TO READ THE LABEL CARDS.  SET FIRST LABEL TO
C   *NOLB* AND PRINT WARNING
150   XL2=4HNOLB
      WRITE(6,2000)
160   IF(NTSV.LE.0) NTSV=6
      WRITE(6,2100) NTSV
      GO TO 180
C   EOF WAS READ WITHOUT COMPLETING THE CLUSTERING
170   K=K-1
      ISTOP=1
      WRITE(6,2400) K
      K=K+1
      NTSV=6
```

```
C   SAVE RESULTS AS DESIRED
180    K=K-1
       IF(NTSV.LE.0.AND.ISTOP.EQ.0) RETURN
       WRITE(NTSV,2300) TITLE
       DO 190 I=1,K
190    WRITE(NTSV,2200) I,II(I),JJ(I),SS(I),IL(I),JL(I),NEXT(I)
       IF(ISTOP.EQ.1) STOP
       RETURN
2000   FORMAT(25H0LABELS CANNOT BE REACHED)
2100   FORMAT(33H CLUSTERING RESULTS SAVED ON UNIT,I5)
2200   FORMAT(3I10,E16.8,3I10)
2300   FORMAT(20A4)
2400   FORMAT(34H0EOF READ BEFORE TREE WAS COMPLETE,/,4H0K =,I10)
       END
```

APPENDIX

F

PROGRAMS FOR NONHIERARCHICAL
CLUSTERING

This appendix includes a computer program for implementing the nearest centroid sorting methods described in Section 7.2. The program incorporates the general design features outlined in Section C.1.

Program DRIVER is a dummy main program; subroutine EXEC includes input specifications, computes storage allocations, and calls other program segments; subroutine KMEAN performs the actual clustering and is discussed further below; subroutine RESULT prints the cluster membership lists and the mean vector for each cluster in the final partition. In addition to these prepared program segments, the user needs to supply two subroutines tailored to the particular problem: subroutine USER is used to read scores on all variables for one data unit; subroutine DIST is used to compute the chosen distance function between a data unit and a seed point. Both of these user supplied subroutines may be used to weight or transform the variables; see the comment cards in subroutine EXEC for illustrative examples.

Six alternative variations of subroutine KMEAN are provided as all possible combinations of two storage modes for the data set: (1) In core, (2) On tape or disk, and three major clustering techniques: (1) Forgy's and Jancey's methods, (2) convergent k-means method, (3) MacQueen's k-means method. Forgy's and Jancey's methods are included in the same subroutine because they differ only in regard to the updating of seed points after each reallocation cycle. In regards to storage the user should prefer the versions of KMEAN which use in-core storage if the data set will fit in core. The versions using tape

306

or disk require a complete pass through the tape or disk file for each reallocation cycle and all this input can be very costly in computer time. However,
the versions using tape or disk storage are capable of handling problems of
virtually unlimited size given enough computer time.

For the MacQueen k-means method the initial seed points are always taken
as the first NC data units in the input sequence. For the other clustering
methods there are three alternatives for specifying the initial partition.

1. Identify specific data units in the data set as seed points. The user can
duplicate the MacQueen k-means start by taking the first NC data units;
alternatively, sequence numbers can be chosen at random, or by any other
method, and the corresponding data units used as seed points.

2. Group the data set into an initial partition so that the first NUMBR(1)
are in the first cluster the next NUMBR(2) are in the second cluster, and so
forth. This alternative is especially useful for improving a given partition
such as one selected from the results of a hierarchical clustering method.

3. Read seed points directly from cards. This alternative may be useful
if previous analyses or other considerations indicate that certain profiles of
scores should be typical of expected clusters.

In the Jancey and Forgy methods clustering can begin with either seed
points or a partition. However, the convergent k-means method proceeds by
improving a complete partition through moving individual data units from
one cluster to another; if the initial configuration is specified using seed
points, the seed points are taken as fixed, and all data units are assigned to
the nearest seed point in order to obtain the initial partition.

Subroutine RESULT gives a fairly abbreviated summary of the clustering
results. However, by saving the cluster membership lists, program POSTDU
in Appendix G may be used to produce more elaborate results for interesting
partitions.

Subroutine EXEC

```
      SUBROUTINE EXEC(X,LIMIT)
C
C  THIS SUBROUTINE READS PARAMETERS, COMPUTES STORAGE AND CALLS MAJOR
C  PROGRAM SEGMENTS NEEDED FOR A NON-HIERARCHICAL CLUSTERING JOB USING
C  ONE OF THE METHODS PROGRAMMED AS A VERSION OF SUBROUTINE *KMEAN*.
C
C  EVERY JOB REQUIRES THREE USER SUPPLIED DECK SEGMENTS.
C
C  1. PROGRAM *DRIVER* PERFORMS THE FOLLOWING TASKS.
C     A. ASSIGNS INPUT/OUTPUT UNITS.
C     B. ESTABLISHES THE DIMENSION OF THE *X* ARRAY AND SETS THIS
C        DIMENSION TO *LIMIT*.
C     C. CALLS SUBROUTINE *EXEC*.
```

```
C  THE FOLLOWING EXAMPLE WILL SUFFICE IN MOST CASES.
C
C      PROGRAM DRIVER(INPUT,OUTPUT,PUNCH,TAPE5=INPUT,TAPE6=OUTPUT,
C     ATAPE7=PUNCH,TAPE1,TAPE2)
C      DIMENSION X(5000)
C      LIMIT=5000
C      CALL EXEC(X,LIMIT)
C      END
C
C  2. SUBROUTINE *USER* IS EMPLOYED TO READ THE COMPLETE SET OF SCORES
C      ON THE VARIABLES FOR ONE DATA UNIT.  THE FOLLOWING EXAMPLE
C      ILLUSTRATES VARIOUS POSSIBILITIES FOR MERGING FILES AND
C      TRANSFORMING VARIABLES AS THEY ARE READ.
C
C      SUBROUTINE USER(X)
C      DIMENSION X(8)
C      READ(1,100) X(7),Y
C      READ(2) (X(I),I=1,6)
C      READ(5,200) X(8),Z
C      X(3)=.5*X(3)
C      X(7)=3.6*X(7)
C      X(8)=.4*X(8)+.35*Y+.25*Z*X(8)
C      RETURN
C 100  FORMAT(2F11.3)
C 200  FORMAT(F8.1,F6.3)
C      END
C
C  3. FUNCTION *DIST* COMPUTES THE DISTANCE BETWEEN TWO DATA UNITS OR
C      BETWEEN A DATA UNIT AND A CLUSTER CENTROID.  THE USER CAN SPECIFY
C      ANY DESIRED DISTANCE FUNCTION AND WEIGHT THE VARIABLES IN ANY
C      MANNER.  THE FOLLOWING EXAMPLE ILLUSTRATES A WEIGHTED SQUARED
C      EUCLIDEAN DISTANCE BETWEEN TWO DATA UNITS DENOTED AS X AND Y.
C      THE PROBLEM INVOLVES 8 VARIABLES AND THE WEIGHTS ARE IN THE
C      *W* ARRAY.
C
C      FUNCTION DIST(X,Y)
C      DIMENSION X(1),Y(1),W(8)
C      DATA (W(I),I=1,8)/3*1.,3.,4.5,2.,2*1./
C      DIST=0.
C      DO 10 I=1,8
C  10  DIST=DIST+W(I)*((X(I)-Y(I))**2)
C      RETURN
C      END
C
C  NOTE THAT SCALING AND TRANSFORMATION OF VARIABLES CAN BE
C  ACCOMPLISHED EITHER IN SUBROUTINE *USER* OR IN SUBROUTINE *DIST*.
C
C-------------------------------------------------------------------
C  INPUT SPECIFICATIONS
C  CARD 1   TITLE
C  CARD 2   PARAMETER CARD
C    COLS  1- 5  NE=NUMBER OF ENTITIES (DATA UNITS)
C    COLS  6-10  NV=NUMBER OF VARIABLES
C    COLS 11-15  NC=NUMBER OF CLUSTERS
C    COLS 16-20  NTIN=INPUT UNIT FOR THE DATA SET
C                NTIN=5, CARD READER
C                NTIN.NE.5, TAPE OR DISK FILE
C    COLS 21-25  NTOUT=OUTPUT UNIT FOR SAVING CLUSTER MEMBERSHIP LISTS
C                NTOUT=7, CARD PUNCH
C                NTOUT.LE.0, DO NOT SAVE MEMBERSHIP LISTS
C    COLS 26-30  MINREL=TERMINATION PARAMETER.  CLUSTERING ENDS WHEN A
C                       CYCLE THROUGH THE DATA SET RESULTS IN *MINREL*
C                       OR FEWER CHANGES IN CLUSTER MEMBERSHIPS
C                MINREL.LE.0, ITERATE TO COMPLETE CONVERGENCE
```

```
C     COLS 31-35  IPART=INITIAL PARTITION PARAMETER
C                 IPART=1, SEED POINTS ARE SELECTED FROM THE DATA UNITS.
C                     READ THE SEQUENCE NUMBERS FOR THE CHOSEN DATA
C                     UNITS FROM CARD(S) 3 IN 20I4 FORMAT. IF THE
C                     DATA SET IS NOT STORED IN CORE, THE LIST OF
C                     OF SEQUENCE NUMBERS MUST BE IN ASCENDING ORDER
C                 IPART=2, THE DATA UNITS ARE GROUPED INTO AN INITIAL
C                     PARTITION IN THE INPUT SEQUENCE WITH THE
C                     FIRST *NUMBR(1)* IN CLUSTER 1, THE NEXT
C                     *NUMBR(2)* IN CLUSTER 2 ETC. READ THE
C                     *NUMBR* ARRAY FROM CARD(S) 3 IN 20I4 FORMAT.
C                 IPART=3, THE SCORE VECTORS FOR THE SEED POINTS ARE
C                     READ FROM CARD(S) 4 IN FORMAT *FMT* WHICH IS
C                     READ FROM CARD 3.
C     COLS 36-40  METHOD=PARAMETER FOR CHOOSING THE ALGORITHM IN ONE
C                     VERSION OF SUBROUTINE *KMEAN*.
C                 METHOD=1, JANCEY ALGORITHM
C                 METHOD.NE.1, FORGY ALGORITHM
C
C***CARDS 3 AND 4 ARE READ IN SUBROUTINE *KMEAN* ACCORDING TO THE
C***PROCEDURE SPECIFIED BY THE CHOSEN VALUE OF *IPART*. NOTE THAT THE
C***BASIC K-MEANS METHOD OF MACQUEEN SIMPLY USES THE FIRST *NC* DATA
C***UNITS AS CLUSTER SEED POINTS AND THEREFORE IGNORES THE *IPART*
C***PARAMETER.
C-------------------------------------------------------------------
C
C     STORAGE ALLOCATIONS IN THE *X* ARRAY
C     X(N1) TO X(N2-1)  NC*NV WORDS--STORAGE OF THE CENTR ARRAY
C     X(N2) TO X(N3-1)  NC WORDS--STORAGE OF THE NUMBR ARRAY
C     X(N3) TO X(N4-1)  NE WORDS--STORAGE OF THE MEMBR ARRAY
C     X(N4) TO X(N5-1)  NC*NV WORDS--STORAGE OF THE TOTAL ARRAY
C     X(N5) TO X(N6)    NV OR NE*NV WORDS-- STORAGE OF THE DATA ARRAY
C     X(N4) TO X(N7)    NE WORDS--STORAGE OF THE LIST ARRAY IN *RESULT*
C
      DIMENSION X(1),TITLE(20)
      READ(5,1000) TITLE
      READ(5,1100) NE,NV,NC,NTIN,NTOUT,MINREL,IPART,METHOD
      WRITE(6,2000) TITLE
      WRITE(6,2100) NE,NV,NC,NTIN,NTOUT,MINREL,IPART,METHOD
      N1=1
      N2=N1+NC*NV
      N3=N2+NC
      N4=N3+NE
      N5=N4+NC*NV
C     *N6* MAY BE INCREASED IN *KMEAN*.
      N6=N5+NV-1
      N7=N4+NE-1
      MAX=N6
      IF(N7.GT.MAX) MAX=N7
      WRITE(6,2200) MAX,LIMIT
      IF(MAX.GT.LIMIT) STOP
      CALL KMEAN(X(N1),X(N2),X(N3),X(N4),X(N5),N5,NE,NV,NC,NTIN,MINREL,
     AIPART,METHOD,LIMIT)
      CALL RESULT(X(N1),X(N2),X(N3),X(N4),TITLE,NE,NV,NC,NTOUT)
      RETURN
1000  FORMAT(20A4)
1100  FORMAT(8I5)
2000  FORMAT(1H1,20A4)
2100  FORMAT(5HONE =,I8,/,5H NV =,I8,/,5H NC =,I8,/,7H NTIN =,I6,/,
     A8H NTOUT =,I5,/,9H MINREL =,I4,/,8H IPART =,I5,/,9H METHOD =,I4)
2200  FORMAT(19HOREQUIRED STORAGE =,I5,6H WORDS,/,
     A        19HOALLOTTED STORAGE =,I5,6H WORDS)
      END
```

Subroutine RESULT

```
      SUBROUTINE RESULT(CENTR,NUMBR,MEMBR,LIST,TITLE,NE,NV,NC,NTOUT)
C THIS SUBROUTINE PRINTS THE RESULTS FROM A CLUSTERING JOB BASED
C ON ANY VERSION OF SUBROUTINE *KMEAN*.
C
      DIMENSION CENTR(1),NUMBR(1),MEMBR(1),LIST(1),TITLE(20)
C
C AS A CONTINGENCY PRECAUTION WRITE OUT THE RAW MEMBERSHIP LIST.
      WRITE(6,2000) TITLE
      WRITE(6,2100) (MEMBR(K),K=1,NE)
      WRITE(6,2200) (NUMBR(J),J=1,NC)
C INVERT THE *MEMBR* ARRAY AND PUT THE RESULT IN THE *LIST* ARRAY.
C FIRST REVISE THE *NUMBR* ARRAY TO CONTAIN START POINTS IN THE
C *LIST* ARRAY FOR EACH CLUSTER
      NUMBR(NC)=NE-NUMBR(NC)+1
      JJ=NC
      JJ1=JJ-1
      DO 10 J=2,NC
      NUMBR(JJ1)=NUMBR(JJ)-NUMBR(JJ1)
      JJ=JJ1
10    JJ1=JJ-1
C BUILD *LIST* ARRAY
      DO 20 K=1,NE
      MEMBRK=MEMBR(K)
      NJ=NUMBR(MEMBRK)
      LIST(NJ)=K
      NUMBR(MEMBRK)=NUMBR(MEMBRK)+1
20    CONTINUE
C SAVE THE SORTED MEMBERSHIP LIST IF DESIRED
      IF(NTOUT.LE.0) GO TO 30
      WRITE(NTOUT,3000) TITLE
      WRITE(NTOUT,3100) (LIST(K),K=1,NE)
C RESTORE THE *NUMBR* ARRAY
30    JJ=NC
      DO 40 J=2,NC
      NUMBR(JJ)=NUMBR(JJ)-NUMBR(JJ-1)
40    JJ=JJ-1
      NUMBR(1)=NUMBR(1)-1
C PRINT RESULTS FOR EACH CLUSTER
      WRITE(6,2000) TITLE
      K1=1
      DO 50 J=1,NC
      WRITE(6,2300) J,NUMBR(J)
      J1=(J-1)*NV
      WRITE(6,2400) (CENTR(J1+I),I=1,NV)
      K2=K1+NUMBR(J)-1
      WRITE(6,2500) (LIST(K),K=K1,K2)
      K1=K2+1
50    CONTINUE
      RETURN
2000  FORMAT(1H1,20A4)
2100  FORMAT(20H0RAW MEMBERSHIP LIST,/,(1X,25I5))
2200  FORMAT(14H0CLUSTER SIZES,/,(1X,25I5))
2300  FORMAT(8H0CLUSTER,I3,9H CONTAINS,I5,11H DATA UNITS)
2400  FORMAT(21H0CENTROID COORDINATES,/,(1X,10E12.4))
2500  FORMAT(16H0MEMBERSHIP LIST,/,(1X,25I5))
3000  FORMAT(20A4)
3100  FORMAT(20I4)
      END
```

Subroutine KMEAN

VERSION 1

```
      SUBROUTINE KMEAN(CENTR,NUMBR,MEMBR,TOTAL,DATA,N5,NE,NV,NC,NTIN,
     AMINREL,IPART,METHOD,LIMIT)
C
C-----------------------------------------------------------------------
C VERSION 1.  THE DATA SET IS STORED IN CENTRAL MEMORY.
C-----------------------------------------------------------------------
C
C THIS SUBROUTINE ITERATIVELY SORTS *NE* DATA UNITS INTO *NC* CLUSTERS
C USING THE ALGORITHM OF (METHOD.NE.1)
C
C FORGY, E.W., CLUSTER ANALYSIS OF MULTIVARIATE DATA, EFFICIENCY
C VERSUS INTERPRETABILITY OF CLASSIFICATIONS, PAPER PRESENTED AT THE
C BIOMETRIC SOCIETY (WNAR) MEETINGS, RIVERSIDE, CALIFORNIA, JUNE
C 1965.  ABSTRACT IN BIOMETRICS, VOLUME 21, NUMBER 3, P 768.
C
C OR THE ALGORITHM OF (METHOD=1)
C
C JANCEY, R.C., MULTIDIMENSIONAL GROUP ANALYSIS, AUSTRALIAN JOURNAL
C OF BOTANY, VOLUME 14, NUMBER 1, APRIL 1966, PP 127-130.
C
C CENTR(NV*(J-1)+I)=SCORE ON I-TH VARIABLE FOR J-TH CLUSTER CENTROID
C TOTAL(NV*(J-1)+I)=TOTAL SCORE ON I-TH VARIABLE FOR DATA UNITS THUS
C                   FAR ALLOCATED TO THE J-TH CLUSTER
C NUMBR(J)=NUMBER OF DATA UNITS THUS FAR ALLOCATED TO THE J-TH CLUSTER
C MEMBR(K)=CLUSTER TO WHICH THE K-TH DATA UNIT CURRENTLY BELONGS
C DATA(NV*(K-1)+I)=SCORE ON I-TH VARIABLE FOR K-TH DATA UNIT
C
      DIMENSION CENTR(1),TOTAL(1),NUMBR(1),MEMBR(1),DATA(1),FMT(20)
     A,NAME(4)
      DATA (NAME(I),I=1,4)/4H     F,4HORGY,4H   JA,4HNCEY/
      I=1
      IF(METHOD.EQ.1) I=3
      WRITE(6,2000) NAME(I),NAME(I+1)
C CHECK FOR SUFFICIENT STORAGE
      N6=N5+NE*NV-1
      WRITE(6,2100) N6,LIMIT
      IF(N6.GT.LIMIT) STOP
C ESTABLISH INITIAL PARTITION
      IF(IPART.NE.3) GO TO 20
C SEED POINTS ARE READ DIRECTLY FROM CARDS
      READ (5,1000) FMT
      WRITE(6,2200) FMT
      WRITE(6,2300)
      J1=0
      DO 10 J=1,NC
      READ(5,FMT) (CENTR(J1+I),I=1,NV)
      WRITE(6,2400) (CENTR(J1+I),I=1,NV)
10    J1=J1+NV
      GO TO 30
C IPART=1 OR 2
20    WRITE(6,2500) IPART
      READ(5,1100) (NUMBR(J),J=1,NC)
      WRITE(6,2600) (NUMBR(J),J=1,NC)
C READ THE DATA SET INTO CENTRAL MEMORY
30    K1=1
      DO 40 K=1,NE
      CALL USER (DATA(K1))
```

```
40      K1=K1+NV
        IF(IPART.EQ.3) GO TO 100
C  IF *IPART* IS 1 OR 2 SET UP THE SEED POINTS
        IF(IPART.EQ.2) GO TO 60
C  IPART=1.  THE DATA UNIT WITH SEQUENCE NUMBER *NUMBR(J)* IS USED AS
C  THE J-TH SEED POINT
        DO 50 J=1,NC
        NJ=(NUMBR(J)-1)*NV
        J1=(J-1)*NV
        DO 50 I=1,NV
        CENTR(J1+I)=DATA(NJ+I)
50      CONTINUE
        GO TO 100
C  IPART=2.  THE DATA UNITS ARE GROUPED INTO CLUSTERS WITH THE J-TH
C  CLUSTER HAVING *NUMBR(J)* MEMBERS.
60      K=0
        J1=0
C  ACCUMULATE THE TOTAL SCORE ON EACH VARIABLE FOR EACH CLUSTER
        DO 80 J=1,NC
        NJ=NUMBR(J)
        DO 70 I=1,NV
70      TOTAL(J1+I)=0.
        DO 80 KJ=1,NJ
        K=K+1
        K1=(K-1)*NV
        DO 80 I=1,NV
        J1=J1+1
        TOTAL(J1)=TOTAL(J1)+DATA(K1+I)
80      CONTINUE
C  COMPUTE THE CENTROIDS
        J1=0
        DO 90 J=1,NC
        DO 90 I=1,NV
        J1=J1+1
        CENTR(J1)=TOTAL(J1)/NUMBR(J)
90      CONTINUE
C  INITIALIZE ARRAYS
100     DO 110 K=1,NE
110     MEMBR(K)=0
        NPASS=1
C  BEGINNING OF MAIN LOOP
120     J1=0
        DO 130 J=1,NC
        NUMBR(J)=0
        DO 130 I=1,NV
        J1=J1+1
130     TOTAL(J1)=0.
        MOVES=0
        TDIST=0
C  ALLOCATE EACH DATA UNIT TO THE NEAREST CLUSTER CENTROID
        K1=0
        DO 160 K=1,NE
        K2=K1+1
        J2=1
C  COMPUTE DISTANCE TO FIRST CLUSTER CENTROID
        DREF=DIST(DATA(K2),CENTR(J2))
        JREF=1
C  TEST DISTANCES TO REMAINING CLUSTER CENTROIDS
        DO 140 J=2,NC
        J2=J2+NV
        DTEST=DIST(DATA(K2),CENTR(J2))
        IF(DTEST.GE.DREF) GO TO 140
        DREF=DTEST
        JREF=J
```

```
140     CONTINUE
C   ALLOCATE DATA UNIT *K* TO CLUSTER *JREF*
        NUMBR(JREF)=NUMBR(JREF)+1
        TDIST=TDIST+DREF
        IF(JREF.EQ.MEMBR(K)) GO TO 150
C   THE DATA UNIT CHANGES ITS MEMBERSHIP
        MOVES=MOVES+1
        MEMBR(K)=JREF
150     J1=(JREF-1)*NV
        DO 160 I=1,NV
        J1=J1+1
        K1=K1+1
        TOTAL(J1)=TOTAL(J1)+DATA(K1)
160     CONTINUE
C   ALL DATA UNITS ALLOCATED.  TEST FOR CONVERGENCE
        WRITE(6,2700) MOVES,NPASS,TDIST
        NPASS=NPASS+1
        JREF=0
        IF(MOVES.GT.MINREL) GO TO 185
        IF(METHOD.NE.1.AND.MOVES.EQ.0) RETURN
        JREF=1
C   COMPUTE TRUE CLUSTER CENTROIDS--FORGY UPDATE
170     J1=0
        DO 180 J=1,NC
        DO 180 I=1,NV
        J1=J1+1
180     CENTR(J1)=TOTAL(J1)/NUMBR(J)
        IF(JREF.EQ.1) RETURN
        GO TO 120
185     IF(METHOD.NE.1) GO TO 170
C   JANCEY UPDATE
190     J1=0
        DO 200 J=1,NC
        DO 200 I=1,NV
        J1=J1+1
200     CENTR(J1)=2.*TOTAL(J1)/NUMBR(J)-CENTR(J1)
        GO TO 120
1000    FORMAT(20A4)
1100    FORMAT(20I4)
2000    FORMAT(1H0,2A4, 53H METHOD OF CLUSTER ANALYSIS.  DATA SET STORED I
        AN CORE)
2100    FORMAT(19H0REQUIRED STORAGE =,I5,6H WORDS,/,
        A        19H0ALLOTTED STORAGE =,I5,6H WORDS)
2200    FORMAT(7H0FORMAT,20A4)
2300    FORMAT( 43H1INITIAL CLUSTER CENTERS READ IN AS FOLLOWS///)
2400    FORMAT(1X,10E12.4)
2500    FORMAT(  9H1 IPART =,I2, 30H,  NUMBR ARRAY READ AS FOLLOWS///)
2600    FORMAT(1X,10I7)
2700    FORMAT(1H0,I5,37H DATA UNITS MOVED ON ITERATION NUMBER,I3,/,
        A38H SUMMED DEVIATIONS ABOUT SEED POINTS =,E16.8)
        END
```

VERSION 2

```
        SUBROUTINE KMEAN(CENTR,NUMBR,MEMBR,TOTAL,DATA,N5,NE,NV,NC,NTIN,
        AMINREL,IPART,METHOD,LIMIT)
C
C------------------------------------------------------------------------
C   VERSION 2.  THE DATA SET IS STORED ON A TAPE OR DISK FILE WHICH IS
C   REWOUND AND READ IN ITS ENTIRETY FOR EACH CYCLE.
C------------------------------------------------------------------------
C
```

```
C    THIS SUBROUTINE ITERATIVELY SORTS *NE* DATA UNITS INTO *NC* CLUSTERS
C    USING THE ALGORITHM OF (METHOD.NE.1)
C
C    FORGY, E.W., CLUSTER ANALYSIS OF MULTIVARIATE DATA, EFFICIENCY
C    VERSUS INTERPRETABILITY OF CLASSIFICATIONS, PAPER PRESENTED AT THE
C    BIOMETRIC SOCIETY (WNAR) MEETINGS, RIVERSIDE, CALIFORNIA, JUNE
C    1965.  ABSTRACT IN BIOMETRICS, VOLUME 21, NUMBER 3, P 768.
C
C    OR THE ALGORITHM OF (METHOD=1)
C
C    JANCEY, R.C., MULTIDIMENSIONAL GROUP ANALYSIS, AUSTRALIAN JOURNAL
C    OF BOTANY, VOLUME 14, NUMBER 1, APRIL 1966, PP 127-130.
C
C    CENTR(NV*(J-1)+I)=SCORE ON I-TH VARIABLE FOR J-TH CLUSTER CENTROID
C    TOTAL(NV*(J-1)+I)=TOTAL SCORE ON I-TH VARIABLE FOR DATA UNITS THUS
C                      FAR ALLOCATED TO THE J-TH CLUSTER
C    NUMBR(J)=NUMBER OF DATA UNITS THUS FAR ALLOCATED TO THE J-TH CLUSTER
C    MEMBR(K)=CLUSTER TO WHICH THE K-TH DATA UNIT CURRENTLY BELONGS
C    DATA(I)=SCORE ON THE I-TH VARIABLE FOR THE CURRENT DATA UNIT
C
      DIMENSION CENTR(1),TOTAL(1),NUMBR(1),MEMBR(1),DATA(1),FMT(20)
     A,NAME(4)
      DATA (NAME(I),I=1,4)/4H    F,4HORGY,4H   JA,4HNCEY/
      REWIND NTIN
      I=1
      IF(METHOD.EQ.1) I=3
      WRITE(6,2000) NAME(I),NAME(I+1)
C    ESTABLISH INITIAL PARTITION
      IF(IPART.NE.3) GO TO 20
C    SEED POINTS ARE READ DIRECTLY FROM CARDS
      READ (5,1000) FMT
      WRITE(6,2200) FMT
      WRITE(6,2300)
      J1=0
      DO 10 J=1,NC
      READ(5,FMT) (CENTR(J1+I),I=1,NV)
      WRITE(6,2400) (CENTR(J1+I),I=1,NV)
10    J1=J1+NV
      GO TO 100
C    IF *IPART* IS 1 OR 2 SET UP THE SEED POINTS
20    WRITE(6,2500) IPART
      READ(5,1100) (NUMBR(J),J=1,NC)
      WRITE(6,2600) (NUMBR(J),J=1,NC)
      IF(IPART.EQ.2) GO TO 50
C    IPART=1.  THE DATA UNIT WITH SEQUENCE NUMBER *NUMBR(J)* IS USED AS
C    THE J-TH SEED POINT
      J1=0
      K2=0
      DO 40 J=1,NC
      K1=K2+1
      K2=NUMBR(J)
      DO 30 K=K1,K2
30    CALL USER(DATA(1))
      DO 40 I=1,NV
      J1=J1+1
      CENTR(J1)=DATA(I)
40    CONTINUE
      GO TO 100
C    IPART=2.  THE DATA UNITS ARE GROUPED INTO CLUSTERS WITH THE J-TH
C    CLUSTER HAVING *NUMBR(J)* MEMBERS.
50    J1=0
      DO 80 J=1,NC
      DO 60 I=1,NV
      J1=J1+1
```

```
60      TOTAL(J1)=0.
        J1=NV*J
        K2=NUMBR(J)
        DO 70 K=1,K2
        CALL USER(DATA(1))
        J1=J1-NV
        DO 70 I=1,NV
        J1=J1+1
        TOTAL(J1)=TOTAL(J1)+DATA(I)
70      CONTINUE
        J1=J1-NV
        DO 80 I=1,NV
        J1=J1+1
        CENTR(J1)=TOTAL(J1)/NUMBR(J)
80      CONTINUE
C  INITIALIZE ARRAYS
100     DO 110 K=1,NE
110     MEMBR(K)=0
        NPASS=1
C  BEGINNING OF MAIN LOOP
120     J1=0
        DO 130 J=1,NC
        NUMBR(J)=0
        DO 130 I=1,NV
        J1=J1+1
130     TOTAL(J1)=0.
        MOVES=0
        TDIST=0
        REWIND NTIN
C  ALLOCATE EACH DATA UNIT TO THE NEAREST CLUSTER CENTROID
        DO 160 K=1,NE
        CALL USER(DATA(1))
        J2=1
C  COMPUTE DISTANCE TO FIRST CLUSTER CENTROID
        DREF=DIST(DATA(1),CENTR(J2))
        JREF=1
C  TEST DISTANCES TO REMAINING CLUSTER CENTROIDS
        DO 140 J=2,NC
        J2=J2+NV
        DTEST=DIST(DATA(1),CENTR(J2))
        IF(DTEST.GE.DREF) GO TO 140
        DREF=DTEST
        JREF=J
140     CONTINUE
C  ALLOCATE DATA UNIT *K* TO CLUSTER *JREF*
        NUMBR(JREF)=NUMBR(JREF)+1
        TDIST=TDIST+DREF
        IF(JREF.EQ.MEMBR(K)) GO TO 150
C  THE DATA UNIT CHANGES ITS MEMBERSHIP
        MOVES=MOVES+1
        MEMBR(K)=JREF
150     J1=(JREF-1)*NV
        DO 160 I=1,NV
        J1=J1+1
        TOTAL(J1)=TOTAL(J1)+DATA(I)
160     CONTINUE
C  ALL DATA UNITS ALLOCATED.  TEST FOR CONVERGENCE
        WRITE(6,2700) MOVES,NPASS,TDIST
        NPASS=NPASS+1
        JREF=0
        IF(MOVES.GT.MINREL) GO TO 185
        IF(METHOD.NE.1.AND.MOVES.EQ.0) RETURN
C  BEFORE RETURNING COMPUTE THE CENTROIDS FOR THE FINAL PARTITION
        JREF=1
```

```
C  COMPUTE TRUE CLUSTER CENTROIDS--FORGY UPDATE
170    J1=0
       DO 180 J=1,NC
       DO 180 I=1,NV
       J1=J1+1
180    CENTR(J1)=TOTAL(J1)/NUMBR(J)
       IF(JREF.EQ.1) RETURN
       GO TO 120
185    IF(METHOD.NE.1) GO TO 170
C  JANCEY UPDATE
190    J1=0
       DO 200 J=1,NC
       DO 200 I=1,NV
       J1=J1+1
200    CENTR(J1)=2.*TOTAL(J1)/NUMBR(J)-CENTR(J1)
       GO TO 120
1000   FORMAT(20A4)
1100   FORMAT(20I4)
2000   FORMAT(1H0,2A4, 61H METHOD OF CLUSTER ANALYSIS.  DATA SET STORED O
      AN TAPE OR DISK)
2200   FORMAT(7H0FORMAT,20A4)
2300   FORMAT( 43H1INITIAL CLUSTER CENTERS READ IN AS FOLLOWS///)
2400   FORMAT(1X,10E12.4)
2500   FORMAT(  9H1 IPART =,I2, 30H,   NUMBR ARRAY READ AS FOLLOWS///)
2600   FORMAT(1X,10I7)
2700   FORMAT(1H0,I5,37H DATA UNITS MOVED ON ITERATION NUMBER,I3,/,
      A38H SUMMED DEVIATIONS ABOUT SEED POINTS =,E16.8)
       END
```

VERSION 3

```
       SUBROUTINE KMEAN(CENTR,NUMBR,MEMBR,TOTAL,DATA,N5,NE,NV,NC,NTIN,
      AMINREL,IPART,METHOD,LIMIT)
C
C------------------------------------------------------------------------
C  VERSION 1.  THE DATA SET IS STORED IN CENTRAL MEMORY.
C------------------------------------------------------------------------
C
C  THIS SUBROUTINE ITERATIVELY SORTS *NE* DATA UNITS INTO *NC* CLUSTERS
C  USING THE CONVERGENT K-MEANS METHOD DESCRIBED IN SECTION 7.2.2.
C
C  CENTR(NV*(J-1)+I)=SCORE ON I-TH VARIABLE FOR J-TH CLUSTER CENTROID
C  TOTAL(NV*(J-1)+I)=TOTAL SCORE ON I-TH VARIABLE FOR DATA UNITS THUS
C                    FAR ALLOCATED TO THE J-TH CLUSTER
C  NUMBR(J)=NUMBER OF DATA UNITS THUS FAR ALLOCATED TO THE J-TH CLUSTER
C  MEMBR(K)=CLUSTER TO WHICH THE K-TH DATA UNIT CURRENTLY BELONGS
C  DATA(NV*(K-1)+I)=SCORE ON I-TH VARIABLE FOR K-TH DATA UNIT
C
       DIMENSION CENTR(1),TOTAL(1),NUMBR(1),MEMBR(1),DATA(1),FMT(20)
       WRITE(6,2000)
C  CHECK FOR SUFFICIENT STORAGE
       N6=N5+NE*NV-1
       WRITE(6,2100) N6,LIMIT
       IF(N6.GT.LIMIT) STOP
C  ESTABLISH INITIAL PARTITION
       IF(IPART.NE.3) GO TO 20
C  SEED POINTS ARE READ DIRECTLY FROM CARDS
       READ (5,1000) FMT
       WRITE(6,2200) FMT
       WRITE(6,2300)
       J1=0
       DO 10 J=1,NC
       READ(5,FMT) (CENTR(J1+I),I=1,NV)
       WRITE(6,2400) (CENTR(J1+I),I=1,NV)
```

```
10      J1=J1+NV
        GO TO 30
C  IPART=1 OR 2
20      WRITE(6,2500) IPART
        READ(5,1100) (NUMBR(J),J=1,NC)
        WRITE(6,2600) (NUMBR(J),J=1,NC)
C  READ THE DATA SET INTO CENTRAL MEMORY
30      K1=1
        DO 40 K=1,NE
        CALL USER (DATA(K1))
40      K1=K1+NV
        IF(IPART.EQ.3) GO TO 51
C  IF *IPART* IS 1 OR 2 SET UP THE SEED POINTS
        IF(IPART.EQ.2) GO TO 60
C  IPART=1.  THE DATA UNIT WITH SEQUENCE NUMBER *NUMBR(J)* IS USED AS
C  THE J-TH SEED POINT
        DO 50 J=1,NC
        NJ=(NUMBR(J)-1)*NV
        J1=(J-1)*NV
        DO 50 I=1,NV
        CENTR(J1+I)=DATA(NJ+I)
50      CONTINUE
C  THE INITIAL CONFIGURATION IS GIVEN IN TERMS OF SEED POINTS.
C  CONSTRUCT AN INITIAL PARTITION BY ASSIGNING EACH DATA UNIT TO THE
C  NEAREST SEED POINT.  SEED POINTS REMAIN FIXED THROUGHOUT ASSIGNMENT
C  OF THE FULL DATA SET.
51      DO 52 K=1,NE
52      MEMBR(K)=0
        J1=0
        DO 53 J=1,NC
        NUMBR(J)=0
        DO 53 I=1,NV
        J1=J1+1
53      TOTAL(J1)=0.
C  ALLOCATE EACH DATA UNIT TO THE NEAREST SEED POINT
        K1=0
        DO 55 K=1,NE
        K2=K1+1
        J2=1
C  COMPUTE DISTANCE TO FIRST SEED POINT
        DREF=DIST(DATA(K2),CENTR(J2))
        JREF=1
C  TEST DISTANCES TO REMAINING SEED POINTS
        DO 54 J=2,NC
        J2=J2+NV
        DTEST=DIST(DATA(K2),CENTR(J2))
        IF(DTEST.GE.DREF) GO TO 54
        DREF=DTEST
        JREF=J
54      CONTINUE
C  ALLOCATE DATA UNIT *K* TO CLUSTER *JREF*
        NUMBR(JREF)=NUMBR(JREF)+1
        MEMBR(K)=JREF
        J1=(JREF-1)*NV
        DO 55 I=1,NV
        J1=J1+1
        K1=K1+1
        TOTAL(J1)=TOTAL(J1)+DATA(K1)
55      CONTINUE
        GO TO 85
C  IPART=2.  THE DATA UNITS ARE GROUPED INTO CLUSTERS WITH THE J-TH
C  CLUSTER HAVING *NUMBR(J)* MEMBERS.
60      K=0
        J1=0
```

```
C   ACCUMULATE THE TOTAL SCORE ON EACH VARIABLE FOR EACH CLUSTER
        DO 80 J=1,NC
        NJ=NUMBR(J)
        DO 70 I=1,NV
70      TOTAL(J1+I)=0.
        DO 80 KJ=1,NJ
        K=K+1
        MEMBR(K)=J
        K1=(K-1)*NV
        DO 80 I=1,NV
        J1=J1+1
        TOTAL(J1)=TOTAL(J1)+DATA(K1+I)
80      CONTINUE
C   COMPUTE THE CENTROIDS
85      J1=0
        DO 90 J=1,NC
        DO 90 I=1,NV
        J1=J1+1
        CENTR(J1)=TOTAL(J1)/NUMBR(J)
90      CONTINUE
C   INITIALIZE ARRAYS
100     NPASS=1
C   BEGINNING OF MAIN LOOP
120     MOVES=0
        TDIST=0
C   ALLOCATE EACH DATA UNIT TO THE NEAREST CLUSTER CENTROID
        K1=0
        DO 160 K=1,NE
        K2=K1+1
        J2=1
C   COMPUTE DISTANCE TO FIRST CLUSTER CENTROID
        DREF=DIST(DATA(K2),CENTR(J2))
        JREF=1
C   TEST DISTANCES TO REMAINING CLUSTER CENTROIDS
        DO 140 J=2,NC
        J2=J2+NV
        DTEST=DIST(DATA(K2),CENTR(J2))
        IF(DTEST.GE.DREF) GO TO 140
        DREF=DTEST
        JREF=J
140     CONTINUE
        TDIST=TDIST+DREF
        IF(JREF.NE.MEMBR(K)) GO TO 155
        K1=K1+NV
        GO TO 160
C   REALLOCATE DATA UNIT*K* FROM CLUSTER *MEMBR(K)* TO CLUSTER *JREF*
155     MOVES=MOVES+1
        J2=MEMBR(K)
        NUMBR(J2)=NUMBR(J2)-1
        NUMBR(JREF)=NUMBR(JREF)+1
        MEMBR(K)=JREF
        J1=(JREF-1)*NV
        J3=(J2-1)*NV
        DO 150 I=1,NV
        J1=J1+1
        J3=J3+1
        K1=K1+1
        TOTAL(J1)=TOTAL(J1)+DATA(K1)
        CENTR(J1)=TOTAL(J1)/NUMBR(JREF)
        TOTAL(J3)=TOTAL(J3)-DATA(K1)
        CENTR(J3)=TOTAL(J3)/NUMBR(J2)
150     CONTINUE
160     CONTINUE
C   ALL DATA UNITS ALLOCATED.  TEST FOR CONVERGENCE
```

```
      WRITE(6,2700) MOVES,NPASS,TDIST
      NPASS=NPASS+1
      IF(MOVES.LE.MINREL) RETURN
      GO TO 120
1000  FORMAT(20A4)
1100  FORMAT(20I4)
2000  FORMAT( 46H0CONVERGENT K-MEANS METHOD OF CLUSTER ANALYSIS,/,
     A 24H DATA SET STORED IN CORE)
     AN CORE)
2100  FORMAT(19H0REQUIRED STORAGE =,I5,6H WORDS,/,
     A       19H0ALLOTTED STORAGE =,I5,6H WORDS)
2200  FORMAT(7H0FORMAT,20A4)
2300  FORMAT( 43H1INITIAL CLUSTER CENTERS READ IN AS FOLLOWS///)
2400  FORMAT(1X,10E12.4)
2500  FORMAT(  9H1 IPART =,I2, 30H,  NUMBR ARRAY READ AS FOLLOWS///)
2600  FORMAT(1X,10I7)
2700  FORMAT(1H0,I5,37H DATA UNITS MOVED ON ITERATION NUMBER,I3,/,
     A38H SUMMED DEVIATIONS ABOUT SEED POINTS =,E16.8)
      END
```

VERSION 4

```
      SUBROUTINE KMEAN(CENTR,NUMBR,MEMBR,TOTAL,DATA,N5,NE,NV,NC,NTIN,
     AMINREL,IPART,METHOD,LIMIT)
C
C----------------------------------------------------------------------
C VERSION 2.  THE DATA SET IS STORED ON A TAPE OR DISK FILE WHICH IS
C REWOUND AND READ IN ITS ENTIRETY FOR EACH CYCLE.
C----------------------------------------------------------------------
C
C THIS SUBROUTINE ITERATIVELY SORTS *NE* DATA UNITS INTO *NC* CLUSTERS
C USING THE CONVERGENT K-MEANS METHOD DESCRIBED IN SECTION 7.2.2.
C
C CENTR(NV*(J-1)+I)=SCORE ON I-TH VARIABLE FOR J-TH CLUSTER CENTROID
C TOTAL(NV*(J-1)+I)=TOTAL SCORE ON I-TH VARIABLE FOR DATA UNITS THUS
C                   FAR ALLOCATED TO THE J-TH CLUSTER
C NUMBR(J)=NUMBER OF DATA UNITS THUS FAR ALLOCATED TO THE J-TH CLUSTER
C MEMBR(K)=CLUSTER TO WHICH THE K-TH DATA UNIT CURRENTLY BELONGS
C DATA(I)=SCORE ON THE I-TH VARIABLE FOR THE CURRENT DATA UNIT
C
      DIMENSION CENTR(1),TOTAL(1),NUMBR(1),MEMBR(1),DATA(1),FMT(20)
      REWIND NTIN
      WRITE(6,2000)
C ESTABLISH INITIAL PARTITION
      IF(IPART.NE.3) GO TO 20
C SEED POINTS ARE READ DIRECTLY FROM CARDS
      READ (5,1000) FMT
      WRITE(6,2200) FMT
      WRITE(6,2300)
      J1=0
      DO 10 J=1,NC
      READ(5,FMT) (CENTR(J1+I),I=1,NV)
      WRITE(6,2400) (CENTR(J1+I),I=1,NV)
10    J1=J1+NV
      GO TO 51
C IF *IPART* IS 1 OR 2 SET UP THE SEED POINTS
20    WRITE(6,2500) IPART
      READ(5,1100) (NUMBR(J),J=1,NC)
      WRITE(6,2600) (NUMBR(J),J=1,NC)
      IF(IPART.EQ.2) GO TO 59
```

```
C     IPART=1.  THE DATA UNIT WITH SEQUENCE NUMBER *NUMBR(J)* IS USED AS
C     THE J-TH SEED POINT
      J1=0
      K2=0
      DO 40 J=1,NC
      K1=K2+1
      K2=NUMBR(J)
      DO 30 K=K1,K2
30    CALL USER(DATA(1))
      DO 40 I=1,NV
      J1=J1+1
      CENTR(J1)=DATA(I)
40    CONTINUE
C     THE INITIAL CONFIGURATION IS GIVEN IN TERMS OF SEED POINTS.
C     CONSTRUCT AN INITIAL PARTITION BY ASSIGNING EACH DATA UNIT TO THE
C     NEAREST SEED POINT.  SEED POINTS REMAIN FIXED THROUGHOUT ASSIGNMENT
C     OF THE FULL DATA SET.
51    DO 52 K=1,NE
52    MEMBR(K)=0
      J1=0
      DO 53 J=1,NC
      NUMBR(J)=0
      DO 53 I=1,NV
      J1=J1+1
53    TOTAL(J1)=0.
C     ALLOCATE EACH DATA UNIT TO THE NEAREST SEED POINT
      K1=0
      REWIND NTIN
      DO 55 K=1,NE
      CALL USER(DATA(1))
      J2=1
C     COMPUTE DISTANCE TO FIRST SEED POINT
      DREF=DIST(DATA(1),CENTR(J2))
      JREF=1
C     TEST DISTANCES TO REMAINING SEED POINTS
      DO 54 J=2,NC
      J2=J2+NV
      DTEST=DIST(DATA(1),CENTR(J2))
      IF(DTEST.GE.DREF) GO TO 54
      DREF=DTEST
      JREF=J
54    CONTINUE
C     ALLOCATE DATA UNIT *K* TO CLUSTER *JREF*
      NUMBR(JREF)=NUMBR(JREF)+1
      MEMBR(K)=JREF
      J1=(JREF-1)*NV
      DO 55 I=1,NV
      J1=J1+1
      TOTAL(J1)=TOTAL(J1)+DATA(I)
55    CONTINUE
      J1=0
      DO 56 J=1,NC
      DO 56 I=1,NV
      J1=J1+1
      CENTR(J1)=TOTAL(J1)/NUMBR(J)
56    CONTINUE
      GO TO 100
C     IPART=2.  THE DATA UNITS ARE GROUPED INTO CLUSTERS WITH THE J-TH
C     CLUSTER HAVING *NUMBR(J)* MEMBERS.
59    J1=0
      K=0
      DO 90 J=1,NC
      DO 60 I=1,NV
      J1=J1+1
```

```
60      TOTAL(J1)=0.
        J1=NV*J
        K2=NUMBR(J)
        DO 70 K=1,K2
        K=K+1
        MEMBR(K)=J
        CALL USER(DATA(1))
        J1=J1-NV
        DO 70 I=1,NV
        J1=J1+1
        TOTAL(J1)=TOTAL(J1)+DATA(I)
70      CONTINUE
        J1=J1-NV
        DO 80 I=1,NV
        J1=J1+1
        CENTR(J1)=TOTAL(J1)/NUMBR(J)
80      CONTINUE
90      CONTINUE
C   INITIALIZE ARRAYS
100     NPASS=1
C   BEGINNING OF MAIN LOOP
120     MOVES=0
        TDIST=0
        REWIND NTIN
C   ALLOCATE EACH DATA UNIT TO THE NEAREST CLUSTER CENTROID
        DO 160 K=1,NE
        J2=1
C   COMPUTE DISTANCE TO FIRST CLUSTER CENTROID
        CALL USER(DATA(1))
        DREF=DIST(DATA(1),CENTR(J2))
        JREF=1
C   TEST DISTANCES TO REMAINING CLUSTER CENTROIDS
        DO 140 J=2,NC
        J2=J2+NV
        DTEST=DIST(DATA(1),CENTR(J2))
        IF(DTEST.GE.DREF) GO TO 140
        DREF=DTEST
        JREF=J
140     CONTINUE
        TDIST=TDIST+DREF
        IF(JREF.EQ.MEMBR(K)) GO TO 160
C   REALLOCATE DATA UNIT*K* FROM CLUSTER *MEMBR(K)* TO CLUSTER *JREF*
        MOVES=MOVES+1
        J2=MEMBR(K)
        NUMBR(J2)=NUMBR(J2)-1
        NUMBR(JREF)=NUMBR(JREF)+1
        MEMBR(K)=JREF
        J1=(JREF-1)*NV
        J3=(J2-1)*NV
        DO 150 I=1,NV
        J1=J1+1
        J3=J3+1
        TOTAL(J1)=TOTAL(J1)+DATA(I)
        CENTR(J1)=TOTAL(J1)/NUMBR(JREF)
        TOTAL(J3)=TOTAL(J3)-DATA(I)
        CENTR(J3)=TOTAL(J3)/NUMBR(J2)
150     CONTINUE
160     CONTINUE
C   ALL DATA UNITS ALLOCATED.  TEST FOR CONVERGENCE
        WRITE(6,2700) MOVES,NPASS,TDIST
        NPASS=NPASS+1
        IF(MOVES.LE.MINREL) RETURN
        GO TO 120
1000    FORMAT(20A4)
```

```
1100   FORMAT(20I4)
2000   FORMAT( 46H0CONVERGENT K-MEANS METHOD OF CLUSTER ANALYSIS,/,
       A 32H DATA SET STORED ON TAPE OR DISK)
2200   FORMAT(7H0FORMAT,20A4)
2300   FORMAT( 43H1INITIAL CLUSTER CENTERS READ IN AS FOLLOWS///)
2400   FORMAT(1X,10E12.4)
2500   FORMAT(  9H1 IPART =,I2, 30H,  NUMBR ARRAY READ AS FOLLOWS///)
2600   FORMAT(1X,10I7)
2700   FORMAT(1H0,I5,37H DATA UNITS MOVED ON ITERATION NUMBER,I3,/,
       A38H SUMMED DEVIATIONS ABOUT SEED POINTS =,E16.8)
       END
```

VERSION 5

```
       SUBROUTINE KMEAN(CENTR,NUMBR,MEMBR,TOTAL,DATA,N5,NE,NV,NC,NTIN,
       AMINREL,IPART,METHOD,LIMIT)
C
C------------------------------------------------------------------------
C   VERSION 1.   THE DATA SET IS STORED IN CENTRAL MEMORY.
C------------------------------------------------------------------------
C
C   THIS SUBROUTINE ITERATIVELY SORTS *NE* DATA UNITS INTO *NC* CLUSTERS
C   USING THE METHOD OF
C   MACQUEEN, J.B., SOME METHODS OF CLASSIFICATION AND ANALYSIS OF
C   MULTIVARIATE OBSERVATIONS, PROCEEDINGS FIFTH BERKELEY SYMPOSIUM
C   ON MATHEMATICAL STATISTICS AND PROBABILITY, VOLUME 1, 1967, PP 281-
C   297, AD669871.
C
C   CENTR(NV*(J-1)+I)=SCORE ON I-TH VARIABLE FOR J-TH CLUSTER CENTROID
C   TOTAL(NV*(J-1)+I)=TOTAL SCORE ON I-TH VARIABLE FOR DATA UNITS THUS
C                     FAR ALLOCATED TO THE J-TH CLUSTER
C   NUMBR(J)=NUMBER OF DATA UNITS THUS FAR ALLOCATED TO THE J-TH CLUSTER
C   MEMBR(K)=CLUSTER TO WHICH THE K-TH DATA UNIT CURRENTLY BELONGS
C   DATA(NV*(K-1)+I)=SCORE ON I-TH VARIABLE FOR K-TH DATA UNIT
C
       DIMENSION CENTR(1),TOTAL(1),NUMBR(1),MEMBR(1),DATA(1),FMT(20)
       WRITE(6,2000)
C   CHECK FOR SUFFICIENT STORAGE
       N6=N5+NE*NV-1
       WRITE(6,2100) N6,LIMIT
       IF(N6.GT.LIMIT) STOP
C   READ THE DATA SET INTO CENTRAL MEMORY
       K1=1
       DO 10 K=1,NE
       CALL USER (DATA(K1))
10     K1=K1+NV
C   SET THE INITIAL SEED POINTS AS THE FIRST *NC* DATA UNITS
       K1=NC*NV
       DO 20 K=1,K1
       CENTR(K)=DATA(K)
20     TOTAL(K)=DATA(K)
C   SET THE NUMBER OF ENTITIES IN EACH CLUSTER TO 1
       DO 30 J=1,NC
30     NUMBR(J)=1
C   ASSIGN EACH DATA UNIT TO NEAREST CENTROID AND UPDATE AFTER EACH
C   ASSIGNMENT
       KK=NC+1
       DO 50 K=KK,NE
       K2=K1+1
       J2=1
C   COMPUTE DISTANCE TO FIRST CLUSTER CENTROID
       DREF=DIST(DATA(K2),CENTR(J2))
       JREF=1
```

```
C   TEST DISTANCES TO REMAINING CLUSTER CENTROIDS
      DO 40 J=2,NC
      J2=J2+NV
      DTEST=DIST(DATA(K2),CENTR(J2))
      IF(DTEST.GE.DREF) GO TO 40
      DREF=DTEST
      JREF=J
40    CONTINUE
C   ALLOCATE DATA UNIT *K* TO CLUSTER *JREF*
      NUMBR(JREF)=NUMBR(JREF)+1
      J1=(JREF-1)*NV
      DO 50 I=1,NV
      J1=J1+1
      K1=K1+1
      TOTAL(J1)=TOTAL(J1)+DATA(K1)
      CENTR(J1)=TOTAL(J1)/NUMBR(JREF)
50    CONTINUE
C   ALL DATA UNITS ALLOCATED TO INITIAL CONFIGURATION
C   REALLOCATE DATA UNITS TO FIXED SEED POINTS
      TDIST=0.
      DO 60 J=1,NC
60    NUMBR(J)=0
      K1=NC*NV
      DO 70 K=1,K1
70    TOTAL(K)=0.
C   REALLOCATE DATA UNITS
      K1=0
      DO 90 K=1,NE
      K2=K1+1
      J2=1
C   COMPUTE DISTANCE TO FIRST CLUSTER CENTROID
      DREF=DIST(DATA(K2),CENTR(J2))
      JREF=1
C   TEST DISTANCES TO REMAINING CENTROIDS
      DO 80 J=2,NC
      J2=J2+NV
      DTEST=DIST(DATA(K2),CENTR(J2))
      IF(DTEST.GE.DREF) GO TO 80
      DREF=DTEST
      JREF=J
80    CONTINUE
C   ALLOCATE DATA UNIT *K* TO CLUSTER *JREF*
      NUMBR(JREF)=NUMBR(JREF)+1
      MEMBR(K)=JREF
      TDIST=TDIST+DREF
      J1=(JREF-1)*NV
      DO 90 I=1,NV
      J1=J1+1
      K1=K1+1
      TOTAL(J1)=TOTAL(J1)+DATA(K1)
90    CONTINUE
C   ALL DATA UNITS ALLOCATED.
      WRITE(6,2700) TDIST
C   COMPUTE FINAL CENTROIDS
      J1=0
      DO 100 J=1,NC
      DO 100 I=1,NV
      J1=J1+1
100   CENTR(J1)=TOTAL(J1)/NUMBR(J)
      RETURN
2000  FORMAT( 44HOMACQUEEN K-MEANS METHOD OF CLUSTER ANALYSIS,/,
     A 24H DATA SET STORED IN CORE)
2100  FORMAT(19HOREQUIRED STORAGE =,I5,6H WORDS,/,
     A          19HOALLOTTED STORAGE =,I5,6H WORDS)
```

```
2700  FORMAT(9H0FINISHED,/,
      A38H SUMMED DEVIATIONS ABOUT SEED POINTS =,E16.8)
      END
```

VERSION 6

```
      SUBROUTINE KMEAN(CENTR,NUMBR,MEMBR,TOTAL,DATA,N5,NE,NV,NC,NTIN,
      AMINREL,IPART,METHOD,LIMIT)
C
C----------------------------------------------------------------------
C  VERSION 2.  THE DATA SET IS STORED ON A TAPE OR DISK FILE WHICH IS
C  REWOUND AND READ IN ITS ENTIRETY FOR EACH CYCLE.
C----------------------------------------------------------------------
C
C  THIS SUBROUTINE ITERATIVELY SORTS *NE* DATA UNITS INTO *NC* CLUSTERS
C  USING THE METHOD OF
C
C  MACQUEEN, J.B., SOME METHODS OF CLASSIFICATION AND ANALYSIS OF
C  MULTIVARIATE OBSERVATIONS, PROCEEDINGS FIFTH BERKELEY SYMPOSIUM
C  ON MATHEMATICAL STATISTICS AND PROBABILITY, VOLUME 1, 1967, PP 281-
C  297, AD669871.
C
C  CENTR(NV*(J-1)+I)=SCORE ON I-TH VARIABLE FOR J-TH CLUSTER CENTROID
C  TOTAL(NV*(J-1)+I)=TOTAL SCORE ON I-TH VARIABLE FOR DATA UNITS THUS
C                    FAR ALLOCATED TO THE J-TH CLUSTER
C  NUMBR(J)=NUMBER OF DATA UNITS THUS FAR ALLOCATED TO THE J-TH CLUSTER
C  MEMBR(K)=CLUSTER TO WHICH THE K-TH DATA UNIT CURRENTLY BELONGS
C  DATA(I)=SCORE ON THE I-TH VARIABLE FOR THE CURRENT DATA UNIT
C
      DIMENSION CENTR(1),TOTAL(1),NUMBR(1),MEMBR(1),DATA(1),FMT(20)
      REWIND NTIN
      WRITE(6,2000)
C  SET UP THE INITIAL SEED POINTS AS THE FIRST *NC* DATA UNITS
      J1=0
      DO 10 J=1,NC
      NUMBR(J)=1
      CALL USER(DATA(1))
      DO 10 I=1,NV
      J1=J1+1
      CENTR(J1)=DATA(I)
      TOTAL(J1)=DATA(I)
10    CONTINUE
C  ASSIGN EACH DATA UNIT TO THE NEAREST CENTROID AND UPDATE THE
C  AFFECTED CENTROID AFTER EACH ASSIGNMENT
      KK=NC+1
      DO 50 K=KK,NE
      CALL USER(DATA(1))
      J2=1
C  COMPUTE DISTANCE TO FIRST CLUSTER CENTROID
      DREF=DIST(DATA(1),CENTR(J2))
      JREF=1
C  TEST DISTANCES TO REMAINING CLUSTER CENTROIDS
      DO 40 J=2,NC
      J2=J2+NV
      DTEST=DIST(DATA(1),CENTR(J2))
      IF(DTEST.GE.DREF) GO TO 40
      DREF=DTEST
      JREF=J
40    CONTINUE
C  ALLOCATE DATA UNIT *K* TO CLUSTER *JREF*
      NUMBR(JREF)=NUMBR(JREF)+1
      J1=(JREF-1)*NV
      DO 50 I=1,NV
```

```
        J1=J1+1
        TOTAL(J1)=TOTAL(J1)+DATA(I)
        CENTR(J1)=TOTAL(J1)/NUMBR(JREF)
50      CONTINUE
C   ALL DATA UNITS ALLOCATED TO INITIAL CONFIGURATION
C   REALLOCATE DATA UNITS TO FIXED SEED POINTS
        REWIND NTIN
        TDIST=0.
        DO 60 J=1,NC
60      NUMBR(J)=0
        K1=NC*NV
        DO 70 K=1,K1
70      TOTAL(K)=0.
C   REALLOCATE DATA UNITS
        DO 90 K=1,NE
        J2=1
        CALL USER(DATA(1))
C   COMPUTE DISTANCE TO FIRST CLUSTER CENTROID
        DREF=DIST(DATA(1),CENTR(J2))
        JREF=1
C   TEST DISTANCES TO REMAINING CENTROIDS
        DO 80 J=2,NC
        J2=J2+NV
        DTEST=DIST(DATA(1),CENTR(J2))
        IF(DTEST.GE.DREF) GO TO 80
        DREF=DTEST
        JREF=J
80      CONTINUE
C   ALLOCATE DATA UNIT *K* TO CLUSTER *JREF*
        NUMBR(JREF)=NUMBR(JREF)+1
        MEMBR(K)=JREF
        TDIST=TDIST+DREF
        J1=(JREF-1)*NV
        DO 90 I=1,NV
        J1=J1+1
        TOTAL(J1)=TOTAL(J1)+DATA(I)
90      CONTINUE
C   ALL DATA UNITS ALLOCATED.
        WRITE(6,2700) TDIST
C   COMPUTE FINAL CENTROIDS
        J1=0
        DO 100 J=1,NC
        DO 100 I=1,NV
        J1=J1+1
100     CENTR(J1)=TOTAL(J1)/NUMBR(J)
        RETURN
2000    FORMAT( 44H0MACQUEEN K-MEANS METHOD OF CLUSTER ANALYSIS,/,
        A 32H DATA SET STORED ON TAPE OR DISK)
2700    FORMAT(9H0FINISHED,/,
        A38H SUMMED DEVIATIONS ABOUT SEED POINTS =,E16.8)
        END
```

G

PROGRAMS TO AID INTERPRETATION OF CLUSTERING RESULTS

Some aids to interpretation of cluster analysis are presented in Chapter 8. This appendix includes computer programs for implementing these aids.

G.1 A Program for Manipulating Hierarchical Trees

Subroutine DETAIL can be used to draw any portion of a hierarchical tree or alter the scale of the criterion against which the tree is plotted. Before considering the particulars of utilizing subroutine DETAIL it is necessary to understand how the tree is drawn on the line printer.

G.1.1 DRAWING A TREE ON THE LINE PRINTER

The hierarchical methods of Chapter 6 and Appendix E build up a merge list which is printed just prior to printing the hierarchical tree. There are eight columns in this merge list which are defined as follows:

    1. K is the stage of clustering, row number in the merge list;

    2. I is the identification number of the lower numbered cluster merged at the Kth stage. A cluster is identified by the lowest numbered entity which it contains; for example, if clusters 3 and 7 are merged, then I = 3, J = 7 (see next definition), and the cluster is henceforth known as cluster number I.

    3. J is the identification number for higher numbered cluster merged at the Kth stage.

4. s is the value of the criterion associated with the merger of clusters I and J.

5. IS is the category number into which s falls. The range of s is divided into 25 segments or categories and IS merely identifies which category contains the value of s for the current stage.

6. IL is the stage number at which cluster I was last involved in a merge. IL = 0 if I consists of a single data unit.

7. JL is the stage number at which cluster J was last involved in a merge.

8. NEXT is the stage number at which the product of merging clusters I and J is next involved in a merge.

The last three pieces of information, IL, JL, and NEXT, provide a complete pathway from any stage to any other stage of clustering. The manner in which this information is used in subroutine TREE can be grasped intuitively through an example but is very difficult to understand from a formal statement of the pertinent operations. Accordingly, a small example involving the clustering of six data units is treated in great detail below. The example merge list is shown in Table G.1.

Beginning with the top row of the merge list, entities 2 and 4 are merged and the result is henceforth known as cluster 2. The value of NEXT is 5 indicating that the next stage where cluster 2 is involved in a merge is stage 5. Going down to stage 5 in the merge list, it is seen that clusters 1 and 2 merge at this stage and IS = 7 for the merge, the last in this small tree. It is necessary to build up cluster 1 before treating the merge of clusters 1 and 2. Noting that cluster 1 corresponds to I at stage 5, the value of IL identifies the previous stage involving cluster 1 as being stage 4. Dropping down to stage 4, both IL and JL are nonzero, so that both clusters 1 and 3 consist of more than one entity; the convention in this instance is to go to the stage indicated by IL which is stage 3 in this case. At stage 3, clusters 1 and 6 consist of single entities as indicated by the 0 values for IL and JL. After the merge at stage 3, the value of NEXT indicates that cluster 1 is next in a merger at stage 4. At stage 4 clusters 1 and 3 are seen to merge, but cluster 3 comes from stage 2 and it is necessary to complete this subtree

TABLE G.1
EXAMPLE OF MERGE LIST

K	I	J	S	IS	IL	JL	NEXT
1	2	4	0.7	1	0	0	5
2	3	5	1.3	2	0	0	4
3	1	6	2.2	3	0	0	4
4	1	3	2.9	4	3	2	5
5	1	2	5.1	7	4	1	0

first. Going back to stage 2, entities 3 and 5 are merged to form cluster 3; the value of NEXT is 4 and therefore transfers attention back to stage 4. At stage 4 clusters 1 and 3 may now be merged (at last) to give a new cluster labeled 1. The value of NEXT is 5 and going to stage 5 finds cluster 2 waiting for the buildup of cluster 1; the merger at stage 5 produces a new cluster which contains all the data units. This process of progressing upward in the tree and backtracking as necessary to build up subtrees exhausts the entire merge list and insures that all entities are included in the tree, unless the process is started and stopped deliberately such that only a subtree is drawn.

This method of drawing a tree on the line printer is by no means the only choice available. Parks (1970) produces much the same kind of tree but uses a much more complicated computing algorithm. Johnson (1967) depicts the tree vertically rather than horizontally on the page and uses a distinctively different way of representing the tree. A third technique is incorporated in a

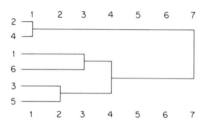

Fig. G.1. Example of hierarchical tree.

program by Ward *et al.* (1967, p. 34) and adopted by Wolfe (1971) for his NORMIX program. An additional alternative is to draw the tree with an off-line plotter; however, programs for this latter approach are always equipment dependent and often incorporate local conventions peculiar to a given computer installation.

For a given sequence of merges, these various tree drawing algorithms may appear to give different trees. At each merger there is an ambiguity as to which cluster should be plotted above the other; referring to Fig. G.1, entities 2 and 4 could swap their positions without actually changing the cluster; likewise at each merge the subtree could be rotated about the axis emanating from the merger of the clusters. Since there are $n - 1$ merges for a data set of n entities, this ambiguity permits construction of 2^{n-1} different but equivalent trees. Subroutine TREE draws a tree which is computationally convenient. McCammon (1968) and McCammon and Wenninger (1970) provide a method for drawing a tree in which the sequence of entities has some significant properties but at some considerable computational cost.

G.1.2 Usage of Subroutine DETAIL

The primary input for subroutine DETAIL is the merge list generated by any of the hierarchical clustering programs in Appendix E. This merge list completely determines the form of the tree; subroutine DETAIL merely provides the facility to draw particular subtrees or vary the scale against which the tree is plotted.

To draw a subtree alone, the analyst needs to specify the beginning and ending stage numbers as KBEG and KEND. KBEG should identify the stage number in the subtree which is nearest the beginning of the merge list, that is, the stage at which the two most similar entities in the subtree are merged. KEND should identify the stage number for the final merge in the subtree, that is, the stage at which the subtree becomes a single cluster. To illustrate, suppose it is desired to draw the subtree consisting of entities 1, 6, 3, and 5 in Fig. G.1. The most similar pair in this subtree consists of entities 3 and 5, whose merger occurs at stage 2; the highest level in the tree occurs at the merger of clusters 1 and 3 at stage 4. Therefore, this subtree would be drawn by setting KBEG = 2 and KEND = 4.

The standard scaling of the criterion values is obtained by finding the largest and smallest criterion values associated with merges in the tree (or subtree) to be drawn and dividing this range into 25 equal segments; this standard scaling is achieved by setting INTRV = 1 on the parameter card. As an alternative, the analyst can specify any desired set of segments by reading in the right endpoint of each segment; this scaling is achieved by setting INTRV = 2 and reading in the EPS array. This latter facility gives the capability to stretch and contract the tree in any desired manner.

G.2 Permuting the Similarity Matrix

The hierarchical clustering approach of Section 6.2 begins with a similarity matrix; when the clustering is finished and the standard tree is drawn, the entities are arranged in a sequence which can be saved on punched cards and used to permute the rows and columns of the similarity matrix, a technique discussed in Section 8.1.2. The program PERMUTE is provided to perform this task with a minimum effort.

The basic input is a sequence list (which may be punched in subroutine TREE) and the similarity matrix which is read using subroutine MTXIN, the same subroutine used in Appendix E to read the similarity matrix for use in subroutines CNTRL and CLSTR. The program is written such that the analyst may deal with any subset of the entities rather than the entire data set; the parameter NSUBT is the number of entities to be represented in the permuted output and IDMAX is the largest sequence number or row number among these entities. The permuted similarity matrix is printed one row at a time with NCOLP columns per page of output; each page is put on a different output unit which the analyst should arrange to copy to the line printer. The analyst has complete control over the output format in that he specifies the number of columns printed per page and the formats for both the page headings and the similarity matrix. See the comment cards at the beginning of the program for additional instruction and explanation.

G.3   Error Sum of Squares Analysis

As explained in Section 8.1.3, the relative influence of each variable in a hierarchical classification of data units may be investigated by examining the fraction of unexplained variance for each variable at each stage. Program ERROR takes as input the original data and the cluster merge list generated by one of the hierarchical clustering programs in Appendix E; then the original sequence of merges is recalculated from the merge list to build up the error sum of squares for each variable. The input specifications are included in comment cards at the beginning of the program. Note that a maximum of 14 variables may be accommodated in one run; if the data set includes more than 14 variables, partition the data set into subsets of variables and run the program for each subset.

G.4   Analysis of a Given Partition

Program POSTDU reorders the original data and computes summary statistics as discussed in Section 8.2. This program takes as input the original data and a sequence list, the latter of which may be punched on cards from either subroutine TREE or subroutine RESULT. The program is limited to partitions of 50 clusters or less, which should be adequate for most cases; the program is also limited to a maximum of 10 variables in each run but this limitation is easily overcome by partitioning the variables into convenient subsets and then running the program for each subset. Each cluster is described by listing the data units it contains, the scores for each data unit on the selected variables, the mean of each variable, and the variance of each variable.

Subroutine DETAIL

```
      SUBROUTINE DETAIL (X,LIMIT)
C
C   THIS SUBPROGRAM IS DESIGNED TO PERMIT DETAILED STUDY OF THE TREE
C   ASSOCIATED WITH A HIERARCHICAL CLUSTERING.  HIERARCHICAL PROGRAMS I
C   THIS SYSTEM AUTOMATICALLY PREPARE THE NECESSARY INPUT.  A SEQUENCE
```

```
C  OF BINARY MERGES FROM ANY HIERARCHICAL METHOD MAY BE PROCESSED IN
C  SUBROUTINES *ALLIN1/PREP* TO GENERATE THE NECESSARY INPUT FOR
C  DRAWING A TREE.
C
C---------------------------------------------------------------------
C  INPUT SPECIFICATIONS
C
C  CARD 1   TITLE CARD
C  CARD 2   INFORMATION FOR SUBROUTINES CLSTR AND TREE
C     COLS   1- 4   KBEG=STAGE NUMBER AT WHICH THE TREE OR SUBTREE BEGINS
C     COLS   5- 8   KEND=STAGE NUMBER AT WHICH THE TREE OR SUBTREE ENDS
C     COLS   9-12   MAXID=HIGHEST CLUSTER ID NUMBER IN THE TREE OR SUBTREE
C     COLS  13-14   NTIN=INPUT UNIT FOR CLUSTER MERGE DATA
C                   NTIN=5, CARD READER
C                   NTIN.NE.5, TAPE OR DISK
C     COL     15    INTRV=OPTION FOR SCALE INTERVALS ON TREE (*EPS* ARRAY)
C                   INTRV=1, DIVIDE RANGE OF CRITERION *SS* INTO 25 EQUAL
C                        INTERVALS
C                   INTRV=2, RIGHT ENDPOINT FOR EACH INTERVAL SPECIFIED ON
C                        CARD 5
C     COLS  16-17   IPRNT=OPTION FOR PRINTING IN *TREE*
C                   IABS(IPRNT)=1, MINIMUM OUTPUT IN ADDITION TO THE TREE
C                   IABS(IPRNT).NE.1, STANDARD OUTPUT INCLUDING THE MERGE
C                        DATA
C                   IPRNT.LE.0, IN ADDITION PUNCH THE SEQUENCE IN WHICH
C                        ENTITIES ARE ARRANGED IN THE TREE
C     COLS  18-19   NTOUT=OUTPUT UNIT FOR SUBROUTINE *TREE*
C                   NTOUT=6, LINE PRINTER (DEFAULT VALUE)
C
C***ANY PREPOSITIONING OF THE I/O UNITS NTSV AND NTIN MUST BE
C  ACCOMPLISHED IN PROGRAM DRIVER OR THROUGH USE OF CONTROL CARDS.
C
C
C  CARD(S) 3  LABEL CARDS FOR ENTITIES.  THERE ARE TWO OPTIONS
C     1.   INCLUDE 1 CARD WITH THE 4 CHARACTERS *NOLB* IN COLUMNS 1-4.
C          UNDER THIS OPTION LABELS ARE NOT PRINTED ON THE TREE OUTPUT.
C
C     2.   INCLUDE*MAXID* CARDS FOR ENTITIES NUMBERED 1 THROUGH *MAXID*.
C          COLUMNS 1 TO 20 CONTAIN THE LABEL FOR THE ASSOCIATED ENTITY.
C
C  CARD 4   FORMAT CARD FOR READING *EPS ARRAY.  INCLUDE ONLY IF INTRV=2
C
C  CARD(S) 5  *EPS* ARRAY OF 25 ELEMENTS READ ACCORDING TO FORMAT ON
C             CARD 4, INCLUDE ONLY IF INTRV=2.
C  CARD(S) 6  CLUSTER MERGE DATA AS PREPARED BY *PREP*, *GROUP*, OR
C             *CLSTR*.  INCLUDE AT LEAST THAT PORTION BETWEEN STAGES
C             *KBEG* AND *KEND* INCLUSIVE.  THE ENTIRE SET OF DATA MAY
C             BE INCLUDED SINCE THAT PRIOR TO *KBEG* IS IGNORED AND
C             THAT AFTER *KEND* IS NOT READ.
C---------------------------------------------------------------------
C
C  DECK SETUP SPECIFICATIONS
C
C  THE USER PROVIDES PROGRAM DRIVER WHICH PERFORMS THE FOLLOWING TASKS.
C     1.   ASSIGNS INPUT/OUTPUT UNITS
C     2.   ESTABLISHES THE DIMENSION OF THE X ARRAY AND SETS THIS
C          DIMENSION EQUAL TO LIMIT.
C     3.   CALLS SUBROUTINE DETAIL.
C  THE FOLLOWING EXAMPLE WILL SUFFICE IN MOST CASES.
C
C     PROGRAM DRIVER (INPUT,OUTPUT,PUNCH,TAPE5=INPUT,TAPE6=OUTPUT,
C     ATAPE7=PUNCH,TAPE1,TAPE2)
C     DIMENSION X(7000)
```

```
C      LIMIT=7000
C      CALL DETAIL(X,LIMIT)
C      END
C----------------------------------------------------------------------
C
C      THE X ARRAY IS PARTITIONED FOR STORAGE AS FOLLOWS
C      X(N1)   TO X(N2-1)    KEND WORDS--STORAGE OF THE II ARRAY
C      X(N2)   TO X(N3-1)    KEND WORDS--STORAGE OF THE JJ ARRAY
C      X(N3)   TO X(N4-1)    KEND WORDS--STORAGE OF THE SS ARRAY

C      X(N4)   TO X(N5-1)    KEND WORDS--STORAGE OF THE IL ARRAY
C      X(N5)   TO X(N6-1)    KEND WORDS--STORAGE OF THE JL ARRAY
C      X(N6)   TO X(N7-1)    KEND WORDS--STORAGE OF THE NEXT ARRAY
C      X(N7)   TO X(N8-1)    25*MAXID WORDS--STORAGE OF THE A ARRAY
C      X(N8)   TO X(N9-1)    5*MAXID WORDS--STORAGE OF THE LABEL ARRAY
C      X(N9)   TO X(N10-1)   MAXID WORDS-- STORAGE OF THE LCLNO ARRAY
C      X(N10)  TO X(N11-1)   MAXID WORDS-- STORAGE OF THE LINE ARRAY
C      X(N11)  TO X(N12-1)   MAXID WORDS-- STORAGE OF THE IS ARRAY
C      X(N12)  TO X(N13)     MAXID WORDS-- STORAGE OF THE LAST ARRAY
C----------------------------------------------------------------------
C
       INTEGER FIRST
       DIMENSION X(1),FMT(20),TITLE(20),EPS(25)
       READ(5,1000) TITLE
       READ(5,1100) KBEG,KEND,MAXID,NTIN,INTRV,IPRNT,NTOUT
       WRITE(6,2000) TITLE
       WRITE(6,2100) KBEG,KEND,MAXID,NTIN,INTRV,IPRNT,NTOUT
C      PARTITION THE STORAGE ARRAY
       N1=1
       N2=N1+KEND
       N3=N2+KEND
       N4=N3+KEND
       N5=N4+KEND
       N6=N5+KEND
       N7=N6+KEND
       N8=N7+25*MAXID
       N9=N8+5*MAXID
       N10=N9+MAXID
       N11=N10+MAXID
       N12=N11+MAXID
       N13=N12+MAXID-1
C      CHECK FOR SUFFICENT STORAGE
       WRITE(6,2300) N13,LIMIT
       IF(N13.GT.LIMIT) STOP
C      READ REMAINING INPUT
C      READ LABELS
       FIRST=N8
       LAST=N8+4
       READ(5,1000) (X(I),I=FIRST,LAST)
       IF(X(FIRST).EQ.4HNOLB) GO TO 20
C      READ REMAINING LABELS
       DO 10 K=2,MAXID
       FIRST=LAST+1
       LAST=LAST+5
10     READ(5,1000) (X(I),I=FIRST,LAST)
20     IF(INTRV.NE.2.AND.INTRV.NE.3) GO TO 30
C      READ EPS ARRAY
       READ (5,1000) FMT
       WRITE(6,2200) FMT
       READ(5,FMT) EPS
C      READ CLUSTER MERGE DATA
30     CALL READCM(X(N1),X(N2),X(N3),X(N4),X(N5),X(N6),KEND,NTIN)
C      DRAW THE TREE
       CALL TREE(X(N1),X(N2),X(N3),X(N4),X(N5),X(N6),X(N7),X(N8),X(N9),
```

```
     AX(N10),X(N11),X(N12),EPS,TITLE,KEND,KBEG,NTOUT,INTRV,IPRNT,MAXID)
     RETURN
1000 FORMAT(20A4)
1100 FORMAT(3I4,I2,I1,2I2)
2000 FORMAT(1H1,20A4)
2100 FORMAT(7H0KBEG =,I6,/,7H KEND =,I6,/,8H MAXID =,I5,/,7H NTIN =,I6,
     A/,8H INTRV =,I5,/,8H IPRNT =,I5,/,8H NTOUT =,I5)
2200 FORMAT(7H0FORMAT,20A4)
2300 FORMAT(19H0REQUIRED STORAGE =,I5,6H WORDS,/,
     A       19H0ALLOTTED STORAGE =,I5,6H WORDS,/)
     END
```

Subroutine READCM

```
     SUBROUTINE READCM(II,JJ,SS,IL,JL,NEXT,KEND,NTIN)
C    THIS SUBROUTINE READS CLUSTER MERGE DATA FOR *DETAIL*
     DIMENSION II(1),JJ(1),SS(1),IL(1),JL(1),NEXT(1)
10   READ(NTIN,2000) K,II(K),JJ(K),SS(K),IL(K),JL(K),NEXT(K)
     IF(K.NE.KEND) GO TO 10
     RETURN
2000 FORMAT(3I10,E16.8,3I10)
     END
```

Subroutine TREE

```
     SUBROUTINE TREE(I,J,S,IL,JL,NEXT,A,LABEL,LCLNO,LINE,IS,LAST,EPS,
     ATITLE,N,KBEG,NT,INTRV,IPRNT,MAXIN)

DATA INPUT THROUGH CALLING SEQUENCE

N=HIGHEST STAGE NUMBER IN THE CLUSTER MERGE DATA (MUST BE EXACT)
KBEG=STAGE NUMBER AT WHICH THE TREE BEGINS, DEFAULT VALUE 1
NT=TAPE NUMBER FOR PRINTED OUTPUT, DEFAULT VALUE = 6
INTRV=INTERVAL OPTION FOR SEGMENTATION
INTRV=1=DEFAULT VALUE, CONSTRUCT EPS BY DIVIDING THE RANGE OF S INTO
        25 EQUAL SEGMENTS
INTRV=2=EPS IS PROVIDED AS PART OF THE ARGUMENT LIST
INTRV=3=THE IS ARRAY IS ALREADY CONSTRUCTED AND EPS IS PROVIDED FOR INFO
IPRNT=PRINT OPTION FOR INPUT INFORMATION
IABS(IPRNT)=1, PRINT ONLY TITLE AND *IS* ARRAY
IABS(IPRNT).NE.1, IN ADDITION PRINT THE CLUSTER MERGE DATA
IPRNT.LE.0, IN ADDITION, PUNCH THE SEQUENCE IN WHICH THE ENTITIES
            APPEAR IN THE TREE (NEEDED FOR POST-ANALYSIS OF DATA
            UNIT CLUSTERING IN SUBROUTINE *POSTDU*).
EPS(M)=RIGHT ENDPOINT FOR THE MTH INTERVAL USED FOR SEGMENTING S
LABEL(M,IJ)=MTH OF 5 WORDS IDENTIFYING THE IJTH OBJECT
TITLE=ARRAY OF 20 WORDS FOR IDENTIFYING THE RUN.
K=INDEX IDENTIFYING STAGE NUMBER IN THE CLUSTERING
I(K)=LOWER NUMBERED CLUSTER IDENTIFICATION NUMBER IN THE MERGE AT THE
     KTH STAGE
J(K)=UPPER NUMBERED CLUSTER IDENTIFICATION NUMBER IN THE MERGE AT THE
     KTH STAGE
S(K)=VALUE OF THE CRITERION FUNCTION FOR THE MERGE AT THE KTH STAGE
IS(K)=CATEGORIZED VALUE OF S= INTEGER IN RANGE 1 TO 25
IL(K)=STAGE NUMBER WHEN I(K) WAS LAST IN A MERGE (0 FOR FIRST MERGE FOR I(K))
JL(K)=STAGE NUMBER WHEN J(K) WAS LAST IN A MERGE (0 FOR FIRST MERGE FOR J(K))
NEXT(K)=STAGE NUMBER WHEN I(K) IS NEXT IN A MERGE
```

```
C    MAXIN=HIGHEST CLUSTER ID NUMBER IN THE CLUSTER MERGE DATA
C
C    OTHER VARIABLES USED IN THE PROGRAM
C
C    LINE(I)=LINE NUMBER IN THE PRINTOUT AT WHICH I(K) IS CARRIED (AFTER
C              MOST RECENT MERGE)
C    LCLNO(L)=THE CLUSTER NUMBER TO BE PRINTED ON LINE L AT THE LEFT OF THE
C    A(M,L)=THE MTH SEGMENT (OF 25) IN THE LTH LINE OF THE PRINTOUT
C    LAST(L)=FARTHEST RIGHT SEGMENT IN LINE L WHICH IS NOT BLANK
           DIMENSION I(N),J(N),S(N),IS(N),IL(N),JL(N),NEXT(N),LCLNO(N),
          AA(25,N),LAST(N)
           DIMENSION LINE(MAXIN),LABEL(5,MAXIN)
           DIMENSION EPS(25),TITLE(20)
           DATA BARI,BLNKI,BARS,BLANK/4H---I,4H    I,4H----,4H
C    DEFAULT VALUES
           IF(KBEG.LT.1) KBEG=1
           IF(INTRV.LT.1.OR.INTRV.GT.3) INTRV=1
           IF(NT.LE.0) NT=6
C    INITIALIZE ARRAYS
           NOBJ=N+1
           DO 10 K=1,NOBJ
           LINE(K)=0
           LCLNO(K)=0
           LAST(K)=0
           DO 10 L=1,25
           A(L,K)=BLANK
10         CONTINUE
C    SEGMENT THE S ARRAY
           GO TO (20,40,120),INTRV
C    CONSTRUCT INTERVALS OF EQUAL LENGTH
20         RANGE=S(N)-S(KBEG)
           DELTA=RANGE/25.
           EPS(1)=S(KBEG)+DELTA
           DO 30 K=2,24
30         EPS(K)=EPS(K-1)+DELTA
           EPS(25)=S(N)
C    CONSTRUCT THE IS ARRAY
40         IF(EPS(1).GT.EPS(2)) GO TO 70
C    S INCREASES WITH DISSIMILARITY (AS DOES A DISTANCE)
           KK=1
           DO 60 K=1,N
50         IF(S(K).LE.EPS(KK)) GO TO 60
           IF(KK.EQ.25) GO TO 60
           KK=KK+1
           GO TO 50
60         IS(K)=KK
           GO TO 120
C    S DECREASES WITH DISSIMILARITY (AS DOES A CORRELATION)
70         KK=24
           KKK=25
           NN=N+1
           DO 90 K=1,N
           KCOMP=NN-K
80         IF(S(KCOMP).LT.EPS(KK)) GO TO 90
           KKK=KK
           KK=KK-1
           IF(KK.EQ.0) GO TO 100
           GO TO 80
90         IS(KCOMP)=KKK
100        DO 110 K=1,KCOMP
110        IS(K)=1
C    PRINT INPUT TO TREE
120        WRITE(NT,2000) TITLE
           WRITE(NT,2100) KBEG,N
```

```
       WRITE(NT,2200)
       WRITE(NT,2300)
       M=1
       WRITE(NT,2400) M,S(KBEG),EPS(M)
       DO 130 M=2,25
       MM=M-1
130    WRITE(NT,2400) M,EPS(MM),EPS(M)
       IF(IABS(IPRNT).EQ.1) GO TO 150
C   PRINT THE CLUSTER MERGE DATA
       WRITE(NT,2000)    TITLE
       WRITE(NT,2500)
       DO 140 K=KBEG,N
       WRITE(NT,2600) K,I(K),J(K),S(K),IS(K),IL(K),JL(K),NEXT(K)
140    CONTINUE
C   START TREE WITH THE MOST SIMILAR PAIR
150    K=KBEG
       LNO=0
C   MERGE CLUSTERS I(K) AND J(K)
160    IK=I(K)
       JK=J(K)
C   SET LINE NUMBERS FOR OUTPUT
       IF(IL(K).NE.0) GO TO 170
       LNO=LNO+1
       LINE(IK)=LNO
       LCLNO(LNO)=IK
170    IF(JL(K).NE.0) GO TO 180
       LNO=LNO+1
       LINE(JK)=LNO
       LCLNO(LNO)=JK
C   FILL IN THE PRINT LINES
180    ISK=IS(K)
       KT=0
       ITEM=IK
190    LITEM=LINE(ITEM)
       IF(ISK-LAST(LITEM)-1) 225,200,210
C   ADD ONLY ONE MORE SEGMENT FOR LINE(ITEM)
200    A(ISK,LITEM)=BARI
       LAST(LITEM)=ISK
       GO TO 225
C   ADD MORE THAN ONE SEGMENT
210    LBEG=LAST(LITEM)+1
       LEND=ISK-1
       DO 220 L=LBEG,LEND
220    A(L,LITEM)=BARS
       GO TO 200
C   REPEAT FOR CLUSTER J(K)
225    KT=KT+1
       IF(KT.NE.1) GO TO 230
       ITEM=JK
       GO TO 190
C   TAKE CARE OF ANY LINES BETWEEN I(K) AND J(K)
230    LIK=LINE(IK)
       LJK=LINE(JK)
       IF(LIK.GT.LJK) GO TO 240
       LBOT=LJK
       LTOP=LIK
       GO TO 250
240    LBOT=LIK
       LTOP=LJK
250    IF(LBOT.EQ.(LTOP+1)) GO TO 270
C   MUST FILL IN SOME VERTICAL CONNECTIONS
       LBEG=LTOP+1
       LEND=LBOT-1
       DO 260 L=LBEG,LEND
```

```
          IF(A(ISK,L).EQ.BARI) GO TO 260
          A(ISK,L)=BLNKI
          LAST(L)=ISK
260       CONTINUE
C   UPDATE LINE NUMBER FOR NEW CLUSTER
270       LINE(IK)=(LINE(IK)+LINE(JK))/2
C   MERGE COMPLETE.  FIND NEXT STAGE
          KLAST=K
          K=NEXT(K)
          IF(K.GT.N.OR.K.LT.KBEG) GO TO 400
          IF(IL(K).LE.0) GO TO 280
          IF(JL(K).LE.0) GO TO 290
          GO TO 300
280       IL(K)=-IL(K)
          GO TO 160
290       JL(K)=-JL(K)
          GO TO 160
C   THIS MERGE INVOLVES CLUSTERS THAT EACH HAVE MORE THAN ONE MEMBER.
C   BACKTRACK TO THE ROOT OF THE TREE ALONG THE UNEXPLORED BRANCH.
300       IF(IL(K).EQ.KLAST) GO TO 310
C   GO DOWN IL(K) BRANCH.  SET JL(K) SO WE KNOW NOT TO GO DOWN THAT BRANCH
          JL(K)=-JL(K)
          K=IL(K)
          GO TO 320
C   GO DOWN JL(K) BRANCH.  SET IL(K) SO WE KNOW NOT TO GO DOWN THAT BRANCH
310       IL(K)=-IL(K)
          K=JL(K)
320       IF(K.LT.1.OR.K.GT.N) GO TO 600
C   TEST TO SEE IF THE END HAS BEEN REACHED.  IL(K)=JL(K) IFF BOTH ZERO.
          IF(IL(K)-JL(K)) 330,160,350
330       IF(IL(K).EQ.0) GO TO 360
340       K=IL(K)
          GO TO 320
350       IF(JL(K).EQ.0) GO TO 340
360       K=JL(K)
          GO TO 320
C   PRINT THE TREE
400       WRITE(NT,2000) TITLE
          WRITE(NT,3000) (K,K=1,25)
          IF(LABEL(1,1).EQ.4HNOLB) GO TO 420
          DO 410 L=1,LNO
          LL=LCLNO(L)
410       WRITE(NT,3100) (LABEL(K,LL),K=1,5),LL,(A(K,L),K=1,25)
          GO TO 440
C   LEAVE LABEL SPACES BLANK
420       DO 430 L=1,LNO
          LL=LCLNO(L)
430       WRITE(NT,3200) LL,(A(K,L),K=1,25)
C   TREE COMPLETE
440       WRITE(NT,3000) (K,K=1,25)
          ENDFILE NT
          IF(IPRNT.GT.0) RETURN
C   PUNCH SEQUENCE LIST
          WRITE(7,3900) TITLE
          WRITE(7,4000) (LCLNO(L),L=1,LNO)
          RETURN
C   ERROR.  PRINT AS MUCH OF THE TREE AS HAS BEEN CONSTRUCTED
600       WRITE(NT,6000) KLAST,K
          GO TO 400
2000      FORMAT(1H1,20X,20A4)
2100      FORMAT(65H0THIS RUN DEPICTS THE PORTION OF THE TREE GENERATED BETW
         AEEN STAGE, I5,10H AND STAGE,,I5,19H OF THE CLUSTERING.)
2200      FORMAT(63H0THE CRITERION VALUES ARE SEGMENTED INTO THE FOLLOWING C
         ALASSES.)
```

```
2300  FORMAT(6H0CLASS,5X,11HLOWER BOUND,5X,11HUPPER BOUND)
2400  FORMAT(1X,I5,2E16.8)
2500  FORMAT(1H0,9X,1HK,9X,1HI,9X,1HJ,15X,1HS,8X,2HIS,8X,2HIL,8X,2HJL,6X
     A,4HNEXT)
2600  FORMAT(1X,3I10,E16.8,4I10)
3000  FORMAT(10H0ITEM NAME,12X,5HID NO,2X,25I4)
C  IF LOCAL CONVENTIONS PERMIT, RECOMMEND THAT THE CARRIAGE CONTROL
C  CHARACTER IN FORMATS 3100 AND 3200 ALLOW 66 LINES OF PRINT PER PAGE.
C  THAT IS, THE MARGINS AT THE TOP AND BOTTOM OF THE PAGE ARE SUPPRESSED
C  AND PRINTING IS SINGLE SPACE.
3100  FORMAT(1HQ,5A4,I6,2X,25A4)
3200  FORMAT(1HQ,20X,I6,2X,25A4)
3900  FORMAT(20A4)
4000  FORMAT(20I4)
6000  FORMAT(36H0ERROR. WHILE BACKTRACKING FROM KLAST,I6,26H K WAS FOUND
     A OUT OF RANGE.,/,1X,3HK =,I20)
      END
```

Program PERMUTE

```
      PROGRAM PERMUTE(INPUT,OUTPUT,TAPE5=INPUT,TAPE6=OUTPUT,
     ATAPE7,TAPE8,TAPE9,TAPE10,TAPE11,
     ZTAPE1)
C  THIS PROGRAM IS WRITTEN TO TAKE A SIMILARITY MATRIX (IN LOWER
C  TRIANGULAR FORM STORED BY ROWS) AND PERMUTE IT TO ANY DESIRED
C  SEQUENCE.
C
C-------------------------------------------------------------------------
C  INPUT SPECIFICATIONS
C
C  CARD 1 TITLE CARD
C  CARD 2
C     COLS  1- 5 NTSEQ=INPUT UNIT FOR THE LIST ARRAY (SEE CARD 7)
C     COLS  6-10 NTMTX=INPUT UNIT FOR SIMILARITY MATRIX, S (SEE CARD 8)
C     COLS 11-15 NCOLP=NUMBER OF COLUMNS OF THE MATRIX TO BE PRINTED
C                      PER PAGE
C     COLS 16-20 NSUBT=TOTAL NUMBER OF SUBJECTS REPRESENTED IN THE
C                      PERMUTED MATRIX
C     COLS 21-25 IDMAX=LARGEST ID NUMBER IN THE LIST ARRAY
C     COLS 26-30 INOPT=INPUT OPTION FOR SIMILARITY MATRIX
C                INOPT.LE.0, EACH RECORD IS ONE ROW OF THE LOWER
C                            TRIANGULAR MATRIX
C                INOPT.GT.0, THE LOWER TRIANGULAR MATRIX IS STORED BY
C                            BY ROWS IN ONE LONG LINEAR ARRAY AND EACH
C                            RECORD IS A BLOCK OF *INOPT* ENTRIES
C                            RECORD
C  *****FORMAT CARDS*****
C  CARD 3 FMTLA=FORMAT FOR READING THE LIST ARRAY
C  CARD 4 FMTSM=FORMAT FOR READING THE SIMILARITY MATRIX
C  CARD 5 FMTHD=FORMAT FOR PRINTING HEADINGS ON EACH OUTPUT UNIT
C  CARD 6 FMTPM=FORMAT FOR PRINTING THE PERMUTED SIMILARITY MATRIX
C  IN THE TWO PRECEDING FORMATS ALLOW FOR ONE COLUMN OF ID NUMBERS
C  AND NCOLP COLUMNS OF THE SIMILARITY MATRIX PER OUTPUT UNIT.
C  *****END OF FORMAT CARDS*****
C  CARD(S) 7 LIST=DESIRED SEQUENCE OF ROWS AND COLUMNS FOR PERMUTED
C                 MATRIX (READ ACCORDING TO FORMAT FMTLA)
C  CARD(S) 8 S=SIMILARITY MATRIX IN LOWER TRIANGULAR FORM STORED
C                 LINEARLY BY ROWS UP THROUGH ROW IDMAX (READ ACCORDING
C                 TO FORMAT FMTSM)
C-------------------------------------------------------------------------
C
```

```
C   ADDITIONAL VARIABLES IN THE PROGRAM
C     NCOLT=TOTAL NUMBER OF COLUMNS IN THE PERMUTED MATRIX=NSUBT-1
C     SROWI(J)=ELEMENT IN COLUMN J AND ROW I OF THE PERMUTED SIMILARITY
C               MATRIX
C
C   THE USER SHOULD INSURE THAT PROGRAM CAPACITY IS SUFFICIENT FOR HIS
C   PROBLEM.  IN PARTICULAR
C     1.  THERE MUST BE ENOUGH TAPE UNITS (ACTUALLY DISK TRACKS)
C   DEFINED TO HANDLE NCOLT COLUMNS OF DATA AT NCOLP COLUMNS PER
C   UNIT.  THE CDC 6600 PERMITS UP TO 50 UNITS.
C
C     2.  ARRAY DIMENSIONS ARE SPECIFIED AS
C     DIMENSION S(M),LIST(N),SROWI(N)
C   WHERE M IS AT LEAST AS LARGE AS IDMAX*(IDMAX-1)/2 AND N IS AT
C   LEAST AS LARGE AS NSUBT.
C
C   AFTER THE CONTROL CARD WHICH LOADS THE PROGRAM, THE USER SHOULD
C   PROVIDE CONTROL CARDS TO REWIND THE DISK FILES AND COPY THEM TO
C   OUTPUT.   FOR EXAMPLE,
C     REWIND(TAPE7)
C     COPYBF(TAPE7,OUTPUT)
C-----------------------------------------------------------------------
C
      DIMENSION S(5000),LIST(100),SROWI(100)
      DIMENSION FMTLA(20),FMTSM(20),FMTHD(20),FMTPM(20),TITLE(20)
      INTEGER FIRST
      READ(5,1100) TITLE
      READ(5,1000) NTSEQ,NTMTX,NCOLP,NSUBT,IDMAX,INOPT
      READ(5,1100) FMTLA
      READ(5,1100) FMTSM
      READ(5,1100) FMTHD
      READ(5,1100) FMTPM
      WRITE(6,2200) TITLE
      WRITE(6,2000) NTSEQ,NTMTX,NCOLP,NSUBT,IDMAX,INOPT
      WRITE(6,2100) FMTLA
      WRITE(6,2100) FMTSM
      WRITE(6,2100) FMTHD
      WRITE(6,2100) FMTPM
      READ(NTSEQ,FMTLA) (LIST(I),I=1,NSUBT)
      CALL MTXIN(S,INOPT,IDMAX,NTMTX,FMTSM)
C   PRINT HEADINGS ON EACH UNIT
15    NCOLT=NSUBT-1
      NTO=6
      FIRST=1
      LAST=NCOLP
20    WRITE(NTO,FMTHD) (LIST(J),J=FIRST,LAST)
      FIRST=LAST+1
      LAST=LAST+NCOLP
      NTO=NTO+1
      REWIND NTO
      IF(LAST.LT.NCOLT) GO TO 20
      WRITE(NTO,FMTHD) (LIST(J),J=FIRST,NCOLT)
C   PERMUTE SIMILARITY MATRIX AND PRINT IN LOWER TRIANGULAR FORM
      DO 40 I=2,NSUBT
      IM1=I-1
C   CONSTRUCT THE LIST(I) ROW
      DO 30 J=1,IM1
      IJ=LFIND(LIST(I),LIST(J))
30    SROWI(J)=S(IJ)
C   PRINT THE ROW
      NTO=6
      FIRST=1
      IF(IM1.LE.NCOLP) GO TO 37
      LAST=NCOLP
```

```
35       WRITE(NTO,FMTPM) LIST(I),(SROWI(J),J=FIRST,LAST)
         FIRST=LAST+1
         LAST=LAST+NCOLP
         NTO=NTO+1
         IF(LAST.LT.IM1) GO TO 35
37       WRITE(NTO,FMTPM) LIST(I),(SROWI(J),J=FIRST,IM1)
40       CONTINUE
         STOP
1000     FORMAT(6I5)
1100     FORMAT(20A4)
2000     FORMAT(8H NTSEQ =,I5,/,8H NTMTX =,I5,/,8H NCOLP =,I5,/,
     A           8H NSUBT =,I5,/,8H IDMAX =,I5,/,8H INOPT =,I5)
2100     FORMAT(7H FORMAT,20A4)
2200     FORMAT(1H1,20A4)
         END
```

Subroutine MTXIN

```
         SUBROUTINE MTXIN(X,IOPT,NE,NTIN,FMT)
C   THIS SUBROUTINE READS A LOWER TRIANGULAR MATRIX *X* REPRESENTING
C   ASSOCIATION AMONG *NE* ENTITIES.  THE MATRIX IS READ FROM UNIT *NTIN*
C   IN FORMAT *FMT*.  THE MODE OF INPUT FOR THE MATRIX IS DETERMINED BY
C   THE *IOPT* PARAMETER AS FOLLOWS.
C     IOPT.LE.0, MATRIX IS READ IN LOWER TRIANGULAR FORM BY ROWS, EACH
C                ROW BEING A NEW RECORD.
C     IOPT.GT.0, MATRIX IS READ IN CONSTANT LENGTH BLOCKS, EACH *IOPT*
C                WORDS LONG.
         DIMENSION FMT(20),X(1)
         INTEGER FIRST
         IF(IOPT.LE.0) GO TO 30
C   READ THE SIMILARITY MATRIX IN BLOCKS IOPT LONG
         FIRST=1
         LAST=IOPT
10       READ(NTIN,FMT) (X(I),I=FIRST,LAST)
C   USE THE END OF RECORD CARD TO SIGNIFY END OF THE SIMILARITY MATRIX
         IF(EOF,NTIN) 60,20
20       FIRST=FIRST+IOPT
         LAST=LAST+IOPT
         GO TO 10
C   READ THE SIMILARITY MATRIX AS ROWS OF A LOWER TRIANGULAR MATRIX,
C   EACH ROW A RECORD.
30       FIRST=1
         LAST=1
         DO 50 K=2,NE
         READ(NTIN,FMT) (X(I),I=FIRST,LAST)
         IF(EOF,NTIN) 200,40
40       FIRST=LAST+1
         LAST=LAST+K
50       CONTINUE
C   PASS THE END OF FILE
         READ(NTIN,FMT) Z
         IF(EOF,NTIN) 60,210
60       RETURN
C   ERROR MESSAGES
200      WRITE(6,2400)
         GO TO 220
210      WRITE(6,2500)
220      WRITE(6,2600)   K,FIRST,LAST,Z,(X(I),I=FIRST,LAST)
         STOP
2400     FORMAT(36H0EOF ENCOUNTERED WHEN NONE EXPECTED.)
2500     FORMAT(30H0NO EOF WHEN ONE WAS EXPECTED.)
2600     FORMAT(1X,3I10,F10.7,/,(1X,12F10.7))
         END
```

Function LFIND

```
      FUNCTION LFIND(I,J)
C  IF THE LOWER TRIANGULAR PORTION OF A SYMMETRIC MATRIX IS STORED BY
C  ROWS IN A ONE-DIMENSIONAL ARRAY, THEN THE ELEMENT (I,J) IN THE FULL
C  MATRIX IS ELEMENT LFIND(I,J) IN THE LINEAR ARRAY
      IF(I.GT.J) GO TO 10
C  ROW J, COLUMN I
      LFIND=((J-1)*(J-2))/2+I
      RETURN
C  ROW I, COLUMN J
10    LFIND=((I-1)*(I-2))/2+J
      RETURN
      END
```

Program ERROR

```
      PROGRAM ERROR(INPUT,OUTPUT,TAPE5=INPUT,TAPE6=OUTPUT,TAPE1,TAPE2)
C
C  THIS PROGRAM IS WRITTEN TO AID IN ANALYZING THE WITHIN GROUP
C  VARIANCE FOR SELECTED VARIABLES.  IF THERE ARE *NS* SUBJECTS AND WE
C  PERMIT *NS* GROUPS, EACH CONSISTING OF ONE SUBJECT, THEN THE GROUPING
C  EXPLAINS ALL OF THE VARIANCE, I.E. EACH SUBJECT IS REPRESENTED
C  WITHOUT ERROR BY THE MEAN VECTOR OF THE CLUSTER TO WHICH IT BELONGS.
C  ON THE OTHER HAND, IF WE COLLECT ALL SUBJECTS INTO A SINGLE GROUP
C  THEN THE GROUPING EXPLAINS NONE OF THE VARIANCE SINCE ALL THE
C  VARIANCE IS WITHIN THE SINGLE CLUSTER.  IN THIS LATTER CASE THE
C  EXPECTED SQUARED ERROR IS MAXIMIZED WHEN WE USE THE CLUSTER MEAN TO
C  PREDICT THE PROFILE OF EACH SUBJECT.  BETWEEN THESE EXTREMES WE TAKE
C  THE SUM OF WITHIN GROUP SQUARED ERRORS AS THE UNEXPLAINED VARIANCE
C  AND THE BETWEEN GROUPS SQUARED ERROR AS THE VARIANCE EXPLAINED BY
C  GROUPING.  THIS PROGRAM CALCULATES THE PROPORTION OF UNEXPLAINED
C  VARIANCE AT SELECTED STAGES OF CLUSTERING FOR EACH VARIABLE.  THIS
C  TECHNIQUE IS DENOTED AS *DECOMPOSITION OF SUM SQUARED ERROR*.  IT
C  IS SUGGESTED IN THE FOLLOWING REFERENCES.
C
C  BALL, G.H., CLASSIFICATION ANALYSIS, TECHNICAL NOTE, STANFORD
C  RESEARCH INTITUTE, MENLO PARK, CALIFORNIA, NOVEMBER 1970, AD 716482
C  SEE PP 73-75.
C
C  HALL, D.J., ET AL, PROMENADE-AN IMPROVED INTERACTIVE-GRAPHICS
C  MAN/MACHINE SYSTEM FOR PATTERN RECOGNITION, REPORT NUMBER
C  RADC-TR-68-572, STANFORD RESEARCH INSTITUTE, MENLO PARK, CALIFORNIA,
C  JUNE 1969, AD 692752, SEE APPENDIX 9G, PP 11-13.
C
C------------------------------------------------------------------------
C  INPUT SPECIFICATIONS
C
C  CARD 1    TITLE CARD
C  CARD 2
C    COLS  1- 5 NTDAT=INPUT UNIT FOR DATA
C    COLS  6-10 NTMRG=INPUT UNIT FOR MERGER SPECIFICATIONS
C    COLS 11-15 NS=NUMBER OF SUBJECTS
C    COLS 16-20 NV=NUMBER OF VARIABLES USED IN THIS RUN (MAX 14)
C  CARD 3    FMTD=FORMAT FOR READING DATA
C  CARD(S) 4   (THERE MUST BE *NV* OF THSE CARDS)
C    COLS  1- 4 LABEL(I)=4 CHARACTER LABEL FOR THE I-TH VARIABLE
C  CARD(S) 5   DATA=ORIGINAL DATA READ ACCORDING TO FORMAT *FMTD*
```

```
C   CARD(S) 6  MERGE DATA AS GENERATED BY ANY HIERARCHICAL GROUPING
C              METHOD IN THIS SYSTEM OR AS CONVERTED BY SUBROUTINE *PREP*
C     COLS  1-10 K=STAGE OF CLUSTERING
C     COLS 11-20 II=LOWER NUMBERED CLUSTER MERGED AT STAGE K
C     COLS 21-30 JJ=UPPER NUMBERED CLUSTER MERGED AT STAGE K
C     COLS 31-46 CR=VALUE OF CRITERION FUNCTION RESULTING FROM THIS MERGE
C-----------------------------------------------------------------------
C
C   OTHER VARIABLES IN THE PROGRAM
C     GSS(I)=GRAND SUM OF SQUARES FOR VARIABLE I (UNCORRECTED FOR MEAN)
C     TSS(I)=TOTAL SUM OF SQUARES FOR VARIABLE I (CORRECTED FOR MEAN)
C     WSS(I)=WITHIN GROUP SUM OF SQUARES FOR VARIABLE I AT CURRENT STAGE
C     NCL=NUMBER OF CLUSTERS AT THE CURRENT STAGE
C     NUMBR(I)=NUMBER OF SUBJECTS CURRENTLY IN CLUSTER I
C     SUM(I)=SUM OF VALUES FOR VARIABLE I
C     UNEXP(I)=UNEXPLAINED PROPORTION OF VARIANCE FOR VARIABLE I
C              =WSS(I)/TSS(I)
C
C   TO CHANGE PROGRAM CAPACITY THE USER NEED CHANGE ONLY THE FOLLOWING
C   DIMENSION CARD
C       DIMENSION NUMBR(M),DATA(N)
C   WHERE M IS AT LEAST AS LARGE AS NS AND N IS AT LEAST AS LARGE AS NS*NV
        DIMENSION NUMBR(255),DATA(2550)
C   THE FOLLOWING ARRAYS ARE DIMENSIONED FOR 14 VARIABLES
        DIMENSION LABEL(14),TSS(14),WSS(14),GSS(14),SUM(14),UNEXP(14)
        DIMENSION FMTD(20),TITLE(20)
        INTEGER FIRST
C   READ IN INFORMATION NEEDED FOR THIS RUN
        READ(5,1100) TITLE
        READ(5,1000) NTDAT,NTMRG,NS,NV
        READ(5,1100) FMTD
        READ(5,1200) (LABEL(I),I=1,NV)
        WRITE(6,2200) TITLE
        WRITE(6,2000) NTDAT,NTMRG,NS,NV
        WRITE(6,2100) FMTD
        WRITE(6,2300) (LABEL(I),I=1,NV)
C   SET INITIAL VALUES
        NCL=NS-1
        DO 20 I=1,NV
        GSS(I)=0.
20      SUM(I)=0.
        LAST=0
        K=0
        DO 30 I=1,NS
        NUMBR(I)=1
        FIRST=LAST+1
        LAST=LAST+NV
        READ(NTDAT,FMTD) (DATA(J),J=FIRST,LAST)
        DO 30 J=1,NV
        K=K+1
        GSS(J)=GSS(J)+DATA(K)*DATA(K)
30      SUM(J)=SUM(J)+DATA(K)
        DO 40 J=1,NV
        TSS(J)=GSS(J)-SUM(J)*SUM(J)/NS
        WSS(J)=0.
40      UNEXP(J)=0.
C   WRITE OUT THE TSS ARRAY
        WRITE(6,3000)
        DO 50 J=1,NV
50      WRITE(6,3100) LABEL(J),TSS(J)
C   WRITE PAGE HEADINGS
        WRITE(6,2200) TITLE
        WRITE(6,2400)
        WRITE(6,2500) (LABEL(I),I=1,NV)
```

```
C   READ MERGE CARDS, COMPUTE WVAR AND PRINT RESULTS.
        NSTGS=NS-1
        DO 90 I=1,NSTGS
        READ(NTMRG,4000) K,II,JJ,CR
C   TEST FOR PROPER SEQUENCE OF MERGE INSTRUCTIONS
        IF(K.NE.I) GO TO 100
C   UPDATE CLUSTER INFORMATION
        IWRD=(II-1)*NV
        JWRD=(JJ-1)*NV
        NTOT=NUMBR(II)+NUMBR(JJ)
        DO 60 J=1,NV
        IWRDJ=IWRD+J
        JWRDJ=JWRD+J
        DIFF=NUMBR(II)*DATA(JWRDJ)-NUMBR(JJ)*DATA(IWRDJ)
        WSS(J)=WSS(J)+DIFF*DIFF/(NUMBR(II)*NUMBR(JJ)*NTOT)
        UNEXP(J)=WSS(J)/TSS(J)
60      DATA(IWRDJ)=DATA(IWRDJ)+DATA(JWRDJ)
        NUMBR(II)=NTOT
70      WRITE(6,2600) K,NCL,II,JJ,CR,(UNEXP(J),J=1,NV)
90      NCL=NCL-1
        STOP
C   ERROR IN SEQUENCE OF MERGE INSTRUCTIONS
100     WRITE(6,5000) I,K
        I=I-1
        WRITE(6,5100) I,(UNEXP(J),J=1,NV)
        STOP
1000    FORMAT(6I5)
1100    FORMAT(20A4)
1200    FORMAT(A4)
2000    FORMAT(8H NTDAT =,I5,/,8H NTMRG =,I5,/,5H NS =,I8,/,5H NV =,I8,
       A/)
2100    FORMAT(7H0FORMAT,20A4)
2200    FORMAT(1H1,20A4)
2300    FORMAT(16H VARIABLE LABELS,/,14(5X,A4))
2400    FORMAT(1H0,34X,48HPROPORTION OF VARIANCE NOT EXPLAINED BY GROUPING
       A)
2500    FORMAT(21H STAGE   NCL    II    JJ,2X,9HCRITERION,1X,14(3X,A4))
2600    FORMAT(1X,4I5,E13.5,14F7.4)
3000    FORMAT(9H0VARIABLE,13X,3HTSS)
3100    FORMAT(5X,A4,E16.8)
5000    FORMAT( 24H1ERROR IN MERGE SEQUENCE ,/,4H I =,I10,/,4H K =,I10)
5100    FORMAT( 28H0RESULTS FOR PRECEDING STAGE,/,1X,I5,29X,14F7.4)
4000    FORMAT(3I10,E16.8)
        END
```

Program POSTDU

```
        PROGRAM POSTDU(INPUT,OUTPUT,TAPE5=INPUT,TAPE6=OUTPUT,TAPE1)
C
C   THIS PROGRAM IS DESIGNED TO ASSIST IN THE INTERPRETATION OF
C   CLUSTERED DATA UNITS.  ORIGINAL DATA IS PERMUTED TO THE SEQUENCE
C   APPEARING IN THE HIERARCHICAL TREE (OR ANY OTHER SEQUENCE THE
C   USER WISHES TO SPECIFY).  CLUSTERS ARE IDENTIFIED BY SIMPLY STATING
C   THE NUMBER OF DATA UNITS IN EACH CLUSTER, SAY N1,N2, ETC.  THEN
C   THE FIRST N1 UNITS IN THE SEQUENCE LIST ARE IN THE FIRST CLUSTER,
C   THE NEXT N2 UNITS IN THE SECOND CLUSTER AND SO FORTH.  EACH CLUSTER
C   IS DESCRIBED BY A LISTING OF ITS DATA UNITS, THEIR SCORES ON
C   SELECTED VARIABLES AND SUMMARY STATISTICS.  THE PRINTED OUTPUT IS
C   LIMITED TO 10 VARIABLES EACH RUN.  IF MORE THAN 10 VARIABLES ARE
C   OF INTEREST, SIMPLY PARTITION THE VARIABLES INTO SUBSETS AND RUN THE
C   PROGRAM FOR EACH SUBSET.
C----------------------------------------------------------------------
```

```
C   INPUT SPECIFICATIONS
C
C   CARD 1   TITLE CARD
C
C   CARD 2   PARAMETER CARD
C      COLS  1- 4   NE=NUMBER OF ENTITIES (DATA UNITS)
C      COLS  5- 6   NV=NUMBER OF VARIABLES (MAX 10)
C      COLS  7- 8   NC=NUMBER OF CLUSTERS (MAX 50)
C      COLS  9-10   NTIN=INPUT UNIT FOR DATA
C
C   CARD 3   LABEL CARD FOR VARIABLES.  A 4 CHARACTER LABEL IS REQUIRED
C   FOR EACH VARIABLE (10A4 FORMAT)
C
C   CARD(S) 4   LABEL CARDS FOR DATA UNITS.  THERE ARE TWO OPTIONS
C      1.   INCLUDE 1 CARD WITH THE 4 CHARACTERS *NOLB* IN COLUMNS 1-4.
C           UNDER THIS OPTION LABELS ARE NOT PRINTED ON THE TREE OUTPUT.
C
C      2.   INCLUDE *NE* CARDS, COLUMNS 1 TO 20 CONTAINING A LABEL FOR ONE
C           DATA UNIT.
C
C   CARD(S) 5   SEQUENCE LIST FOR DATA UNITS (20I4 FORMAT).  USE AS MANY
C   CARDS AS NECESSARY TO LIST *NE* DATA UNITS.  THIS LIST MAY BE
C   PUNCHED IN SUBROUTINE *TREE* AS PART OF A HIERARCHICAL CLUSTERING JOB
C   OR IN SUBROUTINE *RESULT* AS PART OF A NON-HIERARCHICAL CLUSTERING
C   JOB.
C
C   CARD(S) 6   NUMBER OF DATA UNITS IN EACH CLUSTER (20I4 FORMAT).  USE
C   AS MANY CARDS AS NECESSARY TO LIST THE SIZE OF THE *NC* CLUSTERS
C   WHOSE MEMBERS ARE ORDERED IN THE SEQUENCE LIST OF CARD 6.
C
C   CARD 7   FORMAT FOR PRINTING DATA ON OUTPUT.  GIVE FORMAT FOR *NV*
C   FIELDS OF 10 CHARACTERS EACH.  USE ANY COMBINATION OF E, F AND G
C   FIELDS.  THE FORMAT IS LEFT VARIABLE SO THE NUMBER OF SIGNIFICANT
C   DIGITS CAN BE CONTROLLED FOR EACH VARIABLE.  BEGIN THE FORMAT
C   IN COLUMN 1 WITH A LEFT PARENTHESIS AND END WITH A RIGHT PARENTHESIS.
C
C   CARD 8   FORMAT FOR READING DATA
C
C   CARD(S) 9   ORIGINAL DATA (IF ON CARDS)
C-----------------------------------------------------------------------
C   VARIABLES IN THE PROGRAM
C   TITLE=IDENTIFYING TITLE FOR RUN
C   LABLEV(I)=4 CHARACTER LABEL FOR I-TH VARIABLE
C   LABELD(I,J)=J-TH OF 5 WORDS (4 CHARACTERS EACH) LABELLING I-TH DATA
C               UNIT
C   LIST(I)=I-TH DATA UNIT IN THE SEQUENCE LIST
C   NUMBR(I)=NUMBER OF DATA UNITS IN THE I-TH CLUSTER
C   DATA(I,J)=VALUE OF J-TH VARIABLE FOR I-TH DATA UNIT
C   GTOT(I)=TOTAL FOR I-TH VARIABLE OVER ENTIRE DATA SET
C   CTOT(I)=TOTAL FOR I-TH VARIABLE OVER CURRENT CLUSTER
C   GSS(I)=SUM OF SQUARES FOR I-TH VARIABLE OVER ENTIRE DATA SET
C   CSS(I)=SUM OF SQUARES FOR I-TH VARIABLE OVER CURRENT CLUSTER
      DIMENSION TITLE(20),FMT(24),NUMBR(50),FMTD(20)
      DIMENSION LABELV(10),GTOT(10),GSS(10),CTOT(10),CSS(10)
C   THE FOLLOWING DIMENSION STATEMENT IS SET TO HANDLE 1000 DATA UNITS
      DIMENSION LABELD(5,1000),LIST(1000),DATA(10,1000)
      DATA FMT1,FMT2A,FMT2B,FMT3,FMT4A,FMT4B/
     A4H(1X,,4H5A4,,4H20X,,4HI5.2,4HX,    ,4H(28X/
      INTEGER FIRST
      FMT(1)=FMT1
      FMT(2)=FMT2B
      FMT(3)=FMT3
      FMT(4)=FMT4A
      READ(5,1000) TITLE
```

```
          WRITE(6,2000) TITLE
          READ(5,1100)  NE,NV,NC,NTIN
          WRITE(6,2100) NE,NV,NC,NTIN
          READ(5,1000) (LABELV(I),I=1,NV)
          READ(5,1000) (LABELD(I,1),I=1,5)
          IF(LABELD(1,1).EQ.4HNOLB) GO TO 20
C   READ REMAINING LABELS
          DO 10 J=2,NE
10        READ(5,1000) (LABELD(I,J),I=1,5)
          FMT(2)=FMT2A
20        READ(5,1200) (LIST(I),I=1,NE)
          READ(5,1200) (NUMBR(I),I=1,NC)
          WRITE(6,2200) (I,NUMBR(I),I=1,NC)
          READ(5,1300) (FMT(I),I=5,24)
          WRITE(6,2300) (FMT(I),I=1,24)
          READ(5,1000)  (FMTD(I),I=1,20)
          WRITE(6,2300) (FMTD(I),I=1,20)
C   READ DATA SET
          DO 25 J=1,NE
25        READ(NTIN,FMTD) (DATA(I,J),I=1,NV)
C   INITIALIZE GRAND STATISTICS FOR THE ENTIRE DATA SET
          DO 30 I=1,NV
          GTOT(I)=0.
30        GSS(I)=0.
C   COMPUTE STATISTICS FOR EACH CLUSTER AND PRINT RESULTS
          LAST=0
          DO 90 IC=1,NC
          FIRST=LAST+1
          NEC=NUMBR(IC)
          LAST=LAST+NEC
          DO 40 I=1,NV
          CTOT(I)=0.
40        CSS(I)=0.
          WRITE(6,2000) TITLE
          WRITE(6,2400) IC,NEC
          WRITE(6,2500)
          WRITE(6,2600) (LABELV(I),I=1,NV)
          DO 70 J=FIRST,LAST
          JE=LIST(J)
          DO 50 I=1,NV
          CTOT(I)=CTOT(I)+DATA(I,JE)
50        CSS(I)=CSS(I)+DATA(I,JE)**2
          IF(FMT(2).EQ.FLB) GO TO 60
C   NO LABELS
          WRITE(6,FMT) JE,(DATA(I,JE),I=1,NV)
          GO TO 70
C   WITH LABELS
60        WRITE(6,FMT) (LABELD(I,JE),I=1,5),JE,(DATA(I,JE),I=1,NV)
70        CONTINUE
C   UPDATE GRAND STATISTICS AND PRINT CLUSTER STATISTICS
          DO 80 I=1,NV
          GTOT(I)=GTOT(I)+CTOT(I)
          GSS(I)=GSS(I)+CSS(I)
          CTOT(I)=CTOT(I)/NEC
80        CSS(I)=CSS(I)/NEC-CTOT(I)**2
          WRITE(6,2700) (CTOT(I),I=1,NV)
          WRITE(6,2800) (CSS(I),I=1,NV)
          FMT(4)=FMT4A
90        CONTINUE
C   PRINT GRAND STATISTICS
          WRITE(6,2000) TITLE
          WRITE(6,2600) (LABELV(I),I=1,NV)
          DO 100 I=1,NV
          GTOT(I)=GTOT(I)/NE
```

```
100     GSS(I)=GSS(I)/NE-GTOT(I)**2
        WRITE(6,2700) (GTOT(I),I=1,NV)
        WRITE(6,2800) (GSS(I),I=1,NV)
1000    FORMAT(20A4)
1100    FORMAT(I4,3I2)
1200    FORMAT(20I4)
1300    FORMAT(1X,20A4)
2000    FORMAT(1H1,20A4)
2100    FORMAT(5HONE =,I8,/,5H NV =,I8,/,5H NC =,I8,/,7H NTIN =,I6)
2200    FORMAT(21HOSIZE OF EACH CLUSTER,/,(1X,2I10))
2300    FORMAT(7HOFORMAT,24A4)
2400    FORMAT(8HOCLUSTER,I3,11H CONTAINING,I4,12H DATA UNITS.)
2500    FORMAT(11HODATA UNITS,13X,2HID,2X,20H SCORES ON VARIABLES)
2600    FORMAT(28X,10(6X,A4))
2700    FORMAT(6H MEANS,22X,10(E10.3))
2800    FORMAT(10H VARIANCES,18X,10(E10.3))
2900    FORMAT(31HOSTATISTICS FOR ENTIRE DATA SET)
        END
```

H

RELATIONS AMONG CLUSTER ANALYSIS PROGRAMS

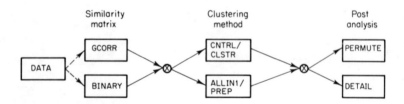

Fig. H.1. Processing flow for clustering variables.

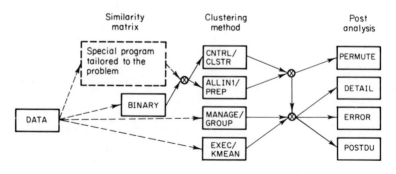

Fig. H.2. Processing flow for clustering data units.

REFERENCES

Abramowtiz, M., and Stegun, I. A. (1968). *Handbook of Mathematical Functions with Formulas, Graphs and Mathematical Tables* (Nat. Bur. of Stand. Appl. Math. Ser., No. 55), 7th printing. US Govt. Printing Office, Washington, D.C.

Aitken, M. A. (1966). "The Correlation between Variate Values and Ranks in a Doubly Truncated Normal Distribution." *Biometrika* **53**, Pts. 1/2, 281–282.

Aitken, M. A., and Hume, M. W. (1966). "Correlation in a Singly Truncated Bivariate Normal Distribution III. Correlation between Ranks and Variate Values." *Biometrika* **53**, Pts. 1/2, 278–281.

Albert, A. (1971). *The Penrose Moore Pseudo Inverse with Statistical Applications. Pt. I. The General Theory and Computational Methods.* Tech. Rep. 173, AD 724752; *Pt. II. Statistical Applications.* Tech. Rep. 176, AD 726692; *Solutions to Problems.* Tech. Rep. 177, AD 728068. Dept. of Statist., Stanford Univ., Stanford, California.

Anderson, T. W. (1958). *An Introduction to Multivariate Statistical Analysis.* Wiley, New York.

Archer, W. B. (1966). *Computation of Group Job Descriptions from Occupational Survey Data.* Rep. No. PRL-TR-66-12, AD 653543. Personnel Res. Lab., Lackland Air Force Base, Texas.

Astrahan, M. M. (1970). *Speech Analysis by Clustering, or the Hyperphoneme Method.* Stanford Artificial Intelligence Proj. Mem. AIM-124, AD 709067. Stanford Univ., Stanford, California.

Baker, F. B. (1965). "Latent Class Analysis as an Association Model for Information Retrieval" *In Statistical Association Methods for Mechanized Documentation* (M. E. Stevens, V. E. Giuliano, and L. B. Heilprin, eds.). *Nat. Bur. Stand. (U.S.) Misc. Publ.* **No. 269**, 149–155.

Ball, G. H. (1966). *A Comparison of Some Cluster-Seeking Techniques.* Rep. No. RADC-TR-66-514, AD 643287. Stanford Res. Inst., Menlo Park, California. (Also available under the title "Data Analysis in Social Sciences: What About the Details?" *Proc. of Fall Joint Comput. Conf.*, **27**, Pt. I, pp. 533–559. Spartan Books, Washington, D.C., 1965.)

Ball, G. H. (1970). *Classification Analysis.* Tech. Note, AD 716482. Stanford Res. Inst., Menlo Park, California.

Ball, G. H., and Hall, D. J. (1965). *ISODATA, A Novel Method of Data Analysis and Pattern Classification*, AD 699616. Stanford Res. Inst., Menlo Park, California.

Ball, G. H., and Hall, D. J. (1967). *PROMENADE—An On-Line Pattern Recognition System.* Rep. No. RADC-TR-67-310, AD 822174. Stanford Res. Inst., Menlo Park, California.

Bartels, P. H., Bahr, G. F., Calhoun, D. W., and Wied, G. L. (1970). "Cell Recognition by Neighborhood Grouping Technique in TICAS." *Acta Cytol.* **14**, No. 6, 313–324. AD 710844.

Borko, H., Blankenship, D. A., and Burket, R. C. (1968). *On-Line Information Retrieval Using Associative Indexing.* RADC-TR-68-100, AD 670195. Systems Develop. Corp., Santa Monica, California.

Bradley, R. A., Katti, S. K., and Coons, I. J. (1962). "Optimal Scaling for Ordered Categories." *Psychometrika* **27**, No. 4, 355–374. AD 407665.

Bryan, J. G. (1961). *Calibration of Qualitative or Quantitative Variables for Use in Multiple-Group Discriminant Analysis,* AD 251139. The Travelers Weather Res. Center, Hartford, Connecticut.

Carroll, J. B. (1961). "The Nature of the Data, or How to Choose a Correlation Coefficient." *Psychometrika* **26**, No. 4, 347–372.

Casetti, E. (1964). *Classificatory and Regional Analysis by Discriminant Iterations.* TR-12, Contract Nonr-1228 (26), AD 608093. Northwestern Univ., Evanston, Illinois.

Castellan, Jr., N. J. (1966). "On the Estimation of the Tetrachoric Correlation Coefficient." *Psychometrika* **31**, No. 1, 67–73.

Cattell, R. B. (1965). "Factor Analysis: An Introduction to Essentials II. The Role of Factor Analysis in Research." *Biometrics* **21**, No. 2, 405–435.

Chernoff, H. (1970). *Metric Considerations in Cluster Analysis.* Tech. Rep. No. 67, AD 714810. Dept. of Statist., Stanford Univ., Stanford, California.

Cochran, W. G., and Hopkins, C. E. (1961). "Some Classification Problems with Multivariate Qualitative Data." *Biometrics* **17**, No. 1, 10–32.

Cole, A. J. (1969). *Numerical Taxonomy.* Academic Press, New York.

Cramér, H. (1946). *The Elements of Probability Theory and Some of Its Applications.* Wiley, New York.

Crawford, R. M. M., and Wishart, D. (1967). "A Rapid Multivariate Method for the Detection and Classification of Groups of Ecologically Related Species." *J. Ecol.* **55**, No. 2, 505–524.

Crawford, R. M. M., and Wishart, D. (1968). "A Rapid Classification and Ordination Method and Its Application to Vegetation Mapping." *J. Ecol.* **56**, No. 2, 385–404.

Cronbach, L. J., and Gleser, G. C. (1953). "Assessing the Similarity between Profiles." *Psychol. Bull.* **50**, No. 6, 456–473.

David, F. N., Barton, D. E., Ganeshalingham, S., Harter, H. L., Kim, P. J., Merrington, M., and Walley, D. (1968). *Normal Centroids, Medians and Scores for Ordinal Data.* Cambridge Univ. Press, London and New York.

Demiremen, F. (1969). *Multivariate Procedures and FORTRAN IV Program for Evaluation and Improvement of Classifications.* Comput. Contrib. 31. State Geol. Survey, Univ. of Kansas, Lawrence, Kansas.

Doyle, L. B. (1966). *Breaking the Cost Barrier in Automatic Classification.* Prof. Pap. SP-2516, AD 636837. Systems Develop. Corp., Santa Monica, California.

Dubes, R. C. (1970). *Information Compression, Structure Analysis and Decision Making with a Correlation Matrix,* AD 720811. Michigan State Univ., East Lansing, Michigan.

Dubin, R. (1971). *Typology of Empirical Attributes: Multi-dimensional Typology Analysis (MTA).* TR-5, AD 728781. Univ. of California, Irvine.

Dubin, R., and Champoux, J. E. (1970). *Typology of Empirical Attributes: Dissimilarity Linkage Analysis (DLA).* Tech. Rep. 3, AD 708678. Univ. of California, Irvine.

Duncan, D. B. (1955). "Multiple Range and Multiple F Tests." *Biometrics* **11**, No. 1, 1–42.

Eades, D. C. (1965). "The Inappropriateness of the Correlation Coefficient as a Measure of Taxonomic Resemblance." *Syst. Zool.* **14**, No. 2, 98–100.

Eddy, R. P. (1968). *Class Membership Criteria and Pattern Recognition.* Rep. 2524, AD 834208. Naval Ship Res. and Develop. Center, Washington, D.C.

Edwards, A. W. F. (1963). "The Measure of Association in a 2 × 2 Table." *J. Roy. Statist. Soc. Ser. A* **126**, Pt. 1, 109–114.

Edwards, A. W. F., and Cavalli-Sforza, L. L. (1965). "A Method for Cluster Analysis." *Biometrics* **21**, No. 2, 362–375.

Ericson, W. A. (1964). "A Note on Partitioning for Maximum between Sum of Squares." *In The Detection of Interaction Effects* (by J. A. Sonquist and J. N. Morgan). Survey Res. Center, Inst. for Soc. Res., Univ. of Michigan, Ann Arbor.

Farris, J. S. (1969). "On the Cophenetic Correlation Coefficient." *Syst. Zool.* **18**, No. 3, 279–285.

Fisher, R. A. (1936). "The Use of Multiple Measurements in Taxonomic Problems." *Ann. Eugenics* **7**, Pt. 2, 179–188.

Fisher, R. A. (1940). "The Precision of Discriminant Functions." *Ann. Eugenics* **10**, 422–429.

Fisher, R. A. (1963). *Statistical Methods for Research Workers*, 13th ed., reprint. Hafner, New York.

Fisher, R. A., and Yates, F. (1953). *Statistical Tables for Biological, Agricultural and Medical Research*. Oliver & Boyd, Edinburgh.

Fisher, W. D. (1958). "On Grouping for Maximum Homogeneity." *J. Amer. Statist. Assoc.* **53**, 789–798.

Fisher, W. D. (1968). *Clustering and Aggregation in Economics*. Johns Hopkins Press, Baltimore, Maryland.

Forgy, E. W. (1965). *Cluster Analysis of Multivariate Data: Efficiency Versus Interpretability of Classifications*. Biometric Soc. Meetings, Riverside, California (Abstract in *Biometrics* **21**, No. 3, 768).

Forgy, E. W. (1966). "Classification So As To Relate to Outside Variables." *Final Rep. Conf. Cluster Analysis of Multivariate Data* (M. Lorr and S. B. Lyerly, eds.), AD 653722. Catholic Univ. of Amer., Washington, D.C. pp. 13.01–13.12.

Friedman, H. P., and Rubin, J. (1967). "On Some Invariant Criteria for Grouping Data." *J. Amer. Statist. Assoc.* **62**, 1159–1178.

Froemel, E. C. (1971). "A Comparison of Computer Routines for the Calculation of the Tetrachoric Correlation Coefficient." *Psychometrika* **36**, No. 2, 165–174.

Garfinkel, R. S., and Nemhauser, G. L. (1970). "Optimal Political Districting by Implicit Enumeration Techniques." *Management Sci.* **16**, No. 8, B495–B508.

Gibbons, J. D. (1971). *Nonparametric Statistical Inference*. McGraw-Hill, New York.

Goodman, L. A., and Kruskal, W. H. (1954). "Measures of Association for Cross Classifications." *J. Amer. Statist. Assoc.* **49**, 732–764.

Gower, J. C. (1966). "Some Distance Properties of Latent Root and Vector Methods Used in Multivariate Analysis." *Biometrika* **53**, No. 3/4, 325–338.

Gower, J. C. (1967). "A Comparison of Some Methods of Cluster Analysis." *Biometrics* **23**, No. 4, 623–637.

Gower, J. C., and Ross, G. J. S. (1969). "Minimum Spanning Trees and Single Linkage Cluster Analysis." *Appl. Statist.* **18**, No. 1, 54–64.

Grason, J. (1970). *Methods for the Computer-Implemented Solution of a Class of "Floor-Plan" Design Problems*, Ph.D. Dissertation, AD 717756. Elect. Eng. Dept., Carnegie-Mellon Univ., Pittsburgh, Pennsylvania.

Green, P. E., and Carmone, F. J. (1970). *Multidimensional Scaling and Related Techniques in Marketing Analysis*. Allyn & Bacon, Rockleigh, New Jersey.

Green, P. E., and Rao, V. R. (1969). "A Note on Proximity Measures and Cluster Analysis." *J. Marketing Res.* **6**, 359–64.

Haggard, E. A. (1958). *Intra-Class Correlation and the Analysis of Variance*. Dryden, New York.

Hall, A. V. (1969). "Avoiding Informational Distortions in Automatic Grouping Programs." *Syst. Zool.* **18**, No. 3, 318–329.

Hall, D. J., Ball, G. H., Wolf, D. E., and Eusebio, J. (1969). *PROMENADE: An Improved Interactive-Graphics Man/Machine System for Pattern Recognition.* Rep. No. RADC-TR-68-572, AD 692752. Stanford Res. Inst., Menlo Park, California.

Hamdan, M. A. (1971a). "Estimation of Class Boundaries in Fitting a Normal Distribution to a Qualitative Multinomial Distribution." *Biometrics* **27**, No. 2, 457–459.

Hamdan, M. A. (1971b). "On the Polychoric Method for Estimation of [Rho] in Contingency Tables." *Psychometrika* **36**, No. 3, 253–259.

Hanson, N. R. (1958). *Patterns of Discovery: An Inquiry into the Conceptual Foundations of Science.* Cambridge Univ. Press, London and New York.

Harding, E. F. (1971). "The Probabilities of Rooted Tree-Shapes Generated by Random Bifurcation." *Advances in Appl. Probability* **3**, No. 1, 44–77.

Harman, H. H. (1960). *Modern Factor Analysis.* Univ. of Chicago Press, Chicago, Illinois.

Harter, H. L. (1961). "Expected Values of Normal Order Statistics." *Biometrika* **48**, Pt. 1/2, 151–165.

Hartigan, J. A. (1967). "Representation of Similarity Matrices by Trees." *J. Amer. Statist. Assoc.* **62**, 1140–1158. AD 674208.

Hartigan, J. A. (1972). "Direct Clustering of a Data Matrix." *J. Amer. Statist. Assoc.* **67**, 123–129.

Henschke, C. I. (1969). *Manpower Systems and Classification Theory,* Ph.D. Dissertation. Univ. of Georgia, Athens, Georgia. Cited in *Dissertation Abstracts* **30**, No. 8, 3911-B.

Himmelblau, D. M. (1972). *Applied Nonlinear Programming.* McGraw-Hill, New York.

Hung, A. Y., and Dubes, R. C. (1970). *An Introduction to Multi-class Pattern Recognition in Unstructured Situations.* Interim Sci. Rep. No. 12, AD 720812. Div. of Eng. Res., Michigan State Univ., East Lansing, Michigan.

Information Please Almanac and Yearbook, 1970. Simon and Schuster, New York.

Jancey, R. C. (1966). "Multidimensional Group Analysis." *Austral. J. Botany* **14**, No. 1, 127–130.

Jardine, C. J., Jardine, N., and Sibson, C. (1967). "The Structure and Construction of Taxonomic Hierarchies." *Math. Biosci.* **1**, No. 2, 173–179.

Jardine, N., and Sibson, R. (1971). *Mathematical Taxonomy.* Wiley, New York.

Johnson, P. O. (1950). "The Quantification of Qualitative Data in Discriminant Analysis." *J. Amer. Statist. Assoc.* **45**, 65–76.

Johnson, S. C. (1967). "Hierarchical Clustering Schemes." *Psychometrika* **32**, No. 3, 241–254.

Kendall, M. G. (1955). *Rank Correlation Methods,* 2nd ed. Griffin, London.

Kendall, M. G. (1968). *A Course in Multivariate Analysis,* 4th printing. Hafner, New York.

Kendall, M. G., and Stuart, A. (1961). *The Advanced Theory of Statistics. Vol. II. Inference and Relationship.* Griffin, London.

Kruskal, Jr., J. B. (1956). "On the Shortest Spanning Subtree of a Graph and the Traveling Salesman Problem." *Proc. Amer. Math. Soc.,* No. 7, pp. 48–50.

Kruskal, W. H. (1958). "Ordinal Measures of Association." *J. Amer. Statist. Assoc.* **53**, 814–861.

Kshirsagar, A. M. (1970). "Goodness of Fit of an Assigned Set of Scores for the Analysis of Association in a Contingency Table." *Ann. Inst. Statist. Math.* **22**, No. 2, 295–306.

Labovitz, S. (1967). "Some Observations on Measurement and Statistics." *Social Forces* **46**, No. 2, 151–160.

Labovitz, S. (1970). "The Assignment of Numbers to Rank Order Categories." *Amer. Sociol. Rev.* **35**, No. 3, 515–524.

Labovitz, S. (1971). "In Defense of Assigning Numbers to Ranks." *Amer. Sociol. Rev.* **36**, No. 3, 520–521.

Lancaster, H. O., and Hamdan, M. A. (1964). "Estimation of the Correlation Coefficient in Contingency Tables with Possibly Nonmetrical Characters." *Psychometrika* **29**, No. 4, 383–391.

Lance, G. N., and Williams, W. T. (1965). "Computer Program for Monothetic Classification ('Association Analysis')." *Comput. J.* **8**, No. 3, 246–249.

Lance, G. N., and Williams, W. T. (1966). "Computer Programs for Hierarchical Polythetic Classification ('Similarity Analyses')." *Comput. J.* **9**, No. 1, 60–64.

Lance, G. N., and Williams, W. T. (1967a). "A General Theory of Classificatory Sorting Strategies. 1. Hierarchical Systems." *Comput. J.* **9**, No. 4, 373–380.

Lance, G. N., and Williams, W. T. (1967b). "A General Theory of Classificatory Sorting Strategies II. Clustering Systems." *Comput. J.* **10**, No. 3, 271–276.

Lazarsfeld, P. F., and Reitz, J. G. (1970). *Toward a Theory of Applied Sociology*, AD 715659. Bur. Appl. Social Res., Columbia Univ., New York.

Linfoot, E. H. (1957). "An Informational Measure of Correlation." *Information and Control* **1**, No. 1, 85–89.

Linscheid, T. R., and Stone, L. A. (1971). "An IBM 360/30 FORTRAN IV Program for the Calculation of Intraclass Correlation Coefficients." *Behavioral Sci.* **16**, No. 4, 413–414.

Litofsky, B. (1969). *Utility of Automatic Classification Systems for Information Storage and Retrieval*, Ph.D. Dissertation. Univ. of Pennsylvania, Philadelphia. Cited in *Dissertation Abstracts* **30**, No. 7, 3264-B (1970). Also available from NTIS as AD 687140.

McCammon, R. B. (1968). "The Dendrograph: A New Tool for Correlation." *Geol. Soc. Amer. Bull.* **79**, 1663–1670.

McCammon, R. B., and Wenninger, G. (1970). *The Dendrograph*. Comput. Contrib. 48. State Geol. Survey, Univ. of Kansas, Lawrence.

MacQueen, J. B. (1967). "Some Methods for Classification and Analysis of Multivariate Observations." *Proc. Symp. Math. Statist. and Probability*, 5th, Berkeley, **1**, 281–297, AD 669871, Univ. of California Press, Berkeley.

McRae, D. J. (1971). "MIKCA: A FORTRAN IV Iterative K-Means Cluster Analysis Program." *Behavioral Sci.* **16**, No. 4, 423–424.

Majone, G., and Sanday, P. R. (1968). *On the Numerical Classification of Nominal Data*. Rep. No. RR-118. AD 665006. Graduate School of Ind. Administration, Carnegie-Mellon Univ., Pittsburgh, Pennsylvania.

Marriott, F. H. C. (1971). "Practical Problems in a Method of Cluster Analysis." *Biometrics* **27**, No. 3, 501–514.

Maung, K. (1941). "Measurement of Association in a Contingency Table with Special Reference to the Pigmentation of Hair and Eye Color of Scottish School Children." *Ann. Eugenics* **11**, 189–223.

Mayer, L. S. (1970). "Comment on the Assignment of Numbers to Rank Order Categories." *Amer. Sociol. Rev.* **35**, No. 5, 916–917.

Mayer, L. S. (1971a). "A Method of Cluster Analysis when There Exist Multiple Indicators of a Theoretical Concept." *Biometrics* **27**, No. 1, 143–155.

Mayer, L. S. (1971b). "A Note on Treating Ordinal Data as Interval Data." *Amer. Sociol. Rev.* **36**, No. 3, 518–519.

Medawar, P. B. (1969). *Induction and Intuition in Scientific Thought*, Jayne Lectures for 1968. Amer. Philos. Soc., Philadelphia, Pennsylvania.

Morrison, D. F. (1967). *Multivariate Statistical Methods*. McGraw-Hill, New York.

Nagy, G. (1968). "State of the Art in Pattern Recognition." *Proc. IEEE* **56**, No. 5, 836–862.

Parks, J. M. (1969). "Classification of Mixed Mode Data by R-Mode Factor Analysis and Q-Mode Cluster Analysis on Distance Functions." *In Numerical Taxonomy* (A. J. Cole, ed.), pp. 216–223. Academic Press, New York.

Parks, J. M. (1970). *FORTRAN IV Program for Q-Mode Cluster Analysis on Distance Function with Printed Dendrogram.* Comput. Contrib. 46. State Geol. Survey, Univ. of Kansas, Lawrence.

Pearson, E. S., and Hartley, H. O. (1954). *Biometrika Tables for Statisticians.* Cambridge Univ. Press, London and New York.

Pearson, W. H. (1966). "Estimation of a Correlation Measure from an Uncertainty Measure." *Psychometrika* 31, No. 3, 421–433. AD 642665.

Rand, W. M. (1971). "Objective Criteria for the Evaluation of Clustering Methods." *J. Amer. Statist. Assoc.* 66, 846–850.

Rao, C. R. (1964). "The Use and Interpretation of Principal Components Analysis in Applied Research." *Sankhyā Ser. A* 26, 329–358.

Robson, D. S., and Whitlock, J. H. (1964). "Estimation of a Truncation Point." *Biometrika* 51, Pts. 1/2, 33–39.

Rohlf, F. J. (1970) "Adaptive Hierarchical Clustering Schemes." *Syst. Zool.* 19, No. 1, 58–83.

Rummel, R. J. (1970). *Applied Factor Analysis.* Northwestern Univ. Press, Evanston, Illinois.

Sammon, Jr., J. W. (1968). *On-Line Pattern Analysis and Recognition System (OLPARS).* Rep. No. RADC-TR-68-263, AD 675212. Rome Air Develop. Center, Griffiss Air Force Base, New York.

Sandon, F. (1961). "The Means of Sections from a Normal Distribution." *British J. Statist. Psychol.* 14, Pt. 2, 117–121.

Sandon, F. (1962). "An Alternative Table for Determining Means of Sections from a Normal Distribution." *British J. Statist. Psychol.* 15, Pt. I, 71–74.

Schweitzer, S., and Schweitzer, D. G. (1971). "Comment on the Pearson r in Random Number and Precise Functional Scale Transformations." *Amer. Sociol. Rev.* 36, No. 3, 517–518.

Scott, A. J., and Symons, M. J. (1971a). "Clustering Methods Based on Likelihood Ratio Criteria." *Biometrics* 27, No. 2, 387–398.

Scott, A. J., and Symons, M. J. (1971b). "On the Edwards and Cavalli-Sforza Method of Cluster Analysis." *Biometrics* 27, No. 1, 217–219.

Shepard, R. N., and Carroll, J. D. (1966). "Parametric Representation of Non Linear Data Structures." *In Multivariate Analysis* (P. R. Krishnaiah, ed.). Academic Press, New York.

Sokal, R. R., and Rohlf, F. J. (1962). "The Comparison of Dendrograms by Objective Methods." *Taxon* 11, 33–40.

Sokal, R. R., and Sneath, P. H. A. (1963). *Principles of Numerical Taxonomy.* Freeman, San Francisco, California.

Solomon, H. (1970). *Numerical Taxonomy.* Tech. Rep. No. 167, AD 715781. Dept. of Statist., Stanford Univ., Stanford, California.

Sonquist, J. A., and Morgan, J. N. (1964). *The Detection of Interaction Effects.* Survey Res. Center, Inst. for Social Res., Univ. of Michigan, Ann Arbor.

Srikantan, K. S. (1970). "Canonical Association between Nominal Measurements." *J. Amer. Statist. Assoc.* 65, No. 329, 284–292.

Stephenson, W. (1935). "Correlating Persons Instead of Tests." *Character and Personality* 4, No. 1, 17–24.

Stephenson, W. (1938). "The Inverted Factor Technique." *British J. Psychol.* 26, No. 4, 344–361.

Stephenson, W. (1953). *The Study of Behavior.* Univ. of Chicago Press, Chicago, Illinois.

Stewart, D., and Love, W. (1968). "A General Canonical Correlation Index." *Psychol. Bull.* **70**, No. 3, 160–163.

Stiles, H. E. (1961). "The Association Factor in Information Retrieval." *Commun. ACM* **8**, No. 1, 271–279.

Stuart, A. (1954). "The Correlation between Variate Values and Ranks in Samples from a Continuous Distribution." *British J. Statist. Psychol.* **7**, Pt. I, 37–44.

Tatsuoka, M. M. (1955). *The Relationship between Canonical Correlation and Discriminant Analysis, and a Proposal for Utilizing Qualitative Data in Discriminant Analysis.* Educ. Res. Corp., Cambridge, Massachusetts.

Tryon, R. C., and Bailey, D. E. (1970). *Cluster Analysis.* McGraw-Hill, New York.

Vargo, L. G. (1971). "Comment on 'The Assignment of Numbers to Rank Order Categories'." *Amer. Sociol. Rev.* **36**, No. 3, 516–517.

Ward, Jr., J. H. (1963). "Hierarchical Grouping to Optimise an Objective Function." *J. Amer. Statist. Assoc.* **58**, No. 301, 236–244.

Ward, Jr., J. H., and Hook, M. E. (1963). "Application of an Hierarchical Grouping Procedure to a Problem of Grouping Profiles." *Educ. and Psychol. Measurement* **23**, No. 1, 69–82.

Ward, Jr., J. H., Hall, K., and Buchhorn, J. (1967). *PERSUB Reference Manual.* Rep. No. PRL-TR-67-3(II), AD 666579. Personnel Res. Lab., Lackland Air Force Base, Texas.

Wherry, Jr., R. J., and Lane, N. E. (1965). *The K-Coefficient, A Pearson-Type Substitute for the Contingency Coefficient.* Rep. No. NSAM-929, AD 622776. Naval School of Aviat. Med., Pensacola, Florida.

Williams, E. J. (1952). "Use of Scores for the Analysis of Association in Contingency Tables." *Biometrika* **39**, Pt. 3/4, 274–289.

Williams, W. T., and Lambert, J. M. (1959). "Multivariate Methods in Plant Ecology I. Association Analysis in Plant Communities." *J. Ecol.* **47**, No. 1, 83–101.

Williams, W. T., Dale, M. B., and Macnaughton-Smith, P. (1964). "An Objective Method of Weighting in Similarity Analysis." *Nature (London)* **201**, 426.

Wishart, D. (1969a). "An Algorithm for Hierarchical Classifications." *Biometrics* **22**, No. 1, 165–170.

Wishart, D. (1969b). *FORTRAN II Programs for 8 Methods of Cluster Analysis (CLUSTAN I).* Comput. Contrib. 38. State Geol. Survey, Univ. of Kansas, Lawrence.

Wolf, D. E. (1968). *PROMENADE: Complete Listing of PROMENADE Programs.* RADC-TR-68-572, Appendix 9d, AD 694114. Stanford Res. Inst., Menlo Park, California (available in microfiche only).

Wolfe, J. H. (1970). "Pattern Clustering by Multivariate Mixture Analysis." *Multivariate Behavioral Res.* **5**, No. 3, 329–350.

Wolfe, J. H. (1971). *NORMIX 360 Computer Program.* Rep. No. SRM-72-4, p. 123, AD 731037. Personnel and Training Res. Lab., San Diego, California.

Yule, G. U. (1912). "On Measuring Associations between Attributes." *J. Roy. Statist. Soc.* **75**, 579–642.

Zahn, C. T. (1971). "Graph-Theoretical Methods for Detecting and Describing Gestalt Clusters." *IEEE Trans. Comput.* **C-20**, No. 1, 68–86.

INDEX

355